Primate ontogeny, cognition and social behaviour

Index

intercourse, frequency 324
interference,
  in aggression 221, 293
  males 384
  (*see also* interventions)
interventions, and dominance 219, 293

Japanese macaque 61, 123
Japanese traditions 61
Java macaque 156, 291, 307
juveniles,
  consciousness 49
  rank 293
  rates of interactions 231
  roles 281
  specialisations 186

kin selection 53, 382
kinship,
  and dominance 351
  and relationships 284
  and social thinking 58

lactation,
  and aggression 345
  allocation of 233, 357
  in humans 319
language development 173
langur 361
leaf eating 109
learning,
  and development 120
  food selection 114
  foraging 39
  general principles 31
  mechanisms 107
linear hierarchies 301
locomotion 155
long-term research,
  cooperation 17
  difficulties 18
  finances 20

Macaca fascicularis 156, 291, 307
Macaca fuscata 61, 123
Macaca mulatta 54, 141, 161, 220, 246, 281, 307
Macaca nemestrina 121, 131, 301
Macaca tonkeana 307
males,
  competition 217
  dominance 207, 383, 334
  mating behaviour 331
  transfer 206, 368, 371, 382
  vigilance 364
mammalian feeding strategies 94
manipulation 155
marmoset 122
maternal behaviour 357

maternal rejection 243
maternal support 291
mating behaviour 333
memory 36
mental maps 75, 87
Microcebus murinus 355
Mikumi Research Project 15
Miopithecus talapoin 329
monogamy 79, 255, 399
morphometry, brain 131
mother–offspring relationship 152, 198, 232, 264
motor development 152
mouse lemur 355

neocortical function 94
neonatal development 137, 147
nesting behaviour 359
nocturnal species 355

oestrus,
  cycles 322, 335, 379
  suppression 338, 345
  synchrony 345
olive baboon 123, 268, 371
omnivory 98
ontogeny 119, 179
optimal design 189
orangutans 119
ovarian cycles 331

Pan troglodytes 32, 39, 75, 120, 124
Papio anubis 123, 268, 371
Papio cynocephalus 13, 205, 268, 344, 381
Papio hamadryas 87
Papio ursinus 124
parenting roles 255
parent–infant interactions 261
parent–offspring conflict 193
paternity,
  infant defence 209
  relationships with mothers 210
peers 231, 284
phenology, plant 106
Piagetian frameworks 155
pigtail macaque 121, 131, 301
plasma cortisol levels 258
play 231
political traits 4, 9
polygyny 399
postpartum,
  hormones 324
  sexual behaviour 321
preference tests 290
pregnancy,
  behaviour 345
  hormones 331
prehension 155

primates,
  consciousness 44
  feeding strategies 96
  minds 27
progesterone 330
psychological processes 56

quality of care 150, 265
quality of resources 231

ranging patterns 84
rank,
  acquisition 219, 291
  and relatedness 351
  stability 303
  (*see also* dominance)
receptivity 336
reconciliation 311
rejections, maternal 247
relationships,
  after aggression 309
  children 121
  male–male 207, 381
  male–infant 201, 257
  mother–offspring 152, 194, 232, 244, 257, 264
reproductive success 356
reproductive suppression 329
reptilian feeding strategies 95
resident males 207
resource,
  distribution 80
  exploitation 79, 105
Resource Holding Potential 395
rhesus macaque 54, 141, 161, 220, 246, 281, 307
rituals 170, 174

Saimiri sciureus 79, 120
siblings,
  interactions 286
  support 294
separation from parents 256
sex differences 277, 284
sex roles 255, 285
sexual behaviour 319
signalling systems 46

social thinking 57
social processes 61
solicitation for care 200
spatial memory 37
specialisations 185
species society 63, 67
squirrel monkey 79, 120
stimulus-response 34
suckling success 233
symbols 169, 173
symmetry in aggression 308
symptoms, communication 172

talapoin monkey 329
task learning 33
territoriality 365
theories, primate minds 27
thinking, evidence for 53
titi monkey 79, 255
tooth eruption 144
Tonkean macaque 307
tradition 47
trait schema 7
transfer,
  female 364
  male 207, 282, 371
travel economy 84
triadic interactions 211, 387
troops, differences between,
  in aggression 351
  infant interactions 275
typological thinking 162

vervet monkey 124, 193, 228
vocal repertoire 161
vocal units 164
vocalisations 161, 169, 200, 265
volume, brain 133, 148
visual system development 134

weaning,
  conflict 202
  time-course 232
weight,
  body 132, 142, 147
  brain 133, 148

yellow baboon 13, 205, 268, 344, 381

Selected Proceedings of the Tenth Congress of the International Primatological Society, held in Nairobi, Kenya, in July 1984

Volume 3

Primate ontogeny, cognition and social behaviour

Edited by

JAMES G. ELSE
Institute of Primate Research, National Museums of Kenya

PHYLLIS C. LEE
*Sub-Department of Animal Behaviour, Department of Zoology
University of Cambridge*

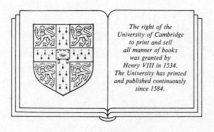

CAMBRIDGE UNIVERSITY PRESS

Cambridge

London   New York   New Rochelle

Melbourne   Sydney

Published by the Press Syndicate of the University of Cambridge
The Pitt Building, Trumpington Street, Cambridge CB2 1RP
32 East 57th Street, New York, NY 10022, USA
10 Stamford Road, Oakleigh, Melbourne 3166, Australia

© Cambridge University Press 1986

First published 1986

Printed in Great Britain at the University Press, Cambridge

British Library cataloguing in publication data

Primate ontogeny, cognition and social
behaviour.—(Selected proceedings of the Tenth
Congress of the International Primatological
Society. . . . ; v. 3)
1. Primates—Behaviour 2. Mammals—Behaviour
I. Else, James G. II. Lee, Phyllis, C. III. Series
599.8'0451   QL737.P9

Library of Congress cataloguing in publication data

Primate ontogeny, cognition, and social behaviour.

(Selected proceedings of the Tenth Congress of the
International Primatological Society; v. 3)
  1. Primates – Behavior – Congresses. 2. Primates –
Development – Congresses. 3. Cognition in animals – Congresses.
4. Social behavior in animals – Congresses. 5. Mammals – Behavior –
Congresses. 6. Mammals – Development – Congresses.
I. Else, James G. II. Lee, Phyllis C. III. Series: International
Primatological Society. Congress (10th : 1984 : Nairobi, Kenya).
Selected proceedings of the Tenth Congress of the International
Primatological Society; v. 3.
QL737.P9I53 1984 Vol. 3   599.8 s [599.8'0451] 86-8288

ISBN 0 521 32452 1 hard covers
ISBN 0 521 31013 X paperback

CONTENTS

List of Contributors
Preface

I Introduction
1 Can primate political traits be identified? *W. A. Mason* — 3
2 Ten years of cooperative baboon research at Mikumi National Park. *R. J. Rhine* — 13

II Primate thinking
1 Introduction: the human primates' theory of the primate mind. *D. K. Candland and R. C. Kyes* — 25
2 Animal thinking – by stimulation or simulation? *D. M. Rumbaugh* — 31
3 Primate consciousness. *M. Singh* — 43
4 Primate social thinking. *V. Reynolds* — 53
5 Anthropomorphism and the Japanese and Western traditions in primatology. *P. J. Asquith* — 61

III Primate behaviour and cognition in nature
Introduction. *D. M. Fragaszy and H. O. Box* — 73
1 Contrasting approaches to spatially distributed resources by *Saimiri* and *Callicebus*. *M. W. Andrews* — 79
2 Ranging patterns in hamadryas baboons: evidence for a mental map. *H. Sigg* — 87
3 Cognition, brain size and the extraction of embedded food resources. *K. R. Gibson* — 93
4 Development of feeding selectivity in mantled howling monkeys, *Alouatta palliata*. *J. M. Whitehead* — 105
5 The development of social behaviour and cognitive abilities. *H. O. Box and D. M. Fragaszy* — 119

IV Perception and performance: growth, manipulation and communication

Editors' introduction — 129

1. Morphometry of the developing brain in *Macaca nemestrina*. J. L. DeVito, J. Graham, G. Schultz, J. W. Sundsten and J. W. Prothero — 131
2. Age-related growth patterns of colony-born rhesus monkeys. D. N. Sharma and K. C. Lal — 141
3. A comparative approach to the question of why human infants develop so slowly. H. Dienske — 147
4. The interaction between prehension and locomotion in macaque, gorilla and child cognitive development. G. Spinozzi and F. Natale — 155
5. Grading in the vocal repertoire of Silver Springs rhesus monkeys. E. H. Peters — 161
6. Symbols: the missing link? J. Liska — 169

V Functional aspects of development

Introduction. P. C. Lee and P. P. G. Bateson — 179

1. Functional approaches to behavioural development. P. P. G. Bateson — 183
2. Parent–offspring conflict: care elicitation behaviour and the 'cry-wolf' syndrome. M. D. Hauser — 193
3. Relationships between adult male and infant baboons. D. A. Collins — 205
4. The role of alliances in the acquisition of rank. S. B. Datta — 219
5. Environmental influences on development: play, weaning and social structure. P. C. Lee — 227

VI Social interactions: development and maintenance

Editors' introduction — 239

1. The use of confidence intervals in arguments from individual cases: maternal rejection and infant social behaviour in small groups of rhesus monkeys. A. Tartabini and M. J. A. Simpson — 243
2. Parenting within a monogamous society. S. P. Mendoza and W. A. Mason — 255
3. Social interactions of free-ranging baboon infants. H. Hendy — 267
4. Juvenile/adolescent role functions in a rhesus monkey troop: an application of household analysis to non-human primate social organization. D. Quiatt — 281

5	Conflict interference and the development of dominance relationships in immature *Macaca fascicularis*. W. J. Netto and J. A. R. A. M. van Hooff	291
6	Dominance rank and related interactions in a captive group of female pigtail macaques. P. Messeri and C. Giacoma	301
7	A comparative study of aggression and response to aggression in three species of macaque. B. Thierry	307

VII Social and reproductive strategies

	Editors' introduction	315
1	Postpartum sexual behavior of American women as a function of the absence or frequency of breast feeding: a preliminary communication. J. M. Stern and S. R. Leiblum	319
2	Social suppression of reproduction in subordinate talapoin monkeys, *Miopithecus talapoin*. D. H. Abbott, E. B. Keverne, G. F. Moore and U. Yodyingyuad	329
3	Reproductive competition among female yellow baboons. S. K. Wasser and A. K. Starling	343
4	The influence of other females on maternal behaviour and breeding success in the lesser mouse lemur (*Microcebus murinus*). A. R. Glatston	355
5	Through the territorial barrier: harem accretion in *Presbytis senex*. G. H. Manley	363
6	Factors affecting intertroop transfer by adult male *Papio anubis*. D. L. Manzolillo	371
7	Lasting alliances among adult male savannah baboons. R. Noë	381
8	Fitness returns from resources and the outcome of contests: some implications for primatology and anthropology. N. G. Blurton Jones	393
	Index	407

CONTRIBUTORS

D. H. Abbott, *Department of Anatomy, University of Cambridge and Institute of Zoology, Regent's Park, London NW1 4RY, UK*

M. W. Andrews, *Animal Behaviour and California Primate Research Center, University of California, Davis, California 95616, USA*

P. J. Asquith, *Department of Anthropology, University of Calgary, Calgary, Alberta, Canada TN2 14N*

P. P. G. Bateson, *Sub-Department of Animal Behaviour, University of Cambridge, Madingley, Cambridge CB3 8AA, UK*

N. G. Blurton Jones, *Graduate School of Education and Department of Anthropology, University of California, Los Angeles, California 90024, USA*

H. O. Box, *Department of Psychology, University of Reading, Whiteknights, Reading RG6 2AL, UK*

D. K. Candland, *Bucknell University, Lewisberg, Pennsylvania 17837, USA*

D. A. Collins, *Zoology Department, University of Edinburgh, Edinburgh EH9 3JT, UK*

S. B. Datta, *Sub-Department of Animal Behaviour, University of Cambridge, Madingley, Cambridge CB3 8AA, UK*

J. L. DeVito, *Regional Primate Research Center, University of Washington, Seattle, Washington 98185, USA*

H. Dienske, *Primate Center TNO, 2280 HV Rijswijk, The Netherlands*

D. Fragaszy, *Veterinary Comparative Anatomy, Pharmacology and Physiology, Washington State University, Pullman, Washington 99164, USA*

C. Giacoma, *Dipartimento di Biologia Animale, University of Torino, Italy*

K. R. Gibson, *Department of Anatomical Sciences, University of Texas Dental Branch, Houston, Texas 77225, USA*

A. R. Glatston, *Biological Research Department, Royal Rotterdam Zoological and Botanical Gardens, Rotterdam, The Netherlands*

J. Graham, *Regional Primate Research Center, University of Washington, Seattle, Washington 98195, USA*

M. D. Hauser, *Department of Anthropology, University of California, Los Angeles, California 90024, USA*

H. Hendy, *Department of Psychology, Pennsylvania State University, Schuylkill Haven, Pennsylvania 17972, USA*

Contributors

J. A. R. A. M. van Hooff, *Laboratory of Comparative Physiology, State University of Utrecht, 3572 LA Utrecht, The Netherlands*

E. B. Keverne, *Department of Anatomy, University of Cambridge, Cambridge CB2 3DY, UK*

R. C. Kyes, *Bucknell University, Lewisberg, Pennsylvania 17837, USA*

K. C. Lal, *Primate Research Facility, All-India Institute of Medical Sciences, Ansari Nagar, New Delhi-29, India*

P. C. Lee, *Sub-Department of Animal Behaviour, University of Cambridge, Madingley, Cambridge CB3 8AA, UK*

S. R. Leiblum, *Department of Psychiatry, Rutgers University Medical School, Piscataway, New Jersey 08854, USA*

J. Liska, *Department of Speech Communication, Indiana University, Bloomington, Indiana 47405, USA*

G. H. Manley, *Department of Anthropology, University of Durham, Durham DH1 3HN, UK*

D. L. Manzolillo, *Institute of Primate Research, P.O. Box 24481, Karen, Kenya*

W. A. Mason, *California Primate Research Center, University of California, Davis, California 95616, USA*

S. P. Mendoza, *California Primate Research Center, University of California, Davis, California 95616, USA*

P. Messeri, *Centro di Studio per la Faunistica ed Ecologia Tropicali del CNR, 50125 Florence, Italy*

G. F. Moore, *Department of Anatomy, University of Cambridge and Institute of Zoology, Regent's Park, London NW1 4RY, UK*

F. Natale, *Istituto de Psicologia del CNR, University of Rome, Roma 00197, Italy*

W. J. Netto, *Laboratory of Comparative Physiology, State University of Utrecht, 3572 LA Utrecht, The Netherlands*

R. Noë, *Laboratory of Comparative Physiology, State University of Utrecht, 3572 LA Utrecht, The Netherlands*

E. H. Peters, *Department of Anthropology, Florida State University, Tallahassee, Florida 32306, USA*

J. W. Prothero, *Department of Biological Structure, University of Washington, Seattle, Washington 98195, USA*

D. Quiatt, *Department of Anthropology, University of Colorado, Denver, Colorado 80202, USA*

V. Reynolds, *Department of Biological Anthropology, Oxford University, Oxford OX 6QS, UK*

R. J. Rhine, *Department of Psychology, University of California, Riverside, California 92521, USA*

D. M. Rumbaugh, *Department of Psychology, Georgia State University, and Yerkes Regional Primate Center, Atlanta, Georgia 30322, USA*

G. Schultz, *Department of Biological Structure, University of Washington, Seattle, Washington 98195, USA*

D. N. Sharma, *Primate Research Facility, All-India Institute of Medical Science, Ansari Nagar, New Delhi-29, India*

H. Sigg, *Biology Zentrallab, Universitatsspital, CH-8901, Zurich, Switzerland*

M. J. A. Simpson, *Sub-Department of Animal Behaviour, University of Cambridge, Madingley, Cambridge CB3 8AA, UK*

M. Singh, *Department of Psychology, University of Mysore, Manasagangotri, Mysore-570 006, India*

G. Spinozzi, *Istituto di Psicologia del CNR, University of Rome, Roma 00157, Italy*

A. K. Starling, *Department of Psychology, University of Washington, Seattle, Washington 98195, USA*

J. M. Stern, *Department of Psychology, Rutgers-State University of New Jersey, New Brunswick, New Jersey 08903, USA*

J. W. Sundsten, *Department of Biological Structure, University of Washington, Seattle, Washington 98195, USA*

A. Tartabini, *Department of Sciences of Education, University of Calabria, Italy 87036*

B. Thierry, *Laboratoire de Psychologie, Universite Louis Pasteur, 67000 Strasbourg, France*

S. K. Wasser, *Department of Psychology, University of Washington, Seattle, Washington 98195, USA*

J. M. Whitehead, *Department of Biology, University of North Carolina, Chapel Hill, North Carolina 27514, USA*

U. Yodyingyuad, *Department of Anatomy, University of Cambridge, Cambridge CB2 3DY and Department of Biology, Chulalongkorn University, Bangkok, Thailand*

PREFACE

Some of the most exciting recent developments in the field of primate behaviour are investigations of the way in which primates perceive and communicate about their physical and social environment. Many of the presentations at the Tenth Congress of the International Primatological Society, held in Nairobi, Kenya in July 1984, considered these topics. The papers included in this volume represent these new ideas and developments. We have included papers on recent approaches to social development, some of which also deal with the development of cognitive abilities in a social context. Others represent the trend towards elucidating the dynamics of relationships; how they are developed, maintained, and their consequences on later behaviour. And finally, we have included a section of papers dealing specifically with the reproductive and social consequences of behaviour among adults. Papers were kept relatively short in order to include as many contributors as possible in each of the sections.

The two papers in Part I discuss very different but equally important issues in the study of primate social behaviour. In Mason's contribution, the problem of defining and classifying complex social behaviour is discussed with reference to the identification of consistent and systematic traits. He reviews the concept of a behavioural trait and its use (and abuse), and presents a new way of using traits to integrate the study of proximate and ultimate functions of behaviour. Rhine covers an issue common to many field primatologists whether they study ecology or social behaviour, the maintenance of long-term field studies of primates. He deals with the problems of financing sites, maintaining continuity of data collection, ensuring harmonious coexistence between researchers, and the importance of positive relations with the governments and local people of the host country. The experiences from Mikumi are examples of ways of dealing with these problems.

The next two sections deal primarily with issues of cognition. The five papers in Part II discuss the ways in which learning and mental abilities are different aspects of primate thinking from those of performance, as it is measured using principles of conditioning. The questions of thinking and performance are viewed from historical, anthropological, ecological and evolutionary perspectives. In Part III, current concepts of cognition are considered by an examination of the ways in which primates exploit their habitats. Principles of learning, processing abilities, the exploitation of spatially dispersed resources, and information gathering and exchange are presented from studies of several different species of wild and captive primates.

Part IV uses the development of the brain as a starting point and moves through questions of growth rates, the development of motor capacities and cognitive abilities in human and non-human primates. The communicative skills of primates are then discussed with reference to the complexity of vocalisations and symbols in primate communication.

Current research on the development of behaviour is presented in Part V using a theoretical framework that emphasises immaturity as an adaptive phase in an individual's life. Issues of the immediate function and of the long-term advantages to behaviour seen during development are presented in papers on mother–infant conflict, paternal behaviour, and rank acquisition. The ways in which social development is constrained by the social or physical environment are also considered.

Further detailed studies on the development and maintenance of social behaviour are presented in Part VI. These papers begin with a new technique for measuring behaviour with small sample sizes and cover different aspects of parental behaviour, relationships with other group members, and the problems of the acquisition, maintenance and function of dominance.

The final section (Part VII) deals with the behaviour of males and females that influences their own and other individuals' reproductive performance. These behavioural tactics or 'strategies' have been a recent focus of studies in primate behavioural ecology. In this section, studies of reproductive physiology and behaviour in the laboratory reveal trends that can also be found in nature and, in the final paper, principles of reproductive success are applied to contests between humans.

Preface

These papers are the results of short presentations at the Nairobi Congress, from a variety of sessions and symposia. They have been reviewed and extensively revised, and we would like to thank the session chairs for their comments and external reviewers for their help. We would like to express our appreciation to the Government of the Republic of Kenya, and to the National Museums of Kenya for hosting the Congress. We are grateful to all those participants who made the Congress and this volume such a success. Dr Robin Pellew of Cambridge University Press has been a constant source of encouragement. PCL is very grateful to Professors R. A. Hinde and P. Bateson for providing facilities while editing.

James G. Else
P. C. Lee
Nairobi and Cambridge
September 1, 1985

Part I

Introduction

I.1

Can primate political traits be identified?

W. A. MASON

Attitudes toward animal behavior have changed much since the time, a few years ago, when scientists somewhat self-consciously proclaimed that their concerns were restricted to the 'objective' study of behavior. The implication was that an adequate and complete science of behavior could be based solely on what was observed. To be sure, it was widely recognized at the time that a satisfactory general account of behavior required the assumption that various covert processes and hypothetical mechanisms played an essential role in the causal network that determines what an organism actually does; hence, the familiar and time-worn concepts of learning, memory, perception, decision-making, needs, motives, emotions, and the like. At the same time, the prevailing attitude toward such hypothetical entities was cautious and conservative. In the name of parsimony and objectivity, the formidable complexities surrounding the ontological and theoretical status of hypothetical processes were avoided by the expedients of explicit operational definitions and a data language that, ideally, was stripped of all mentalistic connotations.

We have lost these inhibitions, it seems, and nowadays freely invoke processes and variables that are covert and entirely hypothetical. As one bit of evidence of this change, I call your attention to the contents of the present *Proceedings*. Even a cursory comparison of the titles of the paper sessions and symposia with those of previous occasions indicates that our attitudes are undergoing a remarkable transformation. Among the topics one notes: primate perception, primate cognition, primate thinking, primate strategies (male and female), and primate political behavior (the title of the symposium in which this contribution was originally included).

Transcending the former constraints on our ways of thinking about primate behavior has been an exhilarating and healthy development. In particular, the explicit recognition that mental processes contribute in important ways to primate behavior has drawn attention to new phenomena and to fresh interpretive possibilities. At the same time, this development has inevitably raised a number of complex and challenging methodological and conceptual issues that need to be addressed if we are to avoid the excesses against which previous generations of scientists reacted so strongly.

One of these issues that I believe is very much in need of critical attention is the concept of behavioral traits. The trait concept is one of those primitive notions that we seem unable to dispense with in describing the behavior of nonhuman primates. It is a commonplace of our everyday experience that individual monkeys and apes not only differ from each other, but appear to do so consistently and in systematic ways. Just as we perceive broad and persistent differences in personality and temperament among our human acquaintances, so too do we describe this particular monkey as being friendly, that one as aggressive, another as timid, placid, and so on. The tendency to apply trait names is probably even more compelling when we describe differences between species – for example, extroverted chimpanzees *versus* introverted gorillas – or between members of the same species which have been raised in different environments – for example, the aggressive rhesus monkeys of the cities *versus* the placid and timid monkeys of the forests. These examples will, I hope, illustrate the point that trait concepts are pertinent and widely used in discussions of primate behavior, whether from the standpoint of phylogeny, evolution and adaptation, or from the standpoint of the proximal determinants of behavior. Indeed, in the sense that traits may be thought of as biologically significant attributes of behavior that are the products of natural selection and also as part of the causal nexus that has an immediate influence on what animals do, they serve as a kind of conceptual bridge between these two complementary orientations.

What are behavioral traits? How do we recognize them and describe them? Obviously, many answers are possible. At one extreme we might say that a behavioral trait is an overt reaction of some sort, which we observe directly and describe as we do any other behavior. Thus, the behavior *is* the trait, or it is a clear and invariant sign of the trait, similar to the stigmata that were once claimed to identify born criminals (Fig. 1). The view that a behavioral trait is a fixed attribute of the individual, fully presented to an observer, is probably not widely accepted. Whatever appeal it has is based on the fact that it resolves most of the conceptual and methodological problems associated with behavioral traits by sweeping them under the rug.

It may be more instructive to consider our actual practices in applying behavioral trait names, the assumptions we make, and their implications. In the first place, identifying and describing a behavioral trait implies a dimension (for example, boldness–timidity) and differences between individuals (or some other focal entity, such as a population) along this dimension. Secondly, behavioral traits imply a certain consistency of reaction across time. One would hardly be willing to ascribe the trait of timidity or aggressiveness to an individual on the basis of a single observation, but might be more inclined to do so if the same behavior was seen on multiple separate occasions. Thirdly, behavioral traits imply some consistency across situations, as well as over time. An aggressive or timid or friendly monkey is expected to show these traits with a variety of different companions, in various contexts. In summary, it is clear that behavioral trait names characteristically refer to hypothetical processes that are inferred from differences between entities along some dimension and that such differences will be consistent across time and situations.

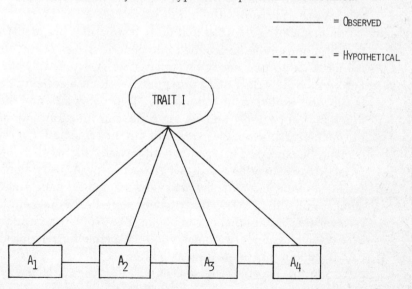

Fig. 1. Trait schema based on the assumption that a trait is directly manifest in one or more attributes (A_1–A_4). The absence of dotted lines is meant to convey that no hypothetical processes are assumed.

To complicate matters further, as we all recognize, traits are influenced by situations. Obviously an animal that we characterize as friendly cannot display this trait in a social vacuum; it is scarcely more likely to do so in a social context if all its friendly overtures are instantly rebuffed. Clearly, some situations will be more relevant to the trait or more conducive to its expression than others. We also know, however, that in its specifics every situation is unique. A person may be regarded as the life of the party, but no two parties are ever precisely the same. Whatever consistency an individual's behavior shows is present in spite of variations among situations. When we refer to traits, therefore, we have in mind not one specific situation, but a class or category of situations. The task of establishing the characteristics or criteria that will be used to determine whether an actual situation fits within a given class or category may be more difficult than it at first appears. For example, is a foraging party a useful category? Is a consort relationship 'a situation'? Is it the same situation for the primary participants as it is for interested bystanders? Beyond the obvious point that some form of 'objective' classification of situations is a necessary component in the use of trait names, there is a need to consider that the effect of any situation will depend to a great extent on the individual's perception of that situation. From this standpoint, situations that are descriptively quite different may be equivalent with respect to their effect on the expression of some individual traits, whereas other situations which are superficially quite similar may in fact have sharply contrasting influences.

Complementing the idea that no two situations are identical, and that situations must be grouped into generic classes or categories, is the fact that no two behavior patterns are ever precisely the same. Furthermore, it is clear that a principal basis for categorizing behavior is function, rather than topography. Two behaviors may be quite different in appearance and yet serve the same function, and therefore be considered as probable expressions of the same trait. For example, fear may be expressed in physical withdrawal, cringing, or stereotyped facial gestures. In other words, we characteristically use trait names to refer to *dispositions* to display behaviors that appear to be functionally similar, rather than to specific behavior patterns. At the methodological level, this implies that beyond being able to obtain accurate records of specific behaviors, we have criteria available for relating them to some common functional dimension (Fig. 2).

An additional complexity needs to be raised, although it may not be as clearly relevant to the behavior of nonhuman primates as are the

Can primate political traits be identified?

preceding points. This is the possibility that different traits may interact with one another and be expressed on the same occasion and in the same behaviors. For example, revealing damaging information about another person may simultaneously express the traits of honesty, cowardice, and hostility, just as in nonhuman primates mounting another animal of the same sex might be construed as expressing the traits of self-assertiveness, friendliness and prurience (Fig. 3).

Finally, there is the question of the dimensions to which trait names refer; this comes close to what is called, in the jargon of psychological tests and measurement, the problem of construct validity. I trust that the particular trait names I have used to illustrate my argument will not

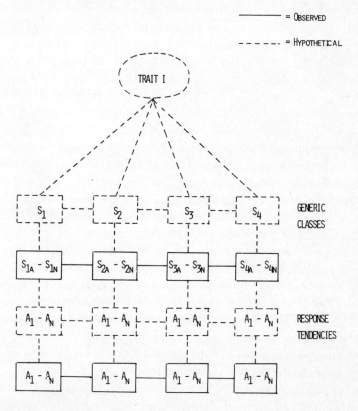

Fig. 2. Trait schema based on the assumption that a trait is inferred from a sampling of situations and behaviors, each of which is a member of a hypothetical set or class. Thus, actual situation S_{1A} falls within the hypothetical set S_1, and the behavior observed in S_{1A} is selected from the hypothetical set A_1-A_N.

be taken seriously. Observers of primate behavior presumably select their trait names with considerably more care than I have shown. Even so, the names they use are taken from the lexicon of everyday language, which virtually guarantees that they will be ambiguous and laden with surplus meaning. There is no need to justify the practice, nor to belabor the difficulties, including the problem of reification. So far as I can see, there is no completely satisfactory solution. Even sophisticated statistical techniques, such as factor analysis, do not finally answer the question of how traits should be labeled and interpreted, although they may help to narrow down the reasonable possibilities.

We are left then with the following conclusions. When behavioral trait names are used a number of complex issues are raised that must

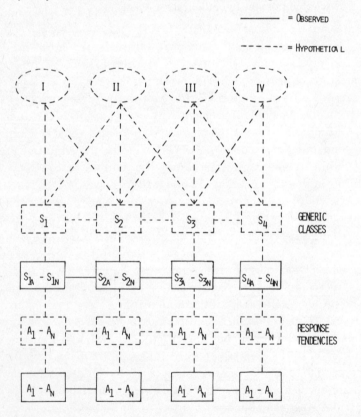

Fig. 3. Trait schema based on the same assumptions as those in Fig. 2, with the additional assumption that one or more traits (I–IV) may jointly contribute to the observed behavior in a given situation.

eventually be confronted. Traits are highly inferential, conceptually abstract and methodologically awkward. Trait names refer to abiding differences among individual entities along some specified dimension, and they allude to aspects of the organization of behavior (e.g. tendencies, dispositions, etc.) that are inherently probabilistic and covert.

This state of affairs is not unique. The status of behavioral traits appears to differ only in degree from that of traditional motivational constructs. Neither trait names nor motivational constructs refer to simple, unitary processes; traits, like drives, are heuristically useful, but they have limited explanatory value; both are fundamentally hypothetical. Furthermore, primatologists are not the first scientists to wrestle with the methodological and conceptual problems associated with behavioral traits. Investigators of human intelligence, personality, and individual differences were long ago forced to confront similar issues, and the majority seemed to have reached conclusions in accord with those presented here (e.g. Anastasi, 1970; Endler & Magnusson, 1976; Epstein, 1983).

In spite of the complexities and uncertainties surrounding the use of trait concepts, I believe that they will continue to find a useful place in discussions of primate behavior. Trait names provide for economy of description. They emphasize the continuity of behavior and the ubiquity of consistent differences with respect to species, gender, age, and individuals. To the extent that trait names refer to genuinely enduring and general dispositions of an entity, they improve prediction, while raising questions about the sources of coherence and stability in the organization of behavior. It is already clear that these sources will operate at many levels (e.g. genetic, epigenetic, physiological, situational), and they will interact in intricate ways. It is also clear that such interactions can be investigated empirically; in fact, this sort of research is commonplace, although it is seldom based on large enough samples of subjects examined over a sufficient range of time and situations to be considered from the standpoint of traits.

The problems and procedures of identifying and analyzing primate 'political' traits are no different in principal from those encountered with any other behavioral trait. To the extent that such phenomena as infanticide, status hierarchies, coalitions, control males, and the like, are established or recurrent features of the social life of one or more primate species, they raise questions about the dispositions within particular entities to behave in certain ways and about the situations with which these dispositions interact to give rise to the outcomes actually observed. This is not to say, of course, that the most influential dispositions or the critical features of the situation will be obvious or conform with our first intuitive expectations; on the contrary, in my experience, this is unlikely to be the case. For example, there is reason to suspect that a cluster of characteristics, including the presence of prominent, agonistically reinforced male dominance hierarchies, frequent coalitions, peripheralization of subadult males, well-defined and relatively exclusive uterine kin-groups, severity and frequency of mother–infant conflict, infrequent infant adoptions, and relatively large inter-animal distances, will differentiate between rhesus and bonnet macaques. It seems unlikely that each of these characteristics constitutes a separate and independent factor in the organization of behavior. At this point we cannot characterize the broader dispositions underlying these phenomena nor describe their interactions with situational variables, although one can readily see how this might be done. Such an approach would only be a first step toward understanding the social dynamics of primate societies, of course, but I am convinced that it would be an important first step. Primate politics – however one chooses to define the term – is thoroughly embedded in the life mode of the species, and in abiding dispositions and tendencies of individual animals. These are key elements in the behavioral traits that have constituted my central theme, in spite of the fact that they are essentially hypothetical, inferential and probabilistic in their expression, and are influenced by a large number of variables such as age, gender, individual history, and present circumstances. From this standpoint it is clear that infanticide, dominance, the formation of coalitions, and many other features of primate social life, are not the expressions of simple traits but the outcomes of complex interactions among individuals and situations (including, of course, the social milieu). If this is a valid position, then it is appropriate to ask at what phenotypic organizational level is natural selection operating to produce these outcomes? With this I return to an earlier point: a more critical and sophisticated approach to the problem of behavioral traits will strengthen the bridge between the two principal orientations toward primate behavior, the one focusing on the proximal organization and prediction of behavior, and the other on its evolution and adaptive functions within a species' total life-history pattern.

Acknowledgements

Preparation of this manuscript was supported in part by the National Institute of Health (Grants RR00169 and R01HD06367). The author would like to thank

Sally P. Mendoza, Michael W. Andrews, and Charles R. Menzel for their comments regarding the manuscript.

References

Anastasi, A. (1970) On the formation of psychological traits. *Am. Psychologist*, **25**, 899–910

Endler, N. S. & Magnusson, D. (1976) Toward an interactional psychology of personality. *Psychol. Bull.* **83**, 956–74

Epstein, S. (1983) A research paradigm for the study of personality and emotions. In *Personality – Current Theory and Research*, ed. R. A. Dienstbier & M. M. Page, pp. 91–154. Lincoln: University of Nebraska Press

I.2

Ten years of cooperative baboon research at Mikumi National Park

R. J. RHINE

Introduction

This paper illustrates the potential advantages and pitfalls of cooperative efforts at a long-term baboon field site, and describes some prerequisites for the effective operation of such sites in a host country. As is typical of many long-term field sites, the Animal Behaviour Research Unit (ABRU) at Mikumi National Park, Tanzania, is distant from its parent institution, the University of California at Riverside (UCR). Baboon research at ABRU has been directed by a Professor with university responsibilities which normally preclude long stays in the field. Consequently, the collection of long-term data and the successful operation of the site depends upon relays of assistants and colleagues, most of whom are unable to devote much more than a year at a time to field work. ABRU, like other such units, is faced with problems of communication and of maintaining the continuity and consistency of data collection. A collaborative balance has to be maintained between the priorities of individual projects and long-term goals. Establishing clear long-term goals and designing a framework for both research and site operations helps to mitigate such problems.

Initial planning for Mikumi grew out of experiences gained from consultation with knowledgeable field workers, prior research at established baboon field stations, and a survey of potential sites. These essential preliminaries provided first-hand information of the main problems and necessities of field research and a basis for project design. They also highlighted the following four central themes which guided the planning of Mikumi research: (1) the importance of understanding primates in their natural habitats for the study of hominid evolution; (2) the tremendous potential of long-term research; (3) the strength of a carefully constructed research design emphasising systematic sampling as a basis for quantitative analyses; and (4) the advantage of collecting data in a manner which facilitates comparisons between troops and sites.

With these themes in mind, research at Mikumi was conceived as a long-term project centered upon the dynamics of changing social and population-demographic events, influenced by year-to-year ecological variation. Initially, a cooperative endeavor was planned, setting the stage for everything that followed. Four students left for Tanzania to begin an overall long-term and comparative project, and at the same time to conduct their own individual research. Their first task was the design and trial of a single set of behavior definitions and procedures for the use of all in the cooperative collection of data. These definitions and procedures were designed also for making systematic comparisons of several troops at different sites.

The early days at Mikumi were the most difficult. The students had to learn to survive on foot in the bush among dangerous animals, and to adjust to the hardship and frustration of shortages and delays. It was also essential to establish with the Park, government and commercial contacts, the relationships needed to work effectively in a foreign country. Perhaps most difficult of all, the first students were unable to start obtaining their main data for many long, exhausting months while attempting to habituate two troops for close-up observation. By April, 1975, one large troop, Viramba troop, was habituated sufficiently and enough animals were individually identified to start data collection. Thereafter, each new generation of researchers and assistants arrived in time to overlap with previous ones and so benefited from accumulating knowledge about the baboons, the site, and appropriate and effective ways of living and operating in Tanzania. At the same time, newly arriving researchers always changed, broadened, and improved the old ways, leaving their own positive imprint on the site.

A major problem of a long-term site is the effort required for the organization and analysis of vast amounts of data collected for several years. Because of the time necessarily devoted to obtaining financial support, manning the site, and keeping it functioning smoothly, continuously collected, long-term data inevitably pile up much faster than they can be analyzed and disseminated in scientific publications. It is desirable to solve this problem without ending or substantially interrupting long-term data collection. The Mikumi solution was a partnership among three experienced Mikumi researchers, with one concentrating upon analysis and writing of the long-term data and the others sharing the main responsibility for site operations and the continuation of Mikumi research.

Table 1 indicates who, among 31 Mikumi researchers and assistants, have undertaken major projects of their own, and illustrates the results

Table 1. *Mikumi Research Projects*

Author	University	Topic
Dissertations		
H. Hendy (1978)	UC, Riverside	Infant–other interactions
D. Rasmussen (1978)	UC, Riverside	Environmental and behavioral correlates of range use
K. Rasmussen (1980)	Cambridge	Consorts and mate selection
H. Klein (1983)	Washington	Paternal care and kin selection
S. Wasser (1981)	Washington	Reproductive competition and cooperation
R. Johnson	Cambridge	Toxic compounds and plants eaten
G. Norton	Cambridge	Leadership in route choice
J. Rogers	Yale	Genetic structure and microevolution
Master's theses		
H. Pennington	Yale	Attitudes toward wildlife conservation
J. Perlsweig	San Francisco	Infant care by multiparous and primiparous mothers
Other research		
R. Rhine (with G. Norton, B. Westlund, C. Kidungho, and others)	UC, Riverside	Long-term: population – demographic, ecological and resource variations, social relationships, feeding–ranging, troop split effects, and rare events
		Determinants of reproductive success
		Spatial organization in progressions
		Troop meetings and movement patterns
S. Wasser	Washington (Seattle)	Coalitions and the effect of stress upon female reproduction
S. Smith	UC, Santa Cruz	Nearest neighbors of juveniles
J. Phillips-Conroy (with J. Rogers)	Washington U. (St Louis)	Genetic structure and microevolution
		Dentition, physical characteristics, and relatedness

of two main Mikumi policies for the enhancement of long-term research: (1) making the site available to people from institutions other than UCR; and (2) encouraging a wide variety of research interests. These policies were adopted because they tend to benefit both the participants and the site itself. People with different backgrounds – those listed in Table 1 are trained in psychology, biology, or anthropology – can usefully share their different perspectives and stores of knowledge. A variety of projects facilitates the accumulation of significant information about the baboons and their environment.

The continued collection of long-term data would not have been possible without volunteers who were willing to pay for their own travel and subsistence in order to obtain experience in field research. Mikumi has been kept operational with the minimal funding needed to provide the core necessities of field vehicles (including petrol and repairs) and furnished housing. A newly arriving researcher need bring only field clothes, personal incidentals, and equipment needed for his/her particular project. While at the site, the individual researcher is additionally responsible for consumables. All else is provided. Some paid assistants were on their own part of the time. For example, Guy Norton, who has played a central role in Mikumi research and who has been associated with Mikumi for 6 years, was a paid assistant for the first 3 years and thereafter was not. Several researchers who were paid assistants were also site leaders. Four of these were Ph.D. students, which is often a very demanding and difficult situation. While many Mikumi assistants were from UCR, others were associated with an additional 10 institutions. In the beginning, assistants were primarily assisting UCR researchers but, more recently, assistants have worked with researchers from other institutions, a trend which is likely to continue.

Long-term data

The benefits of cooperative research go well beyond the passing of field lore from one generation of researchers to the next. Relays of cooperating researchers also provide the means by which site continuity is maintained, and site continuity has been a vital basis of the excellent relationships we have enjoyed with Park and scientific authorities and with many other Tanzanians in all walks of life. These relationships are not merely conveniences; they are essential for productive work.

Each newly arriving researcher must be briefed on local customs, Tanzania and Park requirements, and the importance of tolerating different approaches and values. Cultural diversity is something most students seem to understand in the abstract before leaving their own country, but abstract understanding sometimes rapidly disappears in concrete situations where local customs and styles of work appear to delay or frustrate important goals. Persons unwilling or unable to adjust to different customs or to treat officials and others in the host country with genuine respect should be discouraged from field work. It takes only one thoughtless, discourteous or arrogant researcher to create grave problems for all the rest.

Both researchers and Tanzanian officials have a certain degree of turnover. Nothing can replace personal meetings among people who have learned to respect and trust each other. The advantages of an interlinking chain of introductions of newly informed replacements by researchers already trusted by Tanzanian officials would be largely lost if substantial gaps in site continuity occurred. The primary responsibility for relations with Tanzanian officials rests with the site leader. We have never had a formal application for a research permit denied, though there have sometimes been troublesome delays. Our record in obtaining permits is largely due to the relationship of trust and respect between the site director, Guy Norton, and the Tanzanian scientific authorities.

There is a very real danger that a break in site continuity of, say, one or two years, would end Mikumi research because of the difficulty of essentially starting over. Substantial breaks in site continuity would also have serious effects upon the long-term data. For example, the quality of the long-term data depends upon accurate information about kinship and about population–demographic events, which would be garbled by undue gaps in continuity. Such gaps would not only leave researchers without knowledge of key events, such as the timing of births, transfers, and changes in female reproductive condition, but it would introduce uncertainties in identification, especially among juveniles, and possibly require a long period of rehabituation.

Long-term social, population–demographic, and ecological data are a central feature of a cooperative site. At Mikumi all researchers understand that they are expected to contribute to the maintenance of these data and to a log of rare events. Long-term data are of great interest in their own right; indeed, properly collected and reported, they should be among the most significant data produced by long-term sites. In addition, these data often provide essential background without which many new research projects would not be feasible.

A reasonable amount of core funding to maintain continuity of the site and the long-term data is very difficult to obtain. The more common procedure is to support several uncoordinated, shorter-term studies which may end up producing findings of scientific import and also a considerable amount of good, carefully collected data with puzzling inconsistencies across studies. If different technically excellent snapshots are taken by different people at different places and different points in time, no two of the photos will look alike. Long-term studies provide some of the connecting threads needed to explain variations over time:

> Long-term primate study sites are needed to document the dynamics of change. Ignorance of behavioral variation is a major obstacle to the generality of conclusions about natural primate behavior. Long-term research allows continuous knowledge of well-habituated, individually-identified animals of known kin lines, living in a thoroughly mapped range where major resource areas are identified and located. Lacking such long-term information, a researcher who tried to characterize Mikumi baboons after a study of only one or two years would almost certainly draw an incomplete or distorted picture, as will be seen from a few examples. After several years, a large initial study troop, Viramba troop, split into three distinct parts, including one quite small troop. All of the current most dominant males transferred into the Viramba troops from surrounding ones, and all males who reached young adulthood in a Viramba troop transferred out. In some years the rainfall was double that of others, and foods often eaten in one year were almost excluded from the diet in others. The initial Viramba range greatly expanded before and during the split and has subsequently begun to contract toward the original. After one year, known groves of sleeping trees were numbered to nine, and now there are thirty-four. Recording troop behavior through these and other variations at a long-term site helps separate situation-dependent behaviors from those which are so fundamental they persist over time and under a wide variety of different conditions [Rhine, 1981, p. 8].

Potential difficulties

In our experience, there are four main interpersonal barriers to successful cooperation at field sites. First, situations sometimes develop where one researcher perceives another as interfering with his/her research goals. For example, one researcher may believe that another is not performing reliably or devoting enough time to the collection of data upon which others will depend. Students doing dissertation research at a foreign field site are sometimes under terrible pressure. Usually they have a limited budget which determines the time they can spend in the field. The success of one of the most important undertakings of their life depends upon obtaining adequate amounts of reliable data before time runs out. If some aspect of cooperation seems to interfere with this goal, then a cooperative attitude is difficult to sustain and frustration and anger sometimes occur.

A second barrier to cooperation is a clash in personalities. If researchers dislike each other, the ability to work together toward a common goal may suffer. For example, one researcher was so isolated from the remaining residents that he had almost no social life. In such circumstances, lack of motivation to work long and hard on tasks which largely benefit the other individuals cannot be expected. One of the most devastating events of field sites is irreconcilable personality clashes among people who cannot escape being constantly together and depending upon one another. Probably, the single most important factor, after competence, for a successful stay at a cooperative site is the ability to tolerate the idiosyncrasies of others. Almost everyone applying to work in the field believes he/she has that skill, but some do not. It is a skill which is very difficult to assess in advance because most candidates for field work have never before experienced the kind of isolation and social circumstances that a long stay in the field sometimes entails.

Thirdly, there can be serious disagreements over who originates ideas and who is entitled to use what data and in what ways. Although this problem is not unique to field sites, it is exacerbated for the field researcher who is, in effect, living 24 hours a day in the midst of his/her study, often under difficult circumstances. The resulting powerful sense of commitment tends to intensify territorial urges. For example, individuals who are hired to collect data sometimes forget the terms of their employment, come to think of the data as theirs, and become resentful of their employer's involvement. At Mikumi, after some early misunderstandings about data ownership, we now routinely work out written agreements when overlapping interests are anticipated. Verbal agreements are insufficient because well-meaning people with major life events at stake often remember differently later.

Adjustment to the above three potential problems is largely a matter of character. A person with character has the ability to continue to work effectively under difficult social and environmental conditions without becoming morose or embittered. In our experience, character is not associated with the researcher's sex or physical strength. Professors observing potential field workers in the pleasant confines of the campus often learn little about the student's character.

The fourth potential restraint on cooperative field research is the ability to communicate effectively with other researchers and people of the host country. In the case of Mikumi, communication between UCR and Africa often takes several weeks or more, and in many instances is complicated by letters crossing in transit. Since financial accounting and the setting of broad policy is done in California and the main expenditures and day-to-day site management are done in Tanzania, ample opportunities for misunderstanding cannot be entirely avoided. Problems occur that would be readily alleviated if parties from the two locations could sit down together and talk. It is important to be patient, to recognize the high likelihood that misunderstandings will occur, and to seek ways of keeping a normal amount of disagreement from escalating into irreconcilable differences.

Financial arrangements

If a site is continuously well financed, there is a better chance of selecting researchers and assistants who are both highly competent and suited to field life. Long-term sites are almost never continuously well enough financed to allow volunteers, who contribute significantly to continuity and the long-term data, to be turned down because of hunches about less than perfectly suitable personality traits. At Mikumi, we have usually been fortunate in the choice of researchers, paid assistants, and volunteers, though the record is less than perfect.

Because of the need to keep a long-term site continuously in operation, an inordinate amount of time must be devoted to obtaining financial support. Many of the most common sources of funding for primate field research either do not ordinarily give large enough grants to keep a site in operation for as much as an uninterrupted year or do not regard the advantages of site continuity as a sufficient argument even for granting less than that which the US National Institutes of Health will give under its small grants program. Consequently, much time and energy which could better be devoted to research and site management has to be used writing and rewriting new research proposals to support not the long-term research *per se*, but newly created short-term projects that last a year or two. Fortunately, whenever Mikumi was unsupported by one of these more generously funded shorter-term projects, a bare-bones operation was maintained through smaller, site-saving grants from the Leakey Foundation and/or UCR. With the unavailability of even a small amount of long-term financing, everyone involved in most long-term sites must be prepared to live with ulcer-inducing, never-ending uncertainty.

Conclusions

The history and experience of research at Mikumi National Park supports the view that cooperative research, though not without its difficulties, has many advantages. Cooperation involves relays of researchers who remain at the site for limited periods of time. Effort is needed to ensure comparability of long-term data collected by several observers and to introduce newly arriving researchers to life in a developing country. Conflicts can arise among researchers involved in projects which are major life events. Communication within the host country and between the field and the home unit can sometimes break down. Potential sore points can be avoided or healed if they are anticipated, recognized, and not allowed to fester, in which case the advantages of a cooperative site will far outweigh the disadvantages. Such a site is a practical basis of long-term research from which the connecting threads of the dynamics of change can be revealed. It facilitates broadened training opportunities and the growth of a continually strengthened data base necessary as background information without which many important lines of research could not be effectively pursued.

Acknowledgments

Research at Mikumi National Park evolved from planning started in 1971 at the Gombe Stream Research Centre whose Director, J. Goodall, introduced us to Mikumi and has given invaluable advice and support ever since. Mikumi research is a long-term effort to which many individuals have contributed, including A. Burdick, S. Charnley, H. Doak, L. Eltringham, P. Ender, D. Forthman, H. Hendy, C. Kidungho, H. Klein, R. Johnson, C. Lunn, T. Maple, F. Muroto, G. Norton, H. Pennington, H. Quick, D. and K. Rasmussen, J. Perlsweig, J. Phillips-Conroy, W. Roertgen, J. Rogers, F. Siwezi, S. Smith, A. Starling, E. Sterling, R. Stillwell, S. Wasser, B. and H. Westlund, D. Wilson, and G. and R. Wynn. Over the years, Mikumi research has been supported by grants from the National Institute of Mental Health, the National Science Foundation, the Harry Frank Guggenheim Foundation, the Leakey Foundation, Biomedical Science Support grants to the University of California, Riverside, Intramural and Intercampus Opportunity grants (Riverside campus), and the Center for Social and Behavioral Sciences Research (Riverside campus). For the opportunity to work at Mikumi, we are grateful to the people and leaders of Tanzania, particularly to the late Hon. D. Bryceson, and to the leadership of the Serengeti Research Institute, the Tanzania National Scientific Research Council, and Tanzania National Parks. Suggestions of an anonymous reviewer are appreciated.

References

Hendy, H. (1978) The effects of age, sex, and troop differences on the social interactions of free-ranging baboon infants in their first six months of life. Doctoral thesis, University of California, Riverside

Klein, H. D. (1983) Paternal care and kin selection in yellow baboons (*Papio cynocephalus*). Doctoral thesis, University of Washington, Seattle

Rasmussen, D. R. (1978) Environmental and behavioral correlates of changes in range use in a troop of yellow (*Papio cynocephalus*) and a troop of olive (*P. anubis*) baboons. Doctoral thesis, University of California, Riverside

Rasmussen, K. L. R. (1980) Consort behaviour and mate selection in yellow baboons (*Papio cynocephalus*). Doctoral thesis, University of Cambridge, Cambridge

Rhine, R. J. (1981) The baboons of Mikumi. *L.S.B. Leakey Found. News*, **19**, 7–8

Wasser, S. K. (1981) Reproductive competition and cooperation: general theory and a field study of yellow baboons. Doctoral thesis, University of Washington, Seattle

Part II

Primate thinking

II.1

Introduction: the human primates' theory of the primate mind

D. K. CANDLAND AND R. C. KYES

Introduction

The purpose of this paper is to establish a theme for the section, by analyzing a new development in the history of intellectual thought that has made it now sensible to discuss 'primate thinking'. Some may find audacious the title 'Primate Thinking', but we are united in the belief that nothing is gained and much may be lost by dismissing forever concepts from mentalistic viewpoints regarding how organisms organize their experiences and memories. Although these concepts became shopworn and were removed from analyses of behavior in the earlier part of this century, the contributors are convinced that today it is worthwhile to ask: What can we say about the primates' theory of mind?

During the last three centuries, approximately since the contributions to epistemology of Kant (1781 (1964)) and the contributions to zoology of Cuvier (1863), Geoffroy-Saint-Hilaire (1818), Linnaeus (Frängsmyr, 1983), and Lamarck (1809 (1963)), two models of the mind have guided our thinking about animal behavior. These models derive from the philosophical concept of 'theories of mind'. Said too simply, I have a theory of how your mind works, just as you have a theory of the workings of mine. By theory is meant a set of testable hypotheses. My theory of your mind is that it is very much like mine in the ways in which perceptions are categorized. Otherwise we could not communicate, even imperfectly. Your mind is probably unlike mine in the salience given different aspects of our experience. As human primates, you and I could not act without carrying with us a theory of mind about other persons' minds and, as human primates, we are compelled to force these theories of mind upon our understanding of the behavior of other species.

It is unwise to stray too far from the fact that any theory we have of animal behavior is categorized, filtered, and interpreted through our own ways of knowing. We human beings are unlikely to make substantial progress in understanding animal life until we have first understood the nature of the categories that we impose upon our perceptions of other species.

It is worthwhile to sort the differences between the two models – the two theories of mind – that we have suggested as dominating our thinking, for powerful claims are being made about the theory of mind used by nonhuman primates. Such theories of the primate mind are being made by claims of intra-species and intra-primate communication, by calls for a return to the use of models of the mind that emphasize consciousness and ideation, and by reports of self-awareness in some, but not all, nonhuman primate species.

Two aspects of the human theory of mind need detain us briefly. The first is how categories come to be organized and the second is what we have learned about how we human beings relate categories to one another.

The question of whether categories of perception are organized by experience, or whether experience organizes the categories of perception, is the fundamental question that distinguishes studies of primate thinking. Kant (1781 (1964)), whose analysis of the question set the issue for all studies of behavior that followed, analyzed *a priori* and *a posteriori* categories. The former, characterized by the categories of Time and Space, are *a priori* because they are innate: we cannot imagine anything outside the categories of Time and Space. *A posteriori* categories can be learned: we built the concept 'yellow' or 'length' from experience, but these categories themselves cannot escape the grasp of Time and Space. Any theory of mind meets, first, the constraint imposed by our inability to think beyond *a priori* categories and, secondly, the question of how we judge experiences from the aspects of different categories.

Fechner (1860), working in the century after Kant, saw the significance of this question in the development of what came to be known as the Psychophysical Law. In modern times, Stevens' (1957, 1961) contributions toward deciphering the psychophysical relation between mind and physical environment led to the demonstration that our way of judging experience is to relate categories and events by means of a power function. We human beings make ratio judgements about the relation among objects, events, and categories. We think of such relations in log–log ratios. If we may summarize our basic knowledge about human beings' contemporary theory of the human mind, we find two aspects to be outstanding: the presence of both *a priori* and *a posteriori* categories, and the idea that we human beings relate categories to one another logarithmically.

By examining aspects of the human theory of the human mind, we are now prepared to examine the two models of the animal – and primate – mind that guide our thinking. The views we hold of the animal mind are not alternatives; their principles overlap. It is this overlap that confuses us when we try to sort out the nature of our theory of the primate mind. The views have in common the recognition of *a posteriori* categories as means by which our experiences are organized and thereby learned. I have attempted to demonstrate the relation between the two viewpoints drawn by Table 1.

I have arranged the table with the left-hand column asking questions of an epistemological nature. The first of these concerns the kind of data we select; the second the unit by which we interpret the data; the third how we account for variation, error, or misjudgement within our model of the mind; the fourth is the major hypothetical construct or principle we use in our view of the primate mind; and the last, do we ask of the animal what it can do or what does it know?

It is the final question, that of strategy, that most clearly distinguishes the two theories of mind. When we ask, 'What can the animal

Table 1. *Two Human Theories of the Primate Mind*

Issue	Theory I	Theory II
Data are used to compose a theory of mind	Behavior	Behavior + hypothesis about mental states
The *unit* we use is	a relationship, such as stimulus–response or IRM–FAP	a relationship, such as stimulus–response or IRM–FAP + some mental state, e.g. ideation, representation, intentionality
We explain *variation* by	stimulus or response generalization	the mental state
The major *principle* we use is	examples of learning	examples of performance
The *experimental strategy* is to ask	What can the primate learn to do?	What does the primate know?

learn to do?', we are asking a question about capacities as they are expressed by behavior. A characteristic study is one in which the primate's ability to discriminate hues is determined by its making a behavioral response in the presence of one hue, but not the other. The behavioral response is used to demonstrate the animal's capacity to discriminate.

The second view yields a strategy in which the animal is asked to 'describe what you know'. How can the animal so describe without making some behavioral response? How can we study what the animal knows or has learned without requiring evidence of learned behavior? Seemingly, we arrive at the conclusion upheld by Theory I that we are limited to studying behavior, to describing units of behavior, to placing these units into a framework by investigating how such units are learned – thereby accepting the conclusion that we can only know what the animal has learned, not what it knows.

View II accepts the challenge to determine how to bypass the singular use of learned behaviors, such as pressing a bar, to determine the extent of what the animal knows. It does so in two ways. First, it posits schema or models such as ideation, representation, and intentionality. These schema are regarded as not necessarily beyond the reach of explication through analyses of behavior, for behavior can be used to establish the schema. In this section, both Dr Singh (Chapter II.3) and Dr Rumbaugh (Chapter II.2) demonstrate ways in which such schema can be assessed by attending to the intentional aspects of animal communication. Both papers demonstrate how behavior can be interpreted in ways that help us appreciate the primate's theory of mind.

View II also holds, if not requires, a place for the concepts of consciousness and self-awareness, as the presentations of Drs Asquith (Chapter II.5) and Reynolds (Chapter II.4) demonstrate. When such concepts are used to explain behavior, they may be troublesome if not misleading. When they are used cautiously to set forth the categories of the mind, we may include the findings within our theory of the mind – both our theory of our own mind and our theory of the minds of our fellow primates.

This section, Primate Thinking, explores ways in which some aspects of mentalism can be applied to an understanding of the *a priori* and *a posteriori* nature of the categories. The contributors are aware, as we all are, that misuse of the concepts from mentalism has occurred, but they are convinced that the view that animals know 'nothing but' learned, *a posteriori*, categories may mislead us by encouraging us to overlook what is characteristic of the primate – and other animal – mind; namely, the ability to think within and among categories in patterns that are lawful and regular.

References

Cuvier, G. (1863) *The Animal Kingdom.* (trans. W. B. Carpenter and J. O. Westwood) London: Bohn

Fechner, G. T. (1860) *Elemente der Psychophysik.* Leipzig: Brierkopf & Härtel

Frängsmyr, T. *Lineaus: The Man and His Work.* Berkeley, California: U. California Press, 1983

Geoffroy-Saint-Hilaire, E. (1818) *Philosophie Anatomique.* Paris: Baillière

Kant, I. (1781) *Critik der Reinen Bernunst.* Riga: Hurchnoch (*Critique of Pure Reason*, trans. N. K. Smith, London: Macmillan, 1964)

Lamarck, J. B. P. A. de M. (1809) *Philosophie Zoologique.* (*Zoological Philosphy, an Exposition with regard to the Natural History of Animals*, trans. H. Elliot, New York: Hafner, 1963)

Stevens, S. S. (1957) On the psychophysical law. *Psychol. Rev.*, **64**, 153–81

Stevens, S. S. (1961) To honor Fechner and repeal his law. *Science*, **133**, 80–6.

II.2

Animal thinking – by stimulation or simulation?

D. M. RUMBAUGH

Introduction

The answer to the question, 'Can animals think?', can be found only through thought, and thought is, in a sense, a concept of itself. Little wonder that definite answers about thought, whether it be in human beings or animals, elude us. Nonetheless, it is important that we address the question, 'Can animals think?' It is necessary for us to do so because of our evolutionary perspective. If human beings are said to think, then to the degree that we find life forms that approximate us as do the great apes, we must ask the questions, 'What are the behaviors that we use for concluding that human beings are thinking?', 'Which of these behaviors are found in animals?', and 'In what animals are they found?'

One obvious requisite to 'thinking' as a process is knowledge. To be able to think without 'knowing' about the world seems unlikely at best. 'Knowing' is part of cognition, although cognition might entail more than just 'knowing'. Cognition is generally presumed to include the processes which relate units of knowledge and manipulate them so as to produce new options for behavior; new decisions, as it were.

Cognition as a term is not universally useful to psychologists. It is excluded as a scientific term from the descriptive and experimental work of radical behaviorists. Radical behaviorism (Skinner, 1938, 1950, 1953) rejects all notions of cognition and explains behavior solely in terms of its consequences (i.e. contingencies of the environment). It rejects all notions that the organism is a knowing, thinking agent of initiative. Thus if one is a radical behaviorist, one is unlikely to be concerned with whether or not animals can think in the literal sense of the word.

How is it that a chimpanzee can learn that television portrays scenarios and avails stimuli, or information, as to what is going on elsewhere or otherwise beyond view? Chimpanzees can, for example, readily extract a morsel of food beyond direct sight by watching a television monitor which portrays the food that they are after and the movement of their hand, shielded from direct vision (Menzel & Savage-Rumbaugh, 1985). They are able to respond to televised portrayals of where food is cached, where persons are hiding, which container of several holds prized food, and so on. Also, they prefer to watch football and wrestling rather than the 'soaps' and news. They even have their favorite movies, such as 'Kong'. Why?

Such behaviors would strongly suggest that the chimpanzee knows a great deal about how things in his world work, what causes what to happen and what must be done to accommodate the unknown or unexpected. Thus when Menzel & Savage-Rumbaugh inverted the picture seen by chimpanzees on the monitor, where they had only that picture to direct their hand to the location of an otherwise out-of-sight incentive, what did they do? When the image was inverted with reference to top and bottom, both Sherman and Austin (*Pan troglodytes* subjects) made prompt adjustments. Sherman's was, perhaps, the most concrete. He positioned himself to look at the monitor through his legs, which in effect made the picture 'right-side-up' to him! When previously taped sessions were presented along with the real-time video feedback of hand movement in relation to the location of banana slices (out of sight), the chimps devised a clever method for differentiating them. (The old tape presentation was of no value in the current effort!) They would initiate some novel response, such as waving a hand, then pick the monitor displaying that response for their keen attention as they continued attempting to get the incentive. Despite continuing change, the chimpanzees were able to adjust, even when the images were inverted up–down and right–left. And the task was to them a totally novel one, not obviously related to problems that their ancestors faced in the forests. These behaviors suggest 'thinking' – a problem was perceived, and a novel compensation made to adjust to it.

A review of current perspectives and data on animal cognition is provided by Roitblat, Bever & Terrace (1984), a book resulting from the 1982 Guggenheim conference on Animal Cognition. This book serves to update a previous volume by Hulse, Fowler & Honig (1978). In the introduction, Terrace (1984) makes clear that the assertion of 'cognition' does not imply either consciousness or mentalism. Rather, it is a model that generally posits some form of representation, something

Animal thinking – by stimulation or simulation?

to which the animal can refer so as to behave adaptively and economically.

What kinds of evidence strongly suggest that there is knowing and thinking in animals in the research of those interested in questions of animal cognition and thinking?

Shimp (1976) reported that pigeons could be queried as to which of two keys they had pecked first and which second. A white 'X' was projected on one of three keys in an irregular sequence. The pigeons pecked each 'X', which turned off that key and led to the reappearance of another key, either in the same or in a different position. Pecking the center key initiated the 'question' by turning that key off and turning on the keys on either side. The center key was either red, blue, or white. As it turned off, the side keys came on with the same color which the center key had just had. Red called for the answer as to which key was pecked first in the prior sequence; blue called for the second key pecked; and white center and side keys called for the third key pecked. Shimp reported better than chance performance on all three questions, with the highest accuracy being in the declaration of the most recent key peck. In one sense, then, the birds both knew and remembered which key they had pecked first, second and third.

Olton & Samuelson (1976) and Olton (1978, 1979) demonstrated that in a radial maze, where each arm was baited with food, rats only infrequently re-entered arms at the ends of which they had previously eaten in a given session. Control procedures demonstrated that spatial memory was the basis for this remarkable behavior. Of significance is the fact that the rats did not exhibit repetitive patterns of exhausting the choice of all possible arms in a given session. If not all food was consumed in a given arm, the rats tended to re-enter later, as though they remembered that something remained to be eaten.

Do pigeons 'know' what they must do to get the task done? Do they know what the task requires of them, despite unpredictable changes in the task?

Straub & Terrace (1981) reported intriguing data from pigeons in a simultaneous chaining task. Pigeons learned to peck in order four stimuli (A, B, C, D) presented simultaneously on keys, with the arrangement varied from trial to trial. On test trials where novel arrangements of keys were presented, the pigeons' results were well above chance. This finding was taken to mean that the pigeons had some kind of representation of the sequence in which the keys were to be pecked. Terrace (1984) also has reported that if pigeons are taught to peck three stimuli (A, B, C) in a fixed sequence, regardless of the arrangement on the panel, they apparently learn something about the ordinal position in which a given key should be pecked. This was determined by putting the middle stimulus of the series, B, into new sequences for the pigeons to learn – X, B, Y; B, X, Y; and X, Y, B. The pigeons learned the first of these three tasks very rapidly, an indication according to Terrace that the birds had some 'knowledge' about the stimulus, B, being the one that is pecked second.

Terrace's interest and support for the development of animal cognition is of note because he studied with B. F. Skinner, the father of radical behaviorism, and because of his shift from Skinner's view to something new. The question is, how new, how different is what Terrace says from what Skinner is saying?

Skinner emphasizes stimulus control over operant responses and does not include representations by or in the organism, human or animal, for explaining behavior. Terrace does and makes an interesting statement. He acknowledges the importance of environmental stimuli in the control and generation of behavior. He also acknowledges that to extend '. . . the study of stimulus control to stimuli that the organism generates poses a variety of technical problems. However, from a conceptual point of view, there is no basis at present to distinguish qualitatively between environmental and self-generated stimuli' (1984, p. 20). (Radical behaviorism does not refute the importance of internal stimuli, but it does not allow for self-generated stimuli.)

Allowing for cognition, the question arises, 'Is animal cognition the same as human cognition?' Most assuredly, the answer is that there are some similarities, especially between the great apes and ourselves, but there are surely differences as well. The varied routes of evolution have surely served to select proclivities for both topics of interest to learn and think about, and how they are learned and remembered.

Expectancies

Expectancies were advanced by Tolman (1966) to account for why rats were sensitive to changes in the quality of incentives. One of the more interesting studies that bears upon expectancies, which suggests some form of 'knowing', has been reported by Capaldi & Verry (1981). Rats were given five runs of a maze each 'trial'. On either the first or the last run the rats received 20 pellets of food and no pellets on the remaining runs of a given trial. Thus the rat might obtain 20 pellets on the first run and zero on the next four. Or it might receive zero pellets on the first four runs and then 20 on the fifth run. What was the finding? If the rats received the 20 pellets on the first run, they ran more slowly on the remaining runs of that trial. If they received nothing on the first four trials, they ran more quickly on the last trial. Running speeds were different primarily on the last run of each trial. It would appear that the rats were running in appropriate relation to whether they had been rewarded on the first run or whether the pellets were yet to be obtained on the last. The rats' running time on the last run of each trial suggested that they 'knew' whether or not they would be rewarded.

Bever (1984) seeks to differentiate 'descriptive reductionism' from 'representational reductionism'. The former asserts that science should use weak (simple) rather than strong (complex) descriptions and explanations wherever possible. Representational reductionism asserts that 'an animal organizes each newly acquired behavior with the most concrete mechanisms available' (Bever, 1984, p. 62). Bever proceeds to point out that 'descriptive reductionism is not always the correct move . . .' (p. 62). Animals do not necessarily pay homage to Lloyd Morgan's canon in terms of how and what they learn. In other words, they do not necessarily learn, know, or think in the simplest terms or through the simplest processes.

Is it totally clear that animals have cognitive foundations for that which has been learned? Not necessarily, D'Amato & Salmon (1984) found that the ability of cebus monkeys to execute learned conditional discriminations with auditory sounds (tunes) and visual comparison stimuli was remarkably disrupted by turning the house light in the test apparatus from off to on. Hulse, Cynx & Humpal (1984) reported that although starlings, like humans, generalize frequency patterns within the range used in training, they do not do so when the change is in the order of an octave – a change for which humans would have no problem. Although the perception of 'primitive rhythm and pitch structure' is demonstrable in starlings, it is constrained relative to that of humans. I, too, have noted that the so-called complex learning-set skills of apes and monkeys are remarkably brittle when the nature of the test apparatus is changed; something that one would not expect if there were really a knowledge base of a human type underlying the mastery manifested by the animals. But that is exactly the point. Animals are not humans. Their cognitions, if extant, are surely both different and more circumscribed than in humans.

Categorical learning

Categorical learning can be used to support cognition in animals, although not necessarily so. As Herrnstein (1984) states in his review of categorical learning,

> Human language may depend on categorization, but the evidence shows clearly that categorization does not depend upon language. Generalization of discriminative stimuli shades into categorization as the contingency of reinforcement is associated with actual objects (or representations of them), rather than the usual abstracted lights or tones of psychological research . . . Categories that are fairly easy to describe physically do not seem to be significantly easier for animals, and may even be harder, than categories that are hard to describe in those terms. For example, pigeons appear to find patches of colored light a harder category to form than photographs of trees. (Herrnstein, 1984, pp. 257–8)

Memory

For manifest behavior to be attributed to learning, clearly something must be remembered. The structure and processes of memory are all inferred, just as are those of learning. Studies of animal memory can give us, perhaps, a unique perspective regarding their cognition and thought relative to that of humans.

One of the most robust phenomena to be obtained from humans as they attempt to learn and recall lists of things, such as names, numbers and pictures, is that items at the beginning and at the end of lists are easier than are those in the middle. Why is this so? Reasons for this phenomenon, known as the serial position effect, have been of major theoretical interest.

Sands *et al*. (1984) present an interesting summary of the serial position effect as observed in animal memory research. As they state, the information processing model, advanced by Atkinson & Shiffrin (1968), attempts to account for the effect in terms of the differential sensitivity of ordinal positions in a list in relation to the processes of short-term memory, the rehearsal processes of which are important for rooting items in long-term memory. Items at the beginning and end of lists are sheltered, relatively speaking, in that they are either the first or the last and do not have competition from other items in the limited and brittle limits of short-term memory.

Sands & Wright (1980) devised a clever method for studying the serial position effect in a rhesus monkey. They gave the monkey

10- and 20-item lists of pictures. At the end of each run a 'probe' picture was presented, and the monkey was to declare, through the execution of two different responses, whether it was the SAME as one of the pictures on the list just seen or whether it was DIFFERENT. Both the monkey and a human did appreciably better if the probe picture was one that had been either near the beginning or near the end, rather than in the middle of the list. These findings show remarkable similarity in at least one aspect of memory for two primate forms.

An experiment with the chimpanzee Lana (Buchanan, Gill & Braggio, 1981) revealed that her free recall and reproduction of lists of up to eight words were remarkably similar to those expected of a young child. In addition, she tended to recall words in clusters, based either on word class (such as foods *vs* objects *vs* color names) or on the colored backgrounds of those words (which were faithful within a class; foods and drinks were red, animates were violet, etc.). Human data frequently reveal clustering in memory. (See also Medin, Roberts & Davis (1976) for more on animal memory.)

Spatial memory

Menzel (1978) reported on spatial learning by the chimpanzee and how its search behavior is contingent upon its friends' 'coming along'. His studies demonstrated clearly that the chimpanzee readily learns the locations of food cached in a large compound. The chimpanzees were carried individually about the compound and were shown where the food had been, or was, cached. Subsequently, the chimpanzees were able to go to all of the food sites on the basis of minimum effort; regardless of the specific point of release, they traversed an efficient route to obtain all of the food. It would seem that they knew the field in detail and could conclude (through thought?) how they should travel on each trial, even though quite unique from all previous trials.

Numerical attributes of stimuli

Church & Meck (1984) discuss evidence that indicates '. . . the number of successive events (stimuli or responses) can serve as an effective stimulus for behavior, even when all temporal cues are counterbalanced or held constant' (p. 445). Fernandes & Church (1982), for example, trained rats to press one lever if two sounds were presented and to press another lever if four sounds were presented. The duration of each sound and the time between sounds were empirically discounted as variables which controlled the rats' differential response to the two levers. Accordingly, they concluded that the controlling variable was the number of sequential stimuli, two or four in this study. Other work revealed that there was cross-modal transfer (from hearing to vision) of number. The procedure entailed substituting lights that flashed in lieu of the equivalent number of sounds. Rats that were shown a number of light flashes corresponding to the number of sounds for which they had learned to press different levers did better than those for which there was a reversal of the relation of number of flashes to number of sounds (i.e. they were to press the left lever when four flashes were seen although they pressed that lever in previous training when two sounds were heard and vice versa). This finding serves to support the conclusion that some abstract representation of number served to control the rats' choices of levers on a given trial. (See Ettlinger (1977) for a perspective and review of 'cross-modal' research.)

Given that rats have the capacity to use relative numbers of stimuli to determine which response is appropriate, is there evidence that animals can use specific numbers to represent an appropriate quantity of items? The answer to date is that the evidence is, at best, weak; however, there is some evidence that at least chimpanzees sum two quantities of food items so as to select the greater total when given a choice. The chimpanzee subjects, Sherman and Austin, were given their choice of the left or right pair of food wells on a tray with four wells, each well containing from zero to four candies or chocolate drops (Rumbaugh *et al.*, unpublished). Procedures ensured that their choice of one pair of wells or the other was based on the summation of items in that pair, rather than on whichever pair had the single well with the greatest number of food items. (Choice was indicated by their placement of a finger up to the rim of any well so as to extract the food, whereupon the portion of the tray that held the two adjoining food wells not chosen was withdrawn from reach.) Choice for the greater summed value was better than 90%. Five candies or chocolates were then included in the array of numbers to be assigned to the wells. This provided novel arrays which Sherman and Austin had never seen in prior training and testing (e.g. three and four items in the left two foodwells and one and five in the right two foodwells). Their selection of whichever pair of wells obtained for them the greatest number of food items remained well above 90%. This is not to say, however, that they were adding numbers in the same sense that we add Arabic numbers. Nonetheless, they seemingly showed the capacity for summing pairs of quantities to reach the same conclusion that would be obtained if one were adding in the numeric sense. Whether they can learn to use Arabic numbers in lieu of the physical foods in varying amounts is under study.

Davis & Memmott (1982) reviewed the literature on 'counting behavior' in animals and concluded that although they can discriminate numbers of things, there is no strong evidence that they can show cardination or count. In fact, there is not even any weak evidence of consequence that would support that conclusion. This is not to deny that rodent mothers have ways to 'keep track' of the number of infants in a litter. They do not do so by counting, however. They do so by relying on stimuli (ultrasound in the case of the rat) or simply by exhausting 'the reservoir' by retrieving and transporting the young to a nest until the last check reveals that they are 'all gone'.

It is clear that there is remarkable congruence between biology, behavior and ecological pressures and demands. Consider the plight of Clark's nutcracker (*Nucifraga columbiana*), a bird that may cache up to 33000 seeds! How can the bird subsequently find the seeds, especially when it must find them or die during the long harsh winters endured in its high alpine environment?

The answer is provided, in part, by work reported by Balda & Turek (1984). To be able to retrieve food cached at earlier points in time requires that in some manner the bird 'checks off' sites as food is taken *in toto* from them, else it would return for something no longer there. This is similar to the Olton effect – rats generally do not return to arms of the maze from which they had exhausted the food.

It would seem that inherent to research designs addressing questions of animal cognition and thinking, there are unique test trials for which the animal has no direct experience upon which it can base its efforts to adapt or upon which it can necessarily select the correct and appropriate alternative.

When Clark's nutcracker recovers the thousands of seeds previously cached, is it 'knowing' and 'thinking' in the same sense that Sherman and Austin do when they alter their postures and tactics at the television monitors? Surely not. In the case of the chimpanzee, there must be much more openness. In other words, the chimpanzee can be innovative in its behavioral adaptations over a greater range of problem situations than can birds and other animals. Once again, consonant both with evolutionary theory and with the tenet of comparative psychology, it would appear that there is a continuum from essentially no cognition and/or thinking, to a level where there is some, but which is very delimited by biological or genetically dictated proclivities.

Then there are other levels, such as those of the great apes and humans, levels where there is, by comparison, lesser amounts of biological or genetic constraint. And along this continuum we become somewhat more inclined to talk about animal intelligence.

What animals can be clever about is stimulated by their biology and past experience. The 'simulation' studies, which are favorites of the radical behaviorists, bring forth approximations of behavior that behaviorists with cognitive inclinations hold to suggest the operations of advanced psychological processes, including those of thought. The simulation study typically includes reinforcing the 'target' behavior, that is, the behavior presented by other researchers as indicating advanced cognitive operations. Although such studies have some value in demonstrating the power of reinforcement (generally agreed to) and the cleverness of the experimenter as a teacher, these data serve primarily to cloud the issue in sorting out what animals are able to do by themselves, as it were, because of advanced processes which are theirs because of the animals they have evolved to be. One might, in fact, discern 'cognition', 'thinking' and 'insight', even in pigeons, but one cannot 'put them there' through conditioning them from a null point as one can do with a response, so long as it is in the animal's repertoire.

In closing, I propose that observational learning paradigms will be of great value in studying animals' thinking processes. To the degree that by observation alone they can acquire an arbitrarily selected behavior (such as smoking a cigarette by apes), one has a window to their thought processes not availed through other approaches. (This is not to say, however, that this is the only type of paradigm which should be employed in this area of research, which is only now beginning to quicken its pace and acquire respectability.)

Acknowledgements

Preparation of this paper was supported by NIH grants RR-00165 and HD-06016 to the Yerkes Regional Primate Research Center of Emory University.
Portions of this paper were part of the author's G. Stanley Hall lecture, presented at the American Psychological Association's Convention, Toronto, 1984.

References

Atkinson, R. C. & Shiffrin, R. M. (1968) Human memory: a proposed system and its control processes. In *The Psychology of Learning and Motivation*, Vol. 2, ed. K. W. Spence & J. T. Spence, New York: Academic Press

Balda, R. P. & Turek, R. J. (1984) The cache-recovery system as an example of memory capabilities in Clark's nutcracker. In *Animal Cognition*, ed. H. L. Roitblat, T. G. Bever & H. S. Terrace, pp. 513–32. Hillsdale, NJ: Lawrence Erlbaum Associates

Bever, T. G. (1984) The road from behaviorism to rationalism. In *Animal Cognition*, ed. H. L. Roitblat, T. G. Bever & H. S. Terrace, pp. 61–75. Hillsdale, NJ: Lawrence Erlbaum Associates

Buchanan, J. P., Gill, T. V. & Braggio, J. T. (1981) Serial position and clustering effects in a chimpanzee's 'free recall'. *Memory and Cognition*, 9 (6), 651–60

Capaldi, E. J. & Verry, D. R. (1981) Serial order anticipation learning in rats: memory for multiple hedonic events and their order. *Anim. Learn. Behav.* 9, 441–53

Church, R. M. & Meck, W. H. (1984) The numerical attribute of stimuli. In *Animal Cognition*, ed. H. L. Roitblat, T. G. Bever & H. S. Terrace, pp. 445–64. Hillsdale, NJ: Lawrence Erlbaum Associates

D'Amato, M. R. & Salmon, D. P. (1984) Cognitive processes in cebus monkeys. In *Animal Cognition*, ed. H. L. Roitblat, T. G. Bever & H. S. Terrace, pp. 149–68. Hillsdale, NJ: Lawrence Erlbaum Associates

Davis, H. & Memmott, J. (1982) Counting behavior in animals: a critical evaluation. *Psychol. Bull.* 92 (3), 547–71

Ettlinger, G. (1977) Interactions between sensory modalities in nonhuman primates. In *Primatol.*, vol. 1, ed. A. M. Schrier, pp. 71–104. Hillsdale, New Jersey: Lawrence Erlbaum Associates

Fernandes, D. M. & Church, R. M. (1982) Discrimination of the number of sequential events by rats. *Anim. Learn. Behav.*, 10, 171–6

Herrnstein, R. J. (1984) Objects, categories, and discriminative stimuli. In *Animal Cognition*, ed. H. L. Roitblat, T. G. Bever & H. S. Terrace, pp. 233–61. Hillsdale, NJ: Lawrence Erlbaum Associates

Hulse, S. H., Cynx, J. & Humpal, J. (1984) Cognitive processing of pitch and rhythm structures by birds. In *Animal Cognition*, ed. H. L. Roitblat, T. G. Bever & H. S. Terrace, pp. 183–98. Hillsdale, NJ: Lawrence Erlbaum Associates

Hulse, S. H., Fowler, H. & Honig, W. K. (eds) (1978) *Cognitive Processes in Animal Behavior*. Hillsdale, NJ: Lawrence Erlbaum Associates

Medin, D. L., Roberts, W. A. & Davis, R. T. (eds) (1976) *Processes of Animal Memory*. Hillsdale, NJ: Lawrence Erlbaum Associates

Menzel, E. W., Jr (1978) Cognitive mapping in chimpanzees. In *Cognitive Processes in Animal Behavior*, ed. S. H. Hulse, H. Fowler & W. K. Honig, pp. 375–422. Hillsdale, NJ: Lawrence Erlbaum Associates

Menzel, E. W. & Savage-Rumbaugh, E. S. (1985) Chimpanzee (*Pan troglodytes*) spatial problem solving with the use of mirrors & televised equivalents of mirrors. *J. Comp. Psychol.*, 99, 211–17

Olton, D. S. (1978) Characteristics of spatial memory. In *Cognitive Processes in Animal Behavior*, ed. S. H. Hulse, H. Fowler & W. K. Honig, pp. 341–73. Hillsdale, NJ: Lawrence Erlbaum Associates

Olton, D. S. (1979) Mazes, maps, and memory. *Am. Psychol.* 34, 583–96

Olton, D. S. & Samuelson, R. J. (1976) Remembrance of places passed: spatial memory in rats. *J. Exp. Psychol. Anim. Beh. Processes*, 2, 97–116

Roitblat, H. L., Bever, T. G. & Terrace, H. S. (eds) (1984). *Animal Cognition*. Hillsdale, NJ: Lawrence Erlbaum Associates

Sands, S. F., Urcuioli, P. J., Wright, A. A. & Santiago, H. C. (1984) Serial position effects and rehearsal in primate visual memory. In *Animal Cognition*, ed. H. L. Roitblat, T. G. Bever & H. S. Terrace, pp. 375–88. Hillsdale, NJ: Lawrence Erlbaum Associates

Sands, S. F. & Wright, A. A. (1980) Primate memory: retention of serial list items by a rhesus monkey. *Science*, **209**, 938–40

Shimp, C. P. (1976) Short-term memory in the pigeon: relative recency. *J. Exp. Anal. Behav.*, **25**, 55–61

Skinner, B. F. (1938) *The Behavior of Organisms*. New York: Appleton-Century-Crofts

Skinner, B. F. (1950) Are theories of learning necessary? *Psychol. Rev.* **57**, 193–216

Skinner, B. F. (1953) *Science and Human Behavior*. New York: Macmillan

Straub, R. O. & Terrace, H. S. (1981) Generalization of serial learning in the pigeon. *Anim. Learn. Behav.*, **9**, 454–68

Terrace, H. S. (1984) Animal cognition. In *Animal Cognition*, ed. H. L. Roitblat, T. G. Bever & H. S. Terrace, pp. 7–28. Hillsdale, NJ: Lawrence Erlbaum Associates

Tolman, E. C. (1966) Operational behaviorism and current trends in psychology. In *Behavior and Psychological Man: Essays in Motivation and Learning*, ed. E. C. Tolman, pp. 115–29. Berkeley: University of California Press (Reprinted, originally published in 1936)

II.3

Primate consciousness

M. SINGH

A new perspective

About 8 years ago, Anthony Shafton published a very thought-provoking essay. He formulated certain hypotheses concerning conditions of awareness and the involvement of subjectivity in the social adaptations of man and other primates. Unfortunately, the essay of Shafton has gone unnoticed by many primatologists. I shall here make an attempt to evaluate some of Shafton's hypotheses regarding primate consciousness in the light of field studies on Indian primates. I shall first build up a perspective to talk about primate consciousness, and then make inferences about the cognitive side of their behavior. From here onwards, all citations, unless otherwise mentioned, are from Shafton (1976).

In general, man is considered unique because of his unique psychological characteristics – the states of consciousness. If the other human characteristics are thought to have their phylogenetic homologs in infrahuman primates, human consciousness must also have its own natural history. 'If man's plasticity of action depends upon subjective processes, and if there is a gradient of plasticity among animals, then the possible involvement of subjectivity factors in observable behavior should be studied across species' (p. 13).

Non-human primates, being taxonomically the most evolved Order, are endowed with a very high amount of encoded genetic information. The ecology, the habitats and the niches of primates are highly diverse and complex, which consequently leads to a high and volatile input of environmental information. The social environment of primates is not only complex but also highly diverse. During prolonged socialization, various alternative pathways of development are available. The routine life in a group of primates is so intricate that the inbuilt neuronal codes only serve as nodes for integrating perceptions. Behavior is an integration of information from various sources. Such complexity and diversity of information, both from within and outside, would require a highly plastic perceptual process with a continuation in consciousness and perception.

Consciousness: the concept

Perhaps nothing has been muddled more in psychology than the definition of consciousness. In animal behavior, after an initial spurt by Kohler and Yerkes, the cognitive aspect of primate behavior has remained almost ignored. Now, when we are reviving interest in the subjective side of primate behavior, we have to be very cautious because the reasons for which radical behaviorism rejected the study of consciousness have not yet totally disappeared. Even in the case of the study of human consciousness, Natsoulas (1978) warned, 'When we turn to the psychology of consciousness, subjected as it has been to a recent long period of scientific neglect, we are well advised therefore to begin by exploring the phenomena normally included under consciousness independently of any single theoretical approach or specific scientific ideology'. My point is that at this stage, it will be too much to speculate about categorization and other such abstractions by the primate mind. I propose what we first must try to establish is that non-human primate behavior, especially the day-to-day social interactions, is not simply stimulus–response connections, instinct–reflex events or fixed genetic adaptations. There is involvement of subjectivity, conscious decision making, a relative freedom and intentionality in primate actions.

This brings us again to the definition of these mental states. To quote Griffin (1982a), 'While scientists quite properly attempt to define whatever terms or concepts they use, precise definitions of mental processes and experiences are almost imposssible for the simple reason that we know so little about the phenomena we seek to define'. On the other hand, even if we have a vague idea about a mental process and try to understand it in whatever way possible, the obtained knowledge may help improve the very definition of the process itself.

Since we can speculate about the mental states of non-human primates only by observing their overt behavior, we should care more for what a conscious act would be like rather than consciousness itself. On the basis of prevailing definitions of conscious behavior, and the general phenomena brought under consciousness, without claims of adding anything new, I vaguely define an act of a non-human primate to be a conscious act (of course, with involvement of various degrees of consciousness) if (1) it shows intra- and interindividual variation in everyday social contexts; (2) it apparently involves intentionality, volition and freedom, in spite of its being non-iconic and an indicator of the emotional and motivational states of the actor; and (3) at some level, the action indicates a representation of the self.

Primate consciousness: an evolutionary perspective
Assuming that consciousness, a cognitive capacity, is subject to natural selection, a perspective for its evolution in primates can be constructed as follows:
1. Primates, the most evolved Order, were capable of making adaptations to a wide variety of environments.
2. Differential ecological pressures selected for intraspecific variations in social behavior. At the same time, the 'newly adaptive' factors (Wilson, 1975), such as large size, arboreal life and diurnality, selected for complex but flexible sociality.
3. Intraspecific variations set the pressures and selected for plasticity in the behavior of individuals.
4. Individual plasticity of behavior, and interindividual variation in behavior, coupled with high intelligence of primates, led to a further delicate and intricate social system based on a subtle form of hierarchical organization.
5. Pressures were then set by this kind of social system itself for the genetic selection for a mental ability to judge accurately the actions of others, and to make one's own actions variably, depending upon the requirements of each social situation. This kind of flexibility in overt performance further selected for the PLASTICITY OF PERCEPTION which became *a priori*, and then collateral, to the emergence and evolution of consciousness.

Social communication and consciousness
In a series of writings, Griffin (1976, 1978, 1981, 1982*a,b*) proposed that social communication of a species can help make inferences about their inner cognitive processes. Since communication is one aspect of primate sociality which exhibits the maximum amount of plasticity and flexibility, I intend to look for the involvement of consciousness in social communication. One incidental advantage of this approach is that we can use a large body of data already

available, collected by those scientists who did not consciously look for consciousness in the actions of non-human primates.

Before going to the detailed analysis, I shall first make an attempt upon a different kind of classification of primate signals, which may help the exploration of subjectivity in their actions. The idea behind this classification is to range the signals from purely innate to relatively cognitive, with involvement of more and more plasticity at various levels.

1. Signals that are species-specific action patterns.
2. Signals that may be ritualised behaviors but show interhabit, intraspecific variation.
3. Signal systems that may have been learnt through environmental conditioning by some individuals, but are apparently passed on to others without their undergoing the process of conditioning themselves.
4. Signal systems that apparently involve consciousness.
5. Signal/response systems that apparently involve self-awareness.

The signals in category 1 are fixed and common to all members of a species. Some amount of plasticity is entered in category 2 because of habitat variations. Category 3 refers to 'tradition' in signal systems within a habitat. The signals in category 4 indicate 'subjective self awareness' (Duval & Wicklund, 1972) where the individual focusses its attention on events external to the individual's consciousness, personal history or body. The signal/response system in category 5 indicates 'objective self awareness' where 'Consciousness is focussed exclusively upon the self and consequently the individual attends to his conscious state, his personal history, his body, or any other personal aspect of himself' (Duval & Wicklund, 1972; p. 2).

Primate signalling systems

Let us now search for consciousness in the signalling systems of non-human primates. I want to state here specifically that for this analysis the information from any of the field studies conducted in India by several primatologists could be used, but I will restrict myself to the bonnet macaques of South India, for the following reasons:

1. In most of the field studies, the picture of a species' communication, though concise and insightful, is an overall picture. I need to use more detailed information on individual signals and their sequential combinations – something like Altmann's (1965) analysis of rhesus communication.

2. We have attempted a detailed individual signal study on bonnet macaques (Singh & Prakash, 1981; and other unpublished data) and so I have first-hand information on their communication.

I also wish to point out here that the delineation of messages in the signals is not an easy job, and can many times even be misleading; therefore I shall be talking only about those signals which are simpler in nature and show apparent orientation.

There is no need to talk about the first two categories of signals, and I shall start with the third category, showing some amount of plasticity, and move to the signal systems apparently involving consciousness.

'Tradition' in signals

The bonnet macaques inhabiting roadsides or villages, like other primates, use mostly visual signals in their communication. Since they also inhabit forest areas, I often used to wonder how effective the visual signals could be in thick foliage. But then we found that the bonnets in forests used vocal signals with a much higher frequency. We also soon discovered another adaptation – while the monkeys were on the ground, they would hardly use vocal signals and would communicate through visual signals only. The use of vocal signals on the ground becomes even less than the normal use made by monkeys outside the forest area. Since our study area is infested with tigers, panthers and wild dogs, the avoidance of use of vocal signals is undoubtedly to avoid these ground predators. This process can be easily explained in terms of past learning and environmental conditioning, but the important point is that even the very young individuals follow the same pattern where attracting the predators by making sounds is not a very common and everyday event as far as monkeys are concerned.

(Examples of *tradition*, much more complex than the one mentioned above, can be given for any primate species.)

Signals and consciousness

Assuming that awareness or consciousness is a social adaptation, Shafton argued that the extent of awareness is significantly influenced by the social rank of the individual. Since vision plays a very important role in primate communication, the 'rank contingent freedoms and constraints influence the performance of overt visual orientation at the grade of monkeys' (p. 126). A subordinate must look at a dominant monkey to adjust his own spatial distance and movement accordingly. But in the process of looking at, he also, unwantingly and unavoidably, emits a threat signal. This puts constraints upon the expansion of his field of perceptual experience. On the other hand, the 'dominants have perceptual freedom of space as well as freedom of movement in space' (p. 49). The dominant individual, by overcoming the constraints of the kind imposed on subordinates, has achieved freedom through a high level of perceptual as well as physical force. A major reward of high status is consciousness, achieved by the existential choice involved in a rank reversal when arousal constraints must be overcome by courage to sustain consciousness in spite of fear generated by radical changes in the neuronal model. 'The primitive homolog of existential freedom at the evolutionary grade of monkeys is a rise of status' (p. 128).

The behavior of the subordinate monkey is often emotional and instantaneous as, because of the perceptual constraints, they react to the *social stimuli* without any DELAY. The dominant individual, on the other hand, shows a lot of restraint, and this 'restraint allows him to suspend consummation of integration, assimilate complex information, achieve necessary revision of codes, resolve conflicts between codes, and so to improve his perceptual competence still further and thereby secure his status position' (p. 62). I believe that the presence of DELAY in responding overtly is a very strong indicator of involvement of subjectivity. The dominant monkey *thinks* before it acts.

Take the example of two episodes in macaques – the 'protected threat' and 'interference' in fights.

During protected threat situations, the threatened monkey, crying and screeching, approaches a dominant monkey and then threatens back the aggressor. The threatened monkey makes repeated shifts in elaborate submission signals towards the dominant and threats towards the aggressor. The aggressor stands still or continues to make mild threats. In spite of so much escalated aggression and social stimuli around, the dominant monkey merely self-scratches and only occasionally 'looks at' the aggressor. It is very rare that the dominant monkey actually attacks the aggressor.

Similarly, when a fight erupts between two subordinate monkeys, the behavior of the dominant monkey shows so much restraint and elaboration in making subtle moves that you feel he is very 'thoughtful' about it. The bonnet dominant male, in such situations, first mildly shakes branches from a distance even without looking towards the fighting monkeys. If the fight continues, he makes a slow walk towards them, repeatedly sitting and looking away. He then makes a 'growl' sound, approaches the fighting monkeys, and sits close to them, often looking in the opposite direction. He frequently indulges in self-scratching – apparently showing relaxation and lack of tension. Only on rare occasions does a dominant monkey make an actual attack on one of the fighting monkeys.

In both instances, the behavior of the dominant monkey does not conform to the rule of releasers, S–R conditioning patterns or 'ritualization'. The animal seems to be 'aware' of the whole situation, and behaves in a 'planned' and 'calculated' way, indicating the involvement of thinking.

Consciousness in infants and juveniles

Since infants and juveniles are not bound much by the strict dominance hierarchy of adults, situations can be shown in them which apparently involve consciousness.

In primate communication systems, the context-specific meaning of a signal is a well-known phenomenon. The common view, shared by Shafton, was that in monkeys the grasp of the context is on the part of the receiver. My observation and inference is that even at the monkey level there is a grasp of the context *by the signaller*. In bonnet macaques, as in many other primate species, the play signals are the same as those used during agonistic interactions. However, when an animal goes to induce play in the other, in addition to the usual aggression signals, the initiator adds one of the many discrete signals, such as leaf chewing, bent neck or jerky locomotion, to convey the 'meaning' of his message. But if several animals are already at play, and the initiator approaches a non-playing animal, he uses only the usual play/aggression signals. In the first instance there is no context and so the signaller adds an extra signal; in the second situation there is a context of play and the signaller does not add the extra signal as if he knows that the other animal can now understand the meaning of his signals. The grasp of the context on the part of a signaller, his intentionality of actions and the voluntary emission of a signal are apparent in this interaction. In other words, intentionality in, and voluntary emission of, a signal and the grasp over the context, indicating consciousness, are the aspects of the same cognitive process and are necessarily simultaneous products of evolution.

Sustenance of consciousness and 'neuronal model'

According to Sokolov's (1963) 'neuronal model', 'arousal' or consciousness is produced by a mismatch between moderate 'novelty' and 'significance' of a stimulus and the existing neuronal codes formed earlier by genetic information, developmental impressions and ordinary perceptual integration. The question is 'whether in a given species consciousness must diminish on average as the neuronal model fills out during life span. We intuitively know this not to be the necessary case; as we know that consciousness is often sustained by new additions and integrations of information' (p. 34).

Take for example the case of bonnet monkey communication: While analysing the ontogeny of communication signals, we divided their developmental period into several stages, not on the basis of age alone, but also on the basis of striking changes in overt behavior. The ontogeny of communication indicated three remarkable changes at different stages of development:

1. New signals were being added to the repertoire.
2. Some signals were being dropped.
3. Some signals, already in use for something else, were being given new meanings.

The point I am trying to make is that each developmental stage itself is perceived as 'novel' and capable of producing arousal.

Later in sub-adult and adult stages, the struggle for dominance rank is a continuous process, and a relative rank, as already mentioned, has its own influence on perception. In other words, the neuronal model hardly ever fills out, novelty and significance in social stimuli continue, and arousal or consciousness remain sustained throughout life.

References

Altmann, S. A. (1965) Sociobiology of rhesus monkeys. II. Stochastics of social communication. *J. Theoret. Biol.*, **8**, 490–522

Duval, S. & Wicklund, R. A. (1972) *A Theory of Objective Self Awareness*. New York & London: Academic Press

Griffin, D. R. (1976) A possible window on the minds of animals. *Am. Scientist*, **64**, 530–5

Griffin, D. R. (1978) Prospects for a cognitive ethology. *Behav. Brain Sci.*, **1**, 527–38

Griffin, D. R. (1981) *The Question of Animal Awareness*. New York: Rockfeller University Press

Griffin, D. R. (1982a) Introduction. In *Animal Mind – Human Mind*, ed. D. R. Griffin, pp. 1–12. New York: Springer-Verlag

Griffin, D. R. (1982b) Animal communication as evidence of thinking. In *Language, Mind and Brain*, ed. T. W. Simon & R. J. Scholes, pp. 241–50. London: Lawrence Erlbaum Associates

Natsoulas, T. (1978) Consciousness. *Am. Psychologist*, **38**, 906–14

Shafton, A. (1976) *Conditions of Awareness: Subjective Factors in the Social Adaptations of Man and other Primates*. Portland, Oregon: Riverstone Press

Singh, M. & Prakash, P. (1981) A study of protolinguistic communication system in macaques. Mysore: Central Institute of Indian Languages (Mimeographed)

Sokolov, E. N. (1963) Higher nervous functions: the orienting reflex. *Ann. Rev. Physiol.*, **25**, 545–80

Wilson, E. O. (1975) *Sociobiology: The New Synthesis.* Harvard: Belknap Press

II.4

Primate social thinking

V. REYNOLDS

The most successful explanations of primate social behaviour have been those based on the theory of natural selection. Such theories emphasise the advantage of particular kinds of behaviour for individuals. They emphasise competition between individuals, coupled with cooperation in situations of kin selection as described for instance for the Japanese macaque by Kurland (1977), or reciprocal altruism as described for instance by Packer (1977). It seems likely, therefore, *a priori*, that social thinking in primates which underlies social behaviour will also be best explained in ways which refer the processes of thinking to the phenomenon of natural selection.

A major difference between thinking and behaviour is that the former is much less energetically costly than the latter. Thus if an animal can consciously distinguish at the mental level between a series of alternative courses of action and select the optimal one, it will be saving energy as compared with the animal that tries various alternatives in action and thereby expends energy in the form of various kinds of behaviour. Thus in multiple-choice situations, thinking would be selected for if it could produce the correct situation or better solutions at less cost.

Where do we see evidence of thinking? Quiatt (1984) has referred to a number of such situations. For a start there is the situation in which a female rhesus monkey or some other such species of macaque sits close by another female who has an infant and grooms her, subsequently taking the infant from that mother and holding it herself, perhaps even keeping it for some time. Such baby-snatching and the events preceding it show a considerable amount of forethought and planning on the part of the snatching monkey, together with undoubted evidence of intentionality and perhaps even deceit (Reynolds, 1962).

de Waal (1982) gives numerous examples of carefully planned manoeuvres in chimpanzees. For example, we have the process by which one animal will instigate a quarrel by manoeuvering its opponent into a situation in which it is likely to be attacked by others. Dominant males will studiously ignore the efforts of subordinate males to impress them in some way. Subordinate males wishing to mate and not to incur the hostility of more dominant males will mate clandestinely by positioning themselves so they cannot be seen. A dominant male, in order to obtain the subsequent support of females in a group, will act subordinately towards them. A rival male wanting to achieve dominance will punish females for associating with a dominant male. A socially experienced female, after observing a protracted quarrel between males, will intervene and bring about a reconciliation between them.

Such examples show how, particularly in chimpanzees, social life takes on very complex forms. It is possible, by prolonged careful observation, to see that animals are by no means acting on the spur of the moment but are engaged in quite long-drawn-out social processes in which a great deal of social thinking must be necessary in order for the observed coherence over time to be possible. Such long-drawn-out social relationships are not only seen in chimpanzees but in very many primate species and I have, for example, described such processes in a colony of rhesus monkeys (Reynolds, 1962).

One question that arises concerns the ability of the animals to draw the necessary fine distinctions, for instance to appreciate the status of others in the group. I have argued (Reynolds, 1970) that it is not necessary for monkeys to appreciate each other's status as such. It is only necessary for each to be in some kind of dyadic interaction with others. The resulting status hierarchy will then be a natural outcome of the efforts made by each individual to maximise its own status. I was, however, puzzled by the consistent efforts of one particular female to mate with the dominant male as opposed to the other males in the group. She was unsuccessful in this; nevertheless she persisted despite numerous bites and other unfortunate results. Had such a female been trying to maximise her reproductive success she should surely have mated with one of the other males. It did seem as though she were motivated by a desire to associate with the male of highest status; the phenomenon of association by oestrus females with males of high status is well documented. The question remains, however, whether such females are directly motivated by a desire to associate with a male of high status, or whether they are attracted by some of the overt signs given by males of high status. Such males may, for instance, appear more relaxed, or walk in a particular way, sometimes also holding their tail up while subordinates hold their tails down. Are these the stimuli to which the female responds, or is she responding to some awareness of status *as such*? Before drawing a conclusion, I think more work is needed to establish the relevant criteria for making a decision.

A second important issue here concerns the question of introspection and the ability to impute motives to another monkey. Gallup (1982) has made much of this in his distinction between mindful behaviour and mindless behaviour. He has collected evidence that apes can understand both their own and others' motives, whereas monkeys cannot. Certainly in the case of chimpanzees the work of de Waal (1982) indicates quite a good understanding of the motives of others. Males do seem to know if another male is trying to achieve power and they take steps to make this impossible. Females too rally round a male who is dominant and is having some difficulty with a subordinate rival. They support the dominant male, but if he becomes too aggressive they deny him their support. Clearly in such cases the animals cannot articulate the processes in which they are involved but the ways in which they behave do give an indication that their understanding of the social situation is well developed and that they are by no means acting in a mindless way. Whether Gallup is right in denying the existence of mindful behaviour to monkeys is a subject for further research. Certainly, however, we can see evidence from societies of vervets, baboons and macaques that indicates the existence of similar processes to those found in chimpanzees.

We need constantly to remind ourselves that any kind of linguistic representation is out of the question as regards non-human primates. It is all too easy for us, with our well-developed linguistic ways of thought, to attribute meanings and motives to animals which arise from our own experience of life in human society. As Harris (1984) has pointed out, the meanings in monkey society must arise from the animals themselves, not from any language. How such meanings might appear in the mind of non-human primates is a matter for debate and some people would regard this debate as futile. Leiber (1984) has, however, for the record, distinguished clearly between two forms such thinking might take. First it might be iconic, in which the animal actually has representations of other animals and of particular situations in some kind of form in its brain and it imagines situations in some sort of realistic way. Secondly, there could be abstract representations, not linguistic as in the case of man but in some other form. It seems likely that we shall never know the answer to such questions. However we should bear in mind the point made by Asquith (1984) that when we describe primate behaviour we must always describe it anthropomorphically simply because we have to use our own human language for the description.

Again we should recall that there is no collective rule system in primates and therefore no sociology of primates in the human sense of the word. This does not mean of course that there is no culture in primates: most groups of primates have some cultural features, but behaviour in non-human primates is not bound by social rules in the way it is in humans. There is no book to which monkeys can refer for guidance in social situations, nor is there a body of laws that gives clear indications of when behaviour has to be punished. The dos and don'ts of non-literate societies that are passed on by word of mouth are likewise not found in non-human society. Nevertheless it can be argued that just as much human behaviour is transmitted by imitation learning, so it is in the case of non-human primates. At this psychological level, there is certainly a good deal of overlap between the human and non-human cases.

Thus, what we have in the case of non-human primates are psychological processes consisting of a variety of motivations and their associated behaviours arising out of a particular genetic substrate which is in turn the result of natural selection. For instance, the alliances between two subordinate baboon males described by Packer (1977), by which they can defeat a dominant male and gain access to a female, could be seen as a direct outcome of the enhanced reproductive success such behaviours would bring. The fundamental rule that natural selection would work with in this case is the rule that 'two minds can outwit one'. This should not be construed as any kind of cultural rule, so much as a natural rule, a phenomenon of nature to which we have given the name reciprocal altruism with all the genetic implications of that term.

It is a result of natural selection that we find animals following this rule, but we still need to consider the question: do the animals know the rule? And if they do know it, in what way do they know it? It is perhaps easier to assume that the animals concerned are acting under genetic compulsions which are realised in their brains as psychological motives which they may find easy to express in action. This would imply some kind of facilitated learning, a process well known to animal behaviour students. Nevertheless there is evidence of careful positioning, planning and coordination in any given instance and it seems impossible to accept that all this could be the result of relatively simple compulsions. Clearly there is a lot of cognitive work involved as well and, although the explanation in terms of compulsion has the virtue of simplicity, the true explanation will undoubtedly involve us in attributing to the animals a lot of cognitive mentation of quite complex kinds.

Harré (1984) has contrasted the 'automatic-causal' with the 'cognitive-intentional' framework of analysis. They are quite distinct: in the former, interactants simply react to their perceptions of each other and of the situation; in the latter each has a conscious mental idea of what its goals are and perhaps also what are the goals of others. The latter framework seems appropriate to baboons. For example, Bachman & Kummer (1980) have performed experiments on hamadryas baboons in which a resident pair in a cage is approached by a new incoming male. This work shows that the second male's behaviour depends on his assessment of the female's preference for the first male. This indicates the kind of mental processes involved in making decisions and ultimately in performing various kinds of behaviour by primates. We can only understand such cases in terms of rather careful cognitive assessments of many facets of the situation by the individuals concerned.

It is certainly clear to any student who has observed primates in social situations that primate social life involves many problems and conflicts. For example, a rhesus mother with an infant may be offered grooming by another, more dominant, female but she runs the risk that the latter will subsequently snatch her baby. Likewise a female may chase off a rival female to keep her away from a dominant male, but she runs the risk of being punished by the male for her efforts. A dominant male even runs some risk of being toppled if he becomes too aggressive. These conflicts in primate social life are quite clear to the student who watches a group of primates over a long period. It is possible to see each particular individual caught in a network of relationships often containing problems that are not resolvable. In the case of many animals that leave their natal group, it is quite possibly the fact that they are caught up in unresolvable situations that leads them to make the move. This kind of explanation fits quite well, for instance, with the data on gorillas, chimpanzees and macaques.

Many of the conflicts and problems in a primate society arise precisely because it is a thinking society. In a thinking society all have to think: non-thinkers are outwitted, out-reproduced and thus selected against. But the price of thinking is *worry* and it is quite clear that many primates in their own societies live worried lives. This in turn leads to stress and a number of studies, such as those of Candland & Leshner (1974), have shown different stress hormone levels of primates in different social positions. Such studies bear close comparison to studies of human children, e.g. those of Montagner *et al.* (1978), which show more or less exactly the same thing. In adult humans the situation becomes so much more complex that it is difficult to draw comparisons.

Thus there are common features underlying human and non-human behaviour, even though the characteristic behaviour repertoire of each species is distinct from that of other species. The *form* in which status struggles and other such processes take place is therefore different in each species: chimpanzee facial expressions differ from those of baboons or rhesus monkeys and of course in our own species we have our own particular kinds of non-verbal expressions, gestures and so on. Nevertheless the underlying processes in various species may well be similar. Processes which have been called 'appeasement' (de Waal, 1982), 'equilibration' (Chance & Mead, 1953), 'protected threat' (Kummer, 1957), and 'bandwagon' (Russell & Russell, 1968) may well occur in the societies of many species. All such processes concern the integration of the individual into the group, the pursuit of centrality and ultimately reproductive success. de Waal has demonstrated numerically that the social manoeuvering of males is concerned with gaining and holding the dominant status and that this in turn is correlated with the highest frequency of copulation in the group.

Before ending, a note should be made of the part played by kinship in social thinking. Datta (1983) has produced evidence of a process of careful assessment by younger, high-born rhesus subordinates over their low-born older dominants. That is to say, a younger rhesus monkey who is born in a higher lineage will assess the chance of being able to defeat a more dominant, older member of its social group who was however born in a lower-ranking lineage. These younger high-born monkeys appear to assess the extent of support they can expect to receive from allies in their high-ranking lineage before challenging the older, lower-born monkeys. Support from kin is even more important in the case of dominant males, who need allies to avoid defeat by subordinate monkeys who have allies. Thus it is as necessary for a dominant animal to have allies as it is for a subordinate one, and hence also the formation within groups of certain species of primates such as macaques, of rival sub-groups based on lineage.

We should also note the importance of grooming in monkey social thinking. Besides the support one can expect to receive from one's kind, which can be explained on the basis of kin selection, monkeys endeavour to obtain preferred treatment and support from other members of the group by grooming them. Again we could say that it is simply an automatic outcome of grooming that support is received, and that grooming is not done *with a view* to obtaining benefits. As before, however, the observed data make this unlikely. Monkeys and apes appear to choose very carefully which animals to groom. In rhesus monkeys it has been observed several times that individuals will make efforts to groom more dominant animals, often incurring a certain amount of nervousness which shows in their rather rapid and jerky hand movements. Why should they bother to groom more dominant animals? The answer seems to be that they are prepared to put up with the nervous tension engendered because of the prospect of future support. Smuts (1983) has shown for the baboons at Gilgil that preferred grooming partners will intervene on behalf of each other in any dispute. Where grooming occurs between the sexes, there is also the prospect of a future reproductive reward. As Hinde (1983) has pointed out, the phenomenon of grooming shows also that monkeys can classify one another by rank and by kin.

In conclusion we can see that primates are capable of complex forms of social thinking and that this social thinking tends to revolve around the two central areas of status and reproductive behaviour. These preoccupations are, of course, common ones in the social life of all species of animals. The difference in the case of primates is that they alone appear to have brought the two processes into cognitive awareness and have added a new dimension of refinement to social behaviour as a result of this. There is probably a larger overlap between the processes of thinking in non-human primates and those of man than we have hitherto been inclined to think. Man's uniqueness is constantly being eroded by comparison with non-human primates, and it will be interesting to see how far this process eventually goes.

References

Asquith, P. J. (1984) The inevitability and utility of anthropomorphism in description of primate behaviour. In *The Meaning of Primate Signals*, ed.

R. Harré & V. Reynolds, pp. 138–76. Cambridge: Cambridge University Press

Bachmann, C. & Kummer, H. (1980) Male assessment of female choice in Hamadryas baboons. *Behav. Ecol. Sociobiol.*, **6**, 315–21

Candland, D. K. & Leshner, A. (1974) A model of agonistic behavior: endocrine and autonomic correlates. In *Limbic and Autonomic Nervous Systems Research*, ed. L. V. DiCara. New York: Plenum Press

Chance, M. R. A. & Mead, A. P. (1953) Social behaviour and primate evolution. *Symp. Soc. Exp. Biol.*, **7**, 395–439

Datta, S. J. (1983) Relative power and the acquisition of rank. In *Primate Social Relationships*, ed. R. Hinde, Ch. 6.7. Oxford: Blackwell

Gallup, G. (1982) Self-awareness and emergence of mind in primates. *Am. J. Primatol.*, **2**, 237–48

Harré, R. (1984) Vocabularies and theories. In *The Meaning of Primate Signals*, ed. R. Harré & V. Reynolds, ch. 5. Cambridge: Cambridge University Press

Harris, R. (1984) Must monkeys mean? In *The Meaning of Primate Signals*, ed. R. Harré & V. Reynolds, ch. 7. Cambridge: Cambridge University Press

Hinde, R. A. (1983) Triadic interaction and social sophistication. In *Primate Social Relationships*, ed. R. A. Hinde, ch. 9.1. Oxford: Blackwell

Kummer, H. (1957) Soziales verhalten einer Mantelpaviangruppe. *Schweiz. Z. Psychol.*, **33**, 1–91

Kurland, J. A. (1977) Kin selection in the Japanese monkey. *Contrib. Primatol.*, **12**. Basel: Karger

Leiber, J. (1984) The strange creature. In *The Meaning of Primate Signals*, ed. R. Harré & V. Reynolds, Ch. 4. Cambridge: Cambridge University Press

Montagner, H. *et al.* (1978) Behavioural profiles and corticosteroid excretion rhythms in young children. In *Human Behaviour and Adaptation*, ed. N. Blurton-Jones & V. Reynolds, pp. 209–66. London: Taylor & Francis

Packer, C. (1977) Reciprocal altruism in *Papio anumbis*. *Nature, London*, **265**, 441–3

Quiatt, D. (1984) Devious intentions of monkeys and apes? In *The Meaning of Primate Signals*, ed. R. Harré & V. Reynolds, ch. 7. Cambridge: Cambridge University Press

Reynolds, V. (1962) Social organisation of a colony of rhesus monkeys (*Macaca mulatta*). Ph.D. thesis, London University

Reynolds, V. (1970) Roles and role change in monkey society: the consort relationship of rhesus monkeys. *Man*, **5** (3), 449–65

Russell, W. M. S. & Russell, C. (1968) *Violence, Monkeys and Man*. London: Macmillan

Smuts, B. B. (1983) Dynamics of 'special relationships' between adult male and female olive baboons. In *Primate Social Relationships*, ed. R. Hinde, Ch. 6.9. Oxford: Blackwell

Waal, F. B. M. de (1982) *Chimpanzee Politics*. London: Jonathan Cape

II.5

Anthropomorphism and the Japanese and Western traditions in primatology

P. J. ASQUITH

The papers in this section consider how to conceptualize and describe the nonhuman primate mind. There are profound implications in this of course for reinterpretation and description of social processes in other primates. While there is a growing interest among Western primatologists in the mind of primates, Japanese primatologists by contrast have from the beginning proceeded with the assumption that they are observing thinking animals and there is a singular absence of interest in specifically this question.

In what follows I shall mention briefly those aspects of primate thinking being considered in the West and draw again the distinction between *behaviour*, *action* and *act* and their relationship to anthropomorphism. I will spend the greater amount of time relating some basic assumptions in Japanese primatology that, while not directly stemming from the assumption that they are observing thinking animals, are assumptions that could not perhaps be held *did* they feel the same reserve as many Western primatologists about attributing mind to animals.

Various aspects of how to conceptualize the primate mind being considered in the West can be divided into three categories. One is the justification for and method of inferring mind in animals; a second is suggestions for models of primate thinking, that is, how we will characterize or conceptualize it; and a third is how to describe behaviour, or rather, action (Reynolds, 1980) performed by thinking animals. In the last mentioned, goals and active manipulation of environment and individuals is assumed and it is asked 'How do we talk about them?'.

Excluding the experimental learning and conditioning studies by psychologists who have been working on animal mind for decades, but with a different slant (I wish to focus on naturally occurring behaviour in feral groups), the majority of recent writing on primate thinking falls into category one. The justification for assuming 'mind' in other animals, particularly primates and more particularly the Great Apes, is that given evolution, 'thinking' is not likely to stop at one species barrier and that it is simply 'bad science' to disregard a large area of the life of animals in the interests of strictly quantifiable data.

Experiments by which one can infer thinking in other animals have been devised by, for instance, Menzel (1975, 1979), who postulated that chimpanzees have goals or forethought as evidenced through communication of object-locations in a group of chimpanzees. Gallup (1977) has worked for over a decade testing for self-recognition in species of monkeys and apes with the use of mirrors. de Waal (1982) described the long-term strategies of three adult males of the Arnhem Zoo chimpanzee colony jostling for alpha position through soliciting the help of females and forming and breaking coalitions with one of the other two high-ranking males. Griffin (1976) addressed both justification for inferring that other species think and how to infer it. However, he stopped short of inferring emotions in other animals. This is interesting in view of the fact that Griffin, a specialist in echo-location in bats, is concerned to give objective criteria for inferring mental states, yet de Waal (personal communication) has remarked how easy it is to 'read' emotions in rhesus macaques for instance. Finally, language studies of apes, it was hoped, might offer a direct line to primate thinking. Although it appears to be self-evident to some of the language teachers that the apes are thinking, lying, joking, and so on (e.g. Patterson & Linden, 1981), controversy continues over what exactly the apes are exhibiting (Sebeok & Umiker-Sebeok, 1980; Terrace, 1984).

With reference to the above I want only to make the point that there is growing interest in the *acting*, and not simply the *behaving*, primate (e.g. Harré & Reynolds, 1984). The distinction between behaviour and action that Reynolds (1980) drew is that 'behaviour' refers to movement or physical patterns only, and 'action' to purposeful behaviour. Harré & Secord (1972) added a third dimension to behaviour and action which they named an *act*. The act dimension is concerned with the social context of (human) action: how it is intended, by the actor, to look to others, and how others interpret it. The act refers to the social meaning of a performance so that several different actions may constitute together the act of, for instance, a bid for power or reconciliation.

Comments on inferring thought in other primate species and consequent concern with anthropomorphism in terminology are confined to the behaviour and action aspects as defined above. Anthropomorphism occurs when the behaviour units, defined in terms of physical movements, and the more inclusive behaviour categories, such as 'dominance', 'submission', 'greeting', and so on, are named by ordinary language terms usually used to describe human doings, and become read or understood as action. I have written elsewhere (Asquith, 1984) that the semantic shift occurs as a result of a process parallel to metaphor. However, if *action* is accepted as occurring among other animals, I see no reason why *acts* should not also be attributed to them, especially in the complex social lives of primates.

As mentioned in Japanese studies, there is a notable absence of writing directed to 'the problem of mind in animals'. In a sense, Japanese primatologists have already gone ahead and reaped the fruits of studying social action in primates without feeling they had to edit out of their thinking and writing the 'awareness' of the animals they studied. This allowed, I would say, a far more realistic picture of primate societies, in the sense that from the beginning variability in behaviour among different groups of the same species was expected, in contrast to the species-specific stereotyped behaviour Westerners at first felt was unchanging and uniform among all groups at all times of a species.

Three points are introduced here. One is the bases in the respective intellectual heritages for Western reluctance to attribute mind to animals and for Japanese freedom from that reluctance. The second is the basis for the disagreement by the majority of Japanese primatologists with exclusive use of sociobiological or 'maximizing inclusive fitness' explanations for observed behaviour. The third point is the idea of 'species society' (*shushakai* in Japanese), the explanation or complete description of which is one of the major goals of Japanese primate studies.

On the first point, philosophers such as Midgley (1983) have written that Western scientists' worry over attributing mind to animals rests on a philosophical view that talking about other animals', or people's for that matter, feelings and thought commits us to claiming we know exactly how they feel. This *is* extreme, Midgley points out, for both humans and other animals, but it is a natural sympathy that lets us gauge pretty well what other people feel and which can be extended to other species.

I think, however, there is another source for the Westerner's reluctance. That is that we are still somehow attached to the idea, originating with Plato, of a hierarchy in nature, described as the Chain of Being. In this scheme the ideas of 'continuity' (i.e. in natural history the idea that all organisms can be arranged in one ascending sequence of forms) and 'plenitude' (i.e. in the universe the range of conceivable diversity of kinds of living things is exhaustively exemplified and no genuine potentiality of being can remain unfulfilled) profoundly affected ideas of evolution as late as the eighteenth century. Though our ideas of evolution have changed fundamentally, it is not only in the interests of scientific objectivity that we feel discomfort imagining, for instance, another species' discomfort.

There is nothing like a hierarchical conception of nature in Japanese traditional thought except perhaps in Confucianism, itself an imported philosophy, which stated that those animals that stood nearer to heaven, i.e. were taller, were the more exalted. In such a scheme, the giraffe and elephant are above humans.

Masao Kawai (1969) characterized Western thinking about man's place in nature as vertical, with God above man above animals, and that of Japanese as horizontal and fluid, with animals, man and gods being in some sense interchangeable. In Buddhism, a man can become an animal and an animal can become a god. Darwinian evolution was so offensive to many Westerners because it represented a change in the structure from top/down (God to man, or man made in the image of God) to bottom/up (man evolving from animals). In Japan, evolutionary theory introduced in 1877 was subject to none of the moral implications Westerners attributed to it (Asquith, 1981).

Originally, three points were said to characterize Japanese primate studies and Japanese themselves also cited these points. They were: provisioning, long-term studies and individual identification. Provisioning was first carried out on Koshima with the idea not to control nature but to adjust oneself to nature as it was found. In an early film made in the 1950s called 'Scientists and Monkeys', the scattering of sweet potatoes was first done in the forest along paths the monkeys crossed on Koshima. It was noted that the scientists were both friends and observers of the monkeys. Later, they provisioned at the one clear space on the island, a small beach, where observations were easier. Provisioning in this context is quite natural it was said. They were not controlling, but following nature.

The idea of long-term studies also involves following nature – to look closely and quietly and not to interfere – just to wait. To open your eyes, ears and heart to try to catch the signals released, not to squeeze the information out of nature. The attitude to nature is not that of reason, though reason obviously exists, but of feeling. Intuition plays a fundamental role in the choice of problems to solve in Western science of course, but within some preconceived theoretical framework. For the Japanese, intuition is a means to proceed to gather data, along with more experimental methods. If one looks for the preconceived framework in early Japanese studies, it is society and social structure. This was only possible to study through individual identification of the monkeys.

The aim was and is to grasp what position and what relationships with others each individual has in the monkey society. More than one primatologist has said that parallel phenomena to what occurs in Japanese society was sought in monkey society. When you come to Japan you get a *meishi* or calling card. It states your name, title, business or university and position in that place of work. When you meet someone you exchange *meishi*. On the professional level, what one does or what position one occupies in the society is more important than one's individuality or personality. Without exchanging *meishi* one does not know how to get along with the new acquaintance. One's identity is to a great extent one's group identity.

Status, relationship and kinship are given much attention. Therefore it was natural to look for similar phenomena in the monkey groups. That was the first question to be asked of monkey society – is there anything like status? The Japanese approach primate studies from a cultural rather than a biological point of view. After successfully provisioning monkeys, sociological studies blossomed and they sought to elucidate the diachronic and synchronic aspects of society in each group (and of each individual monkey). For Western primatologists, the approach has been biological; behaviour has been explained in terms of how it maximizes reproductive advantage and social structure in terms of adaptation to environmental conditions. This has been a single unifying theme in Western studies, though it is now being questioned by some people such as Richard (1981) and Gould & Lewontin (1979). Richard (1981) for instance, has written that the concept of adaptation has provided a flimsy theoretical base which seems successfully to have blocked serious consideration of any other theoretical propositions. In Japanese studies, although evolution of social structure is an important basic question, explanation in terms of reproductive advantage is rarely given. On the ecological side, of course, Japanese are also studying monkeys' adaptation to different environments; their endemic species after all lives in ecological zones ranging from sub-tropical Yakushima to the subfrigid forests at the northern limit of non-human primates in the world in northern Honshu. However, one primatologist pointed out that citing correspondence between a certain environment and a certain social structure is not an acceptable explanation. We are not saying that environmental factors never affect social life, he said, but that they are only one part of the story. Over long-term studies the Japanese have witnessed intertroop relations changing, choice of food species changing – in other words, an ecological use change occurring that can equally be said to be based on behaviour, not the other way around.

But there is more to primate behaviour than that. Most theories proposed in the West are considered too logical and simplistic by the Japanese. What else is there? In the words of Sugiyama (1980) in his book *Kogoroshi no kōdogaku* [Infant Killing Behaviour], in translation,

> On the one hand ethology has so far been working on the mechanism of the behaviour patterns (so-called instinct) inherent in man and animals. To the question 'Why do they do this?', ethology would answer that they are genetically programmed to do it and 'got the right of existence' in evolutionary history.
>
> Further, ecologists use such words as strategy. It explains in this instance (infant killing) that to prevent an excess of population even by killing infants is a strategy for survival and development of the species (and a strategy by an overtaking male to maximize his reproductive potential). I think both are correct, but in these explanations there is no room for living things. At least in mammals including primates each has its own motivation, thought and feelings and soul in its own behaviour.

Sugiyama is not entirely against sociobiological explanations for behaviour but they do not provide a sufficient explanation for the complexities of primate life.

The idea of soul or something like individuality in monkeys is by definition almost impossible to make objective. But that it is thought to exist is reflected in the proliferating monkey memorial services held each year at the Japan Monkey Centre, the Primate Research Institute in Inuyama, Osaka University, and, most recently, at the Awajishima Monkey Centre, Awajishima, where both experimental and field primatologists go to pray for the souls of monkeys that have died in the course of their researches. There is no parallel to this in the West (see Asquith, 1983).

I have mentioned the early recognition of variability or cultural behaviour among different groups of the same species by Japanese. That is the synchronic aspect of social behaviour studies. The diachronic or historical aspect of both the individual and the group is equally important and was almost wholly neglected by Western primatologists in their shorter-term studies. To the Japanese, current variability among groups and change in individuals and groups over time must be included for a complete picture of the dynamics of a society. That complete picture is the 'species society' which brings us to the third point.

The pioneer of Japanese primate studies, Kinji Imanishi, called the species society (*shushakai* in Japanese) the *specia*. This term appeared in a paper written by Imanishi (1960) for *Current Anthropology* but was not explained by him and was largely ignored as an unnecessary neologism (Reynolds, 1976). However, the third and now fourth generation of Japanese primatologists still refer to *specia* and it seems to me to be necessary to an understanding of their theme. *Specia* is not to be confused with species. The definition of a biological species is very simple: 'a group, the members of which can mate and produce fertile offspring'. Reproduction is the important element, as is evident in various explanations of primate social behaviour by Western primatologists. The *specia*, by contrast, is 'the society of a species'. It includes all the members of a species. It is not just a container, however; it has its own subjectivity. Each individual member is a member of society and contributes to maintain *shushakai* (species society). Thus the species itself makes a system consisting of its member individuals which support the system through their interactions.

Imanishi's theory really starts at the level of *specia*, not with members of the species. He said in his book *What is Evolution?* (1976) that 'a *specia* consists of more than one individual. I call a *specia* a society because it consists of many like individuals. The *specia* and its individuals are two separate entities, but inseparable. A *specia* has a particular identity which it bestows on each member. This identity is unchanging.'

Japanese primatologists study the *specias* of primates; Westerners study *species* of primates. This gives rise to very different assumptions about behaviour. A *specia* has a permanent unchanging identity over and above the synchronic and diachronic changes of individuals and groups. All the variability over time and place of these cultural individualistic animals must be identified before one can understand the overall system, hence the long-term 20–30-year studies. What the identity or species society is, has not been described yet in any one species of primate.

In a species, too, a social norm is expected (species-specific behaviour) but there is an unresolved tension between that and individual competition for reproductive success. I think a source of the holistic, societal and cultural orientation of Japanese studies is the idea of interindividual relationships supporting a *specia* identity.

Westerners have based many explanations for a species' behaviour on single group studies. Japanese study several single groups of a species in an effort to understand how they make a larger society of a species. Itani, Professor of the Laboratory of Human Evolution Studies at Kyoto University, has said that the individual's relationship to the group is not so important as his relationship to the entire *specia*. There are always some individuals (in Japanese macaques) living alone. The individual has been in a group, then left it, and then after 2 or 3 years will perhaps belong to another group. We think of *specia* and also the individual's life history. The *species* concept is concerned with elements only. The *specia* concept is concerned with interaction among individuals and the individual's life history. Westerners attempt to generalize from the individual to the species. Japanese primatologists are always thinking of the individual as a member of the *specia*. If we think about the individual we must always bear in mind the social structure because every individual belongs to the *specia*.

In earlier papers, Itani (1967) described the personality of boss monkeys of the Takasakiyama group. Although for 16 years he has not mentioned personality in English publications, he says one of his basic ideas is that personal variability becomes wider as you move from simpler to more complex primate species. In prosimians, the individual variability is very small. In Japanese macaques it becomes wider, in *Pan* even wider, and in man the individuality is even larger. Thus the variability range, how it has changed or transformed through these grades, is of interest to trace the origin of humanistic behaviour.

On the question of what is humanness there are many factors to consider – language, family, culture. Until now we have been searching for these things in other species. For instance, in monkeys we found culture; in chimpanzees, language (perhaps); nowhere have we found family. An important question is why is individual variability there? How does society support or allow it to exist? (Itani, personal communication.)

This year (1984) Professor Itani will receive the Royal Anthropologic Society of London's highest award – the Thomas Henry Huxley

Memorial Award. He is the first Japanese scientist to receive it since its inception in 1901. It is in recognition of his work by the world community of anthropologists. I would like to present Itani's views (stated at seminars at Kyoto University in September, 1982 and June, 1983) which have particular reference to primate thinking. He says our methodology (that is Kyoto University's Laboratory of Human Evolution primatologists) is the anthropomorphic method. It is the only method by which to solve sociological problems of primates. As long as study is limited to very high primates, the analogy should be effective. You make your own subjective interpretation when you see some primate behaviour and anthropomorphic inference can be made only through the individual identification method. By other methods you cannot get the depth applicable to that type of analysis. Many important problems about the Great Apes, especially sociological aspects, have not been answered. If you adhere to objective observation you have to avoid those important problems (e.g. suppression, domination, permission, tolerance, etc.). These can be better described by the anthropomorphic method. The same method as used in Japanese monkey studies has been applied to chimpanzee studies for almost 20 years now. To look for single units of behaviour is meaningless. The history of the individual in the group and the group history is necessary to understand behaviour. Westerners have in the past treated data as typical behaviour and tried to find patterns or species-specific behaviour. This is not the idea which Japanese primatologists have. We accumulate individual data to analyse the whole society of *specia*.

In conclusion, I would like to make the point that what has been thought to be a less rigorous approach in Japanese primate field studies is in fact not simply a cultural phenomenon of being 'close to nature' or 'not traditionally objectifying nature'. Rather, the Japanese approach is a carefully considered methodology. Four generations of primatologists, while being completely *au fait* with theories proposed by Western colleagues, still feel a need to include what Sugiyama called *living things* in the study. I think the fact that two symposia on primate thinking are included in the 1984 IPS Congress shows that that idea is not entirely foreign to Western primatologists.

Acknowledgements

Grateful acknowledgement is made to Professor Douglas Candland of Bucknell University for inviting me to participate in this symposium, to primatologists throughout Japan for making the research there possible and in particular Dr Y. Sugiyama and Professor J. Itani of Kyoto University, and to the Social Sciences and Humanities Research Council of Canada for finances to attend this Congress. The Japanese Ministry of Education (Monbusho) supported the research in Japan.

References

Asquith, P. J. (1981) Some aspects of anthropomorphism in the terminology and philosophy underlying Western and Japanese studies of the social behaviour of non-human primates. D.Phil. thesis, University of Oxford

Asquith, P. J. (1983) The Monkey Memorial Service of Japanese Primatologists. *RAIN*, Feb., No. 54, 3–4

Asquith, P. J. (1984) The inevitability and utility of anthropomorphism in description of primate behaviour. In *The Meaning of Primate Signals*, ed. R. Harré & V. Reynolds, pp. 138–76. Cambridge: Cambridge University Press

Gallup, G. R. (1977) Self-recognition in primates, a comparative approach to the bidirectional properties of consciousness. *American Psychologist*, **32**, 329–38

Gould, S. J. & Lewontin, R. C. (1979) The spandrels of San Marco and the panglossian paradigm: a critique of the adaptationist programme. *Proc. R. Soc. London*, **205**, 581–98

Griffin, D. (1976) *The Question of Animal Awareness. Evolutionary Continuity of Mental Experience*. New York: Rockefeller University Press

Harré, R. & Reynolds, V. (eds) (1984). *The Meaning of Primate Signals*. Cambridge: Cambridge University Press

Harré, R. & Secord, P. F. (1972) *The Explanation of Social Behaviour*. Oxford: Basil Blackwell

Imanishi, K. (1960) Social organization of subhuman primates in their natural habitat. *Curr. Anthropol.*, **1** (5–6), 393–407

Imanishi, K. (1976) *Shinka to wa nani ka*. [*What is Evolution?*] Tokyo: Kodansha (in Japanese)

Itani, J. (1967) Editorial. *Primates*, **8** (1), 89–90; **8** (2), 187–8

Kawai, M. (1969) *Nihonzaru no Seitai*. [Life of Japanese Monkeys.] Tokyo: Kawade-Shohoshinsha (in Japanese)

Menzel, E. W. (1975) Natural language of chimpanzees. *New Scientist*, Jan. 16, 127–30

Menzel, E. W. (1979) Communication of object-locations in a group of young chimpanzees. In *The Great Apes*, ed. D. A. Hamburg & E. R. McCown, pp. 359–71. California: Benjamin/Cummings

Midgley, M. (1983) *Animals and Why They Matter*. Harmondsworth: Penguin

Patterson, F. & Linden, E. (1981) *The Education of Koko*. New York: Russell & Volkening Inc.

Reynolds, V. (1976) The origins of a behavioural vocabulary: the case of the rhesus monkey. *J. Theoret. Soc. Behav.*, **6**, 105–42

Reynolds, V. (1980) *The Biology of Human Action*, 2nd edn. San Francisco & Oxford: W. H. Freeman & Co.

Richard, A. (1981) Changing assumptions in primate ecology. *Am. Anthropologist*, **83**, 517–33

Sebeok, T. A. & Umiker-Sebeok, J. (eds) (1980) *Speaking of Apes. A Critical Anthology of Two-Way Communication with Man*. New York: Plenum Press

Sugiyama, Y. (1980) *Kogoroshi no kōdogaku*. [Infant Killing Behaviour.] Hokuto shuppan. (In Japanese)

Terrace, H. S. (1984) 'Language' in apes. In *The Meaning of Primate Signals*, ed. R. Harré & V. Reynolds, pp. 179–207. Cambridge: Cambridge University Press

Waal, Frans de (1982) *Chimpanzee Politics*. London: Jonathan Cape

Part III

Primate Behaviour and Cognition in Nature

Introduction *D. M. Fragaszy and H. O. Box*

There has been a tremendous growth of interest in, and information about, nonhuman cognition in recent years (e.g. Hulse, Fowler & Honig, 1978; Griffin, 1982; Roitblat, Bever & Terrace, 1984). This symposium has afforded the participants an opportunity to consider the usefulness of cognitive concepts in understanding normal behavior in nonhuman primates. Such an enterprise must be interdisciplinary in outlook. Cognitive concepts have been defined and explored primarily by experimental psychologists working in laboratory settings. Yet many of the questions asked by behavioral scientists studying animals, particularly primates, in nature implicitly involve cognition. The use of space, dietary selection, and a myriad of other behavioral activities all depend to a large extent upon what individuals know about their physical and social environments. To quote Alison Richard (1981), 'Until we know more about the decision-making capacity of an animal confronting changed conditions, models predicting [behavior] in any but the most stable environments will be useless' (p. 526).

In these opening remarks, our purpose is to touch upon some general issues in cognitive research which are relevant to an interdisciplinary concern with natural behavior. These include the proper domains of cognitive research, some general characteristics of cognitive concepts, and the present and potential value of such concepts to behavioral primatologists. We approach this task in the belief, shared with Hinde (1982), that some of the most interesting developments in animal behavior take place at the boundaries between established disciplines. Fostering communication among scientists of different backgrounds is an essential step in achieving interdisciplinary advance.

Historically, inquiry into cognition in nonhumans has generally paralleled the study of human cognition. The study of representations of the external environment has been central (Roitblat, 1982). The computer metaphor of information processing is currently a dominant feature, and there has been a strong emphasis on studies of memory in a variety of forms, such as working, serial, and long-term memory. Researchers in animal cognition have also studied concept formation, judgements of similarity and difference, and representations of space, time, and quantity (Roitblat *et al.*, 1984). Most of the work on non-human primates, as well as on humans, has taken place in laboratory settings with only minimal recognition of the relation between the abilities under study and the demands placed on the individual in the real world (Neisser, 1976). However, there are some exemplary laboratory studies which have clear relevance to the behavior of the study species in nature (e.g. Menzel, 1973; Olton, Handelmann & Walker, 1981; Balda & Turek, 1984; Kamil, 1984). These studies show us that ecological validity in laboratory studies of cognition is a matter of perspective. The same point can be made for the notion of 'cognitive validity', if you will, in studies made on subjects in natural, unrestrained conditions. Among nonhuman primates, Menzel's (1966, 1969) studies of responsiveness to objects by free-ranging rhesus and Japanese macaques stand out as well-known examples of this point; others come from the literature on tool-using among wild chimpanzees (e.g. Teleki, 1975; McGrew, 1977).

A cognitive orientation is appropriate for the study of any behavior involving voluntary choice, decision-making, or purposeful action – all synonyms for behavior in which what the individual knows about its world plays an important role. This, of course, covers a tremendous range of behavior in primates; too great a range for the scientist reasonably to expect a precise statement of governing processes relevant to the entire range. At this point, a cognitive orientation provides a general heuristic framework, not a monolithic well-developed model, in which to consider behavior. Consequently, work in cognitive phenomena in nature is scattered across journals and disciplines. We hoped to remedy that situation in a small way by the collection of several reports on relevant topics in one place. We have singled out four topics for inclusion in this symposium in which we believe a cognitive orientation is particularly useful. These areas are: patterns of resource exploitation (Andrews: Chapter III.1); temporal and spatial mapping abilities (Sigg: Chapter III.2); instrumental activities (Gibson: Chapter III.3); and attentional capacities and social learning (Whitehead: Chapter III.4; Box & Fragaszy: Chapter III.5). Research in these areas seems to be on the verge of providing some new directions in behavioral primatology.

General characteristics of cognitive concepts

Like all conceptual entities, cognitive terms can be used in a variety of ways (Honig, 1978). It is best to clarify at the outset what their functions can be. One common use of cognitive terms is descriptive, i.e. to summarize in an abstract way the behaviors which need explanation. For example, in Olton's work with rats (e.g. Olton & Samuelson, 1976), a rat is placed on a central platform of a maze which has eight arms, each baited with one piece of food, radiating from the center in a sunburst pattern. The rat is free to move in the maze, entering or re-entering any arm in any sequence. The general outcome is that the rat enters each arm once before it re-enters any arm, much more systematically than expected on a random basis. Furthermore, the rat is equally efficient on each trial when it is placed in the maze several times in succession. Olton uses the term 'working memory' to summarize the rat's ability to search the arms of the maze systematically on each trial. Note that this term does not explain the behavior; it merely labels an inferred process underlying it.

At a more theoretical level, cognitive terms can describe a hypothetical structure which organizes an observed behavior. The term 'cognitive map' is generally used in this way. When Menzel's chimpanzees searched a field for hidden objects, they followed a route which was shorter than the route they had experienced as they watched the objects being hidden (Menzel, 1973). The term 'cognitive map' describes the cognitive entity the use of which would produce this outcome. Note that the term 'cognitive map' labels the entity but does not tell you about its characteristics.

The main purpose of the use of these terms in current laboratory research is not to affirm the presence of purely cognitive entities, which most researchers now acknowledge, but rather to determine the functional characteristics of the states or processes to which they refer. For example, in Olton's radial maze studies, the characteristics of working memory are studied by examining the effects of increasing the number of arms in the maze. Characteristics of a cognitive map can be determined by altering features of the environment, such as the number of objects, or the size of the area to be mapped. For example, in one variation of this approach, Menzel, Premack & Woodruff (1978) modified the visual image upon which the map was based, in an effort to determine parameters of map-using ability.

Cognitive concepts can also contribute to prediction of new observations, serving as explanatory rather than descriptive or heuristic concepts (Honig, 1978). For example, in Cole et al.'s (1982) work with nectar-feeding birds, predictions about performance based on cognitive characteristics of the subjects were made. Specifically, Cole et al. predicted that the birds would exhibit win-shift learning more readily than win-stay learning; that is, after finding food at one location, they would preferentially visit a different location on their next opportunity, rather than return to the location in which they had successfully fed. This pattern runs counter to the conventional wisdom about reinforcement (in this case, food) increasing the likelihood of the individual repeating an action. The authors had an ecological basis for this prediction concerning replenishment rates of typical food resources in nature. It takes a nectar-producing flower several hours to produce a new crop of nectar; visits to a particular flower more often than once every few hours are inefficient.

Present and potential value of cognitive concepts

In behavioral field research on primates, descriptive use of cognitive terms still seems to be the dominant mode. In some cases, this involves translation of ecological concepts into cognitive equivalents. For example, theories of foraging behavior generally assume the subject 'knows' more or less about certain aspects of the foraging environment, such as the abundance, distribution, or location of resources, in the sense that these environmental parameters co-vary with feeding behavior. Examinations of foraging patterns are then made to test behavioral hypotheses drawn from these models. At the conclusion of the study, a statement can be made about the type of information which appears to be most important to the animal in the determination of feeding behavior. John Robinson's (unpublished) recent work with *Cebus olivaceous* is an elegant example of this approach. He found that his subjects were in fact moving about their home range primarily on the basis of known information about the location of particular fruit resources (i.e. a 'map'). This finding does not preclude the monkeys' knowing about the overall abundance or distribution of other fruit or invertebrate foods in their home range, but these factors did not regulate their behavior to the same extent. This approach is not directly concerned with cognition, but it does confirm the regulatory role of particular knowledge in a fundamental behavioral task.

In our view, a cognitive orientation towards behavior in nature could do much more for us than to provide alternative phrasing of ecological models. It could invigorate laboratory research on cognitive phenomena (Neisser, 1976); after all, many important cognitive phenomena are not even recognized except in natural conditions – the use of tools by chimpanzees during feeding is one good example. It could also enrich studies of behavior in nature. Field researchers from many disciplines need to be aware of the cognitive characteristics of their subjects which may be critical to an understanding of their behavior, such as the sources of individual variability, or the basis for less-than-optimal foraging or use of space, at least as predicted from application of optimality models. These are thorny problems in behavioral ecology which will not, we believe, be resolved without consideration of cognitive factors.

With these concerns in mind, we invite you to read the papers which follow.

References

Balda, R. P. & Turek, R. J. (1984). The cache-recovery system as an example of memory capabilities in Clark's nutcracker. In *Animal Cognition*, ed. H. L. Roitblat, T. G. Bever & H. S. Terrace, pp. 513–32. Hillsdale, NJ: Lawrence Erlbaum Associates

Cole, S., Hainsworth, F. R., Kamil, A. C., Mercier, T. & Wolf, L. L. (1982) Spatial learning as an adaptation in humming birds. *Science*, **217**, 656–7

Griffin, D. R. (1982) *Animal Mind – Human Mind*. (Dahlem Konferenzen, 1982) Berlin: Springer-Verlag

Hinde, R. A. (1982) *Ethology*. Oxford: Oxford University Press

Honig, W. K. (1978) On the conceptual nature of cognitive terms: an initial essay. In *Cognitive Processes in Animal Behavior*, ed. S. Hulse, H. Fowler & W. Honig, pp. 1–14. Hillsdale, NJ: Lawrence Erlbaum Associates

Hulse, S., Fowler, H. & Honig, W. (1978) *Cognitive Processes in Animal Behavior*. Hillsdale, N.J.: Lawrence Erlbaum Associates

Kamil, A. (1978) Systematic foraging by a nectar-feeding bird, the Amakihi (*Loxops virens*). *J. Comp. Physiol. Psychiatry*, **92**, 388–96

Kamil, A. C. (1984) Adaptation and cognition: knowing what comes naturally. In *Animal Cognition*, ed. H. L. Roitblat, T. G. Bever & H. S. Terrace, pp. 533–44. Hillsdale, NJ: Lawrence Erlbaum Associates

McGrew, W. C. (1977) Socialization and object manipulation of wild chimpanzees. In *Primate Biosocial Development*, ed. F. E. Poirier & S. Chevalier-Skolnikoff, pp. 216–88. New York: Garland

Menzel, E. (1966) Responsiveness to objects in free-ranging Japanese monkeys. *Behaviour*, **26**, 130–49

Menzel, E. (1969) Naturalistic and experimental approaches to primate behavior. In *Naturalistic Viewpoints in Psychological Research*, ed. E. P. Williams & A. L. Raush, pp. 78–121. New York: Holt, Rinehart, and Winston

Menzel, E. (1973) Chimpanzee spatial memory organization. *Science*, **182**, 943–5

Menzel, E., Premack, D. & Woodruff, G. (1978) Map reading by chimpanzees. *Folia Primatol.*, **29**, 241–9

Neisser, U. (1976) *Cognition and Reality*. San Francisco: Freeman

Olton, D. & Samuelson, R. J. (1976) Remembrance of places passed: spatial memory in rats. *J. Exp. Psychol.: Anim. Behav. Processes*, **2**, 97–116

Olton, D. S., Handelmann, G. E. & Walker, J. A. (1981) Spatial memory and food searching strategies. In *Foraging Behavior*, ed. A. C. Kamil & T. D. Sargent. New York: Garland

Richard, A. (1981) Changing assumptions in primate ecology. *Am. Anthropologist*, **83** (3), 517–33

Roitblat, H. L. (1982) The meaning of representation in animal memory. *Behav. Brain Sci.*, **5**, 353–406

Roitblat, H. L., Bever, T. G. & Terrace, H. S. (1984) *Animal Cognition*. Hillsdale, NJ: Lawrence Erlbaum Associates

Teleki, G. (1975) Chimpanzee subsistence technology: materials and skills. *J. Hum. Evol.*, **4**, 575–94

III.1

Contrasting approaches to spatially distributed resources by *Saimiri* and *Callicebus*

M. W. ANDREWS

Introduction

Two species of the family Cebidae, *Saimiri sciureus* (squirrel monkey) and *Callicebus moloch* (titi monkey), were compared with respect to foraging behavior. These species are sympatric over much of their ranges (Hershkovitz, 1963) and may be found in the same forest area (Thorington, 1968). Based upon weights at the California Primate Research Center, individuals of both species average approximately 1 kg; in *Saimiri* only, the males are a little larger than the females. Both species are omnivorous frugivores, consuming a range of food items including fruits, leaves, insects and nuts (compare reviews in Baldwin & Baldwin, 1981; Kinzey, 1981).

Despite similarities in habitat, individual size, and diet, the species differ markedly in their social behavior, ranging activities and modes of travel. *Callicebus* typically live in monogamous family groups in which the members remain in close association during travel. The adult male and female cooperate in caring for immature offspring (Mason, 1966, 1968; Robinson, 1979). In contrast, *Saimiri* live in large groups consisting of multiple males and females in which the individuals appear to travel in a much more fluid association than do *Callicebus* family groups. They show little evidence of strong attachment between opposite-sexed individuals, and offspring are cared for exclusively by females (Thorington, 1968; Baldwin, 1971; Baldwin & Baldwin, 1971, 1981). Ranging patterns of the two species also differ. In comparsion to *Saimiri*, *Callicebus* travel much shorter distances in a day within a substantially smaller home range (Mason, 1968; Baldwin & Baldwin, 1972, 1981; Kinzey, 1978). Differences in both social and ranging behaviors are reflected in the characterization of *Callicebus* as territorial (Mason, 1966), whereas the evidence suggests that *Saimiri* are not territorial (Baldwin & Baldwin, 1972). Finally, the species differ in their modes of travel. *Callicebus* family groups generally travel together along habitual pathways, using primarily the larger limbs in the lower and middle regions of the canopy (Mason, 1966; summarized in Fragaszy, 1980). When traveling and foraging, *Saimiri* individuals travel on different routes between fruiting trees and they use most levels of the canopy, traveling in trees and on branches of various sizes (Moynihan, 1976; Thorington, 1968).

The differences between *Callicebus* and *Saimiri* in ranging patterns and modes of travel suggest that they may also differ in their approaches to the general problems of finding and making effective use of food resources. The primary objective of this study was to compare the behavior of heterosexual pairs of *Callicebus* and *Saimiri* in an environment designed to model a patchy distribution of trees bearing ripe fruit. Since both species forage in trees containing ripe fruits, it is reasonable to assume that this task is relevant to both and that any differences in their performance would illuminate psychological differences in the organization of foraging behavior.

Methods
Subjects

All subjects were pair-housed adults; eight heterosexual pairs of *Callicebus moloch* and eight heterosexual pairs of *Saimiri sciureus* were tested. At the time the study began, *Callicebus* pairmates had been housed together from a minimum of 7 months to a maximum of 3 years and 3 months (median = 1 year and 6 months). For *Saimiri* pairmates, one pair had been housed together for only 1 month, but the other pairs had been housed together from a minimum of 8 months to a maximum of 7 years and 11 months (median = 4 years and 8 months).

Testing

Eight pairs of each species were each given six test sessions over a period of 10 months. In each session a pair was released for 1 hour in an outdoor enclosure (18 m × 36 m), set apart from the living area, in which 10 artificial 'trees' were spatially distributed as indicated in Fig. 1. Each tree was constructed of wood and PVC pipe and contained four opaque food cups. For all but two pairs, each cup contained one half of a small, soft candy (miniature marshmallow). For two pairs in which one of the pairmates did not readily consume half pieces of miniature marshmallow in quantity, each cup contained one quarter of a miniature marshmallow for one pair or one quarter of a piece of sugared breakfast cereal (Fruit Loop) for the other, because these items were readily consumed in quantity by both pairmates. The trees were interconnected by an elevated (0.85 m) runway system (Fig. 1).

Six rows of posts ran the entire length of the enclosure with a post located every 3.6 m; each post was designated the center of an 'area'. To facilitate data collection, posts were labeled with numbers following the coordinate system indicated in Fig. 1 and trees were clearly identified by letters. Pairmates were observed simultaneously by two trained observers using focal animal sampling (Altmann, 1974). Data were obtained on both social and non-social aspects of the use of space.

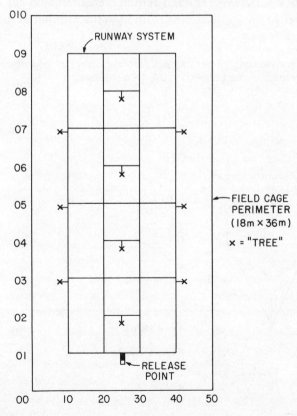

Fig. 1. Diagram of the test environment. The runway system and locations of artificial trees are shown. Numbers indicate the coordinate system used to identify 'areas'.

In addition to the six test sessions in the outdoor enclosure, all pairs were habituated to taking food from a 'tree' in a special training cage (0.75 m wide × 1.8 m high × 3.7 m long) before testing began. Furthermore, individual animals were given food acceptance trials in their living areas. Single food items identical to those used in the test sessions were offered to an animal until 40 pieces (the total number available during a test session) had been consumed, or 1 hour had elapsed, whichever came first.

Results

Saimiri were clearly superior to *Callicebus* at obtaining the food items from the artificial trees (Fig. 2). They obtained approximately 86% of the available food, in contrast to 64% obtained by *Callicebus* ($F(1,14) = 15.5$, $p < 0.01$). This indicates that relative to *Callicebus*, *Saimiri* were more efficient at obtaining the food from the trees which were entered and/or they entered more of the available trees.

With regard to the first possibility, it appears that obtaining the food from those trees which were entered is not a major reason for the species difference in foraging success. Although the number of items

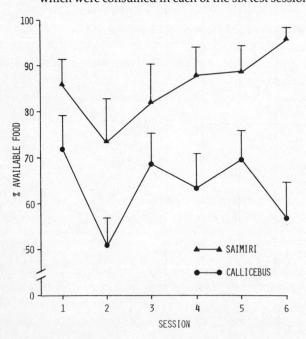

Fig. 2. Mean percentage per pair of the total number of food items which were consumed in each of the six test sessions.

taken from entered trees was greater in *Saimiri* than in *Callicebus* (percentage empty cups in trees entered in a session, $\bar{X} = 95.8$ for *Saimiri* vs 91.0 for *Callicebus*; $F(1,14) = 5.22, p < 0.05$), the magnitude of the difference was small and hence cannot account for the interspecies difference in foraging success. *Callicebus* were 95% as efficient as *Saimiri* in those trees which were entered, but they obtained only 74% as much of the food which was available in all trees.

With respect to the second possibility, namely that the superior foraging success of *Saimiri* is due to the number of available trees entered by the pairs, the evidence is more persuasive. *Callicebus* pairs entered only 78% as many of the available trees as did *Saimiri* pairs. Of the total of 10 trees available in the test area, *Saimiri* pairs entered an average of 8.9 different trees (89% of available trees), as compared to 6.9 (69% of available trees) by *Callicebus* ($F(1,14) = 14.08, p < 0.01$). Behavioral measures obtained permit evaluation of several explanations of this difference.

One possible explanation of the interspecies difference in percentage of available trees entered is that the species differed in food motivation. The results of the food acceptance trials, however, indicated that there were no interspecies differences in acceptability of the food items used in the study or in satiety relative to the maximum number of food items available during a test session. Out of a possible total of 40 pieces of food, *Callicebus* consumed an average of 36.6 food pieces per individual and *Saimiri* consumed an average of 38.9 food pieces ($F(1,30) = 2.35$, n.s.). On average, each individual of a pair was capable, by itself, of consuming nearly all the food which was available to both animals during a test session.

A second possible explanation of the interspecies difference in percentage of available trees entered is that, relative to the travel of *Callicebus* individuals, the travel of *Saimiri* individuals may have brought them into proximity to a greater number of available trees which were then exploited. A tree approach was scored if an animal entered an area in which a tree was located. When the species were compared on percentage of available trees approached in a session, no species differences were found. *Callicebus* individuals approached an average of 76% of available trees and *Saimiri* individuals approached an average of 78% ($F(1,30) = 0.27$, n.s.).

A third possible explanation of the interspecies difference in number of available trees entered is that the percentage of foraging areas entered by either pairmate may have been greater for *Saimiri* pairs than for *Callicebus* pairs. If two animals tend to travel independently to different areas, they may cover more combined area than two animals traveling to the same areas. Restricting analysis to those areas interconnected by the runway system, including all those areas in which a tree was located, no statistically significant species difference in the percentage of areas entered by either one or both members of a pair was found. *Callicebus* entered an average of 77% of these areas and *Saimiri* entered an average of 84% ($F(1,14) = 1.00$, n.s.).

A fourth possible explanation of the difference in percentage of trees entered is that *Saimiri* may have had a greater economy of travel among those trees which were entered than did *Callicebus*. An animal that moves efficiently from one tree to the next, taking the food from each in turn, would have a good chance of exploiting a majority of the available trees. The data indicate that *Saimiri* did have a greater economy of travel than did *Callicebus*. *Callicebus* were only about 70% as likely to enter a tree per unit of travel as were *Saimiri*. Although this difference was not statistically significant (tree entries per area entry, $\bar{X} = 0.14$ for *Saimiri* vs 0.19 for *Callicebus*; $F(1,30) = 3.78$, n.s.), two additional measures of travel economy did reveal a significant difference between the species. *Saimiri* were more likely than were *Callicebus* to enter the first tree encountered as they entered the test area (percentage of sessions in which first tree entry was made to one of the three trees closest to the release point, $\bar{X} = 89.6$ for *Saimiri* vs 68.3 for *Callicebus*; $F(1,30) = 5.60$, $p < 0.05$). They were also less likely to enter a tree, leave it, and return to that same tree before moving into another tree. Analyzing consecutive re-entries based on the first 10 tree entries in a session, it was found that the percentage of consecutive re-entries was higher for *Callicebus* than for *Saimiri* (percentage of repeat tree entries that were consecutive, $\bar{X} = 46.2$ for *Callicebus* vs 26.4 for *Saimiri*; $F(1,30) = 8.70$, $p < 0.01$). In general, the travel of *Saimiri* appeared to be much better organized for the efficient exploitation of the available trees than was that of *Callicebus*.

Two determinants of the reduced travel economy of *Callicebus* relative to *Saimiri* are suggested by further analysis of the data. First, *Callicebus* were more disposed than were *Saimiri* to maintain proximity to the pairmate. Each time an animal entered a tree or an area, an approach to the pairmate was scored if its pairmate was already present. On 17% of all entries into trees or areas for *Callicebus* the pairmate was present, whereas this was the case on only 5% of such occasions for *Saimiri*; this difference was statistically significant ($F(1,30) = 30.99$, $p < 0.001$).

The interspecies difference in tendency to approach the pairmate suggests that in circumstances in which approach to the pairmate conflicted with entry into a tree, the outcome was more likely to result in reduced economy of travel for *Callicebus* than for *Saimiri*, because they had a greater tendency than did *Saimiri* to approach the pairmate, thereby foregoing the opportunity to enter a tree.

Secondly, *Callicebus* appeared more disposed than did *Saimiri* to establish familiarity with the environment at the cost of immediate foraging efficiency. Travel between tree entries by *Callicebus* was often broken by long pauses which were spent scanning the surrounding environment. In contrast, *Saimiri* traveled rapidly from tree to tree, usually without stopping between tree entries. The observation that *Callicebus* were more likely than *Saimiri* to return to a tree that they had just left also suggests an animal somewhat occupied with establishing familiarity with the environment rather than solely concerned with finding all the food in that environment. Furthermore, *Callicebus* were more likely than *Saimiri* to pass up an opportunity to enter a tree as they proceeded into the test environment, suggesting that their attention was more strongly directed to details of the environment other than the locations of available food resources.

Summary and conclusion

Callicebus and *Saimiri* were compared with respect to foraging behavior in an environment designed to model a patchy distribution of food. The species clearly differed in foraging efficiency in this situation. Despite the apparent absence of interspecies differences in food motivation, *Saimiri* were clearly superior to *Callicebus* at obtaining the food items available in the artificial trees and approached more closely an optimum foraging pattern. *Saimiri* obtained only slightly more of the food from the trees which they entered. They entered more trees, however, and they exhibited a much greater economy of travel relative to those trees which were entered, as compared to *Callicebus*. The data suggest that the reduced foraging efficiency of *Callicebus*, particularly as reflected in reduced economy of travel, relates to two dispositions that support the monogamous and territorial aspects of the *Callicebus* lifestyle. These dispositions are maintenance of proximity to the pairmate and establishment of familiarity with the environment.

Acknowledgments

I am indebted to Dr William A. Mason whose guidance and support made this work possible. I am also grateful to Dr Sally P. Mendoza, Charles R. Menzel and Dr Gustl Anzenberger for important comments and discussions regarding this research. US Public Health Service Grant RR-00169 supported the animals with which I worked.

References

Altmann, J. (1974) Observational study of behavior: sampling methods. *Behaviour*, **49**, 227–67

Baldwin, J. D. (1971) The social organization of a semifree-ranging troop of squirrel monkeys (*Saimiri sciureus*). *Folia Primatol.*, **14**, 23–50

Baldwin, J. D. & Baldwin, J. I. (1971) Squirrel monkeys (*Saimiri*) in natural habitats in Panama, Colombia, Brazil, and Peru. *Primates*, **12**, 45–61

Baldwin, J. D. & Baldwin, J. I. (1972) The ecology and behavior of squirrel monkeys (*Saimiri oerstedi*) in a natural forest in western Panama. *Folia Primatol.*, **18**, 161–84

Baldwin, J. D. & Baldwin, J. I. (1981) The squirrel monkeys, genus *Saimiri*. In *Ecology and Behavior of Neotropical Primates*, Vol. 1, ed. A. F. Coimbra-Filho & R. A. Mittermeier, pp. 241–76. Rio de Janeiro: Academia Brasileira de Ciencias

Fragaszy, D. M. (1980) Comparative studies of squirrel monkeys (*Saimiri*) and titi monkeys (*Callicebus*) in travel tasks. *Z. Tierpsychol.*, **54**, 1–36

Hershkovitz, P. A. (1963) A systematic and zoogeographic account of the monkeys of the genus *Callicebus* (Cebidae) of the Amazonas and Orinoco River basins. *Mammalia*, **27**, 1–79

Kinzey, W. G. (1978) Feeding behaviour and molar features in two species of titi monkeys. In *Recent Advances in Primatology*, Vol. 1, *Behaviour*, ed. D. J. Chivers & J. Herbert, pp. 373–85. London: Academic Press

Kinzey, W. G. (1981) The titi monkeys, genus *Callicebus*. In *Ecology and Behavior of Neotropical Primates*, Vol. 1, ed. A. F. Coimbra-Filho & R. A. Mittermeier, pp. 241–76. Rio de Janeiro: Academia Brasileira de Ciencias

Mason, W. A. (1966) Social organization of the South American monkey, *Callicebus moloch*: a preliminary report. *Tulane Stud. Zool.*, **13**, 23–8

Mason, W. A. (1968) Use of space by *Callicebus* groups. In *Primates: Studies in Adaptation and Variability*, ed. P. C. Jay, pp. 200–16. New York: Holt, Rinehart & Winston

Moynihan, M. (1976) *The New World Primates*. Princeton, NJ: Princeton University Press

Robinson, J. G. (1979) Vocal regulation of use of space by groups of titi monkeys *Callicebus moloch*. *Behav. Ecol. Sociobiol.*, **5**, 1–15

Thorington, R. W., Jr (1968) Observations of squirrel monkeys in a Colombian forest. In *The Squirrel Monkey*, ed. L. A. Rosenblum & R. W. Cooper, pp. 69–85. New York: Academic Press

III.2

Ranging patterns in hamadryas baboons: evidence for a mental map

H. SIGG

Introduction

The concept of a mental map has previously been examined by Tolmann (1948). Menzel (1973, 1978) studied the spatial memory organisation of young chimpanzees in terms of the optimal route taken in an experimental setting. Boesch & Boesch (1984) proved this cognitive ability on the basis of tool transportation in free-ranging chimps. Unlike these studies, which are based on the relatively simple paradigm of the shortest connections between various places, it is almost impossible to test the mental map hypothesis in free-ranging baboons. However, most observers of free-ranging baboons are convinced that the baboons forage on the basis of a mental representation of the environment (e.g. Altmann & Altmann, 1970). Why is it so hard to test this mental capacity? There are three reasons. First, the actual route of a foraging baboon may optimize many different needs, such as (1) sufficient intake of water and food, (2) avoiding predators by selecting safe routes thus reducing the risk of contact, and by preparing for flight and defence, (3) regulation of body temperature by seeking out shady places and regulation of marching speed, (4) saving energy, (5) training for optimal physical strength, and (6) retaining an accurate knowledge of the condition of the resources. Secondly, the observer does not know the actual goal of a travelling baboon. Thirdly, the route may result from a compromise made between the various members of the group (Stolba, 1979). Nevertheless, there is some striking evidence to support the concept of a mental map in free-ranging hamadryas baboons (Sigg & Stolba, 1981) which will be summarized in the following section.

Methods

During our field study from December 1973 to May 1975 at Erer-Gota, Ethiopia, we recorded 75 dayranges of Cone-Rock-Band I (66 individuals) with an average route length of 8.6 km, and 13 dayranges of Band II (95 individuals) with an average route length of 10.4 km. For every dayrange, the route of a particular clan was drawn on a map (1:50000). Every 15 min our exact position was marked on the map. The richly structured topography allowed the mapping of each position with an exactitude of ±100 m.

The most important aim of the baboons on the daily march is to find drinking water and, at night, sleeping cliffs. As waterholes and cliffs are quite scarce in the region, these are most likely to be important goals. Since most cliffs are easily visible from a distance, only the position of waterholes can be used to test the mental map hypothesis.

Results

As a first hypothesis, we presumed that the baboons left the sleeping cliff in the direction of the waterhole they were intending to reach. We compared the departure directions for 75 dayranges, i.e. the directions in which the baboons left the cliff region, with the location of the first waterhole reached after 09.30h.

In 32% of the dayranges, the midday waterhole was situated in the same sector of 45° as the departure direction. In a further 43% it was in the adjacent sector. The correspondence between the departure direction and the location of the midday waterhole is called goal-certainty. This goal-certainty varies depending on the sector of departure. Two explanations for this variation come to mind. (A) Goal-certainty is mainly affected by the condition of the habitat in each sector. The local topography may make it easy or difficult to keep to one particular course; or encounters with nomads, dogs or predators, which turn the baboons away from their course, may occur more often in some sectors than in others. This hypothesis does not need a mental-map capacity. (B) Goal-certainty is mainly determined by the attractiveness of the midday area. The probability that the initial direction is maintained in spite of obstacles and hazards and by walking around them, is a function of the baboons' appetence for the goal. This hypothesis requires a mental map.

We assumed that the frequency with which a particular place is visited is a fair measure of its attractiveness. We found goal-certainty to be positively correlated with attractiveness. At first, this correlation seems to be trivial, since higher goal-certainty means fewer deviations and therefore more routes reaching the goal. However, looking at the proportion of direct routes leading to a certain goal, a negative correlation occurs, i.e. the deviated routes contribute more to the total number of visits to attractive goals than to less attractive ones. But the probability of a change in direction is smaller if the departure direction lies in the sector of an attractive goal. Hypothesis A can therefore be rejected.

We may therefore suggest that the departing baboons have an idea of where to go. This suggestion is strongly supported by the fact that the band normally splits up on its way to the waterhole. The different parties often lose contact. In five cases separated parties were pursued by two observers. The different parties followed quite different routes. Nonetheless they came together again near the waterhole.

While the waterhole is obviously an important goal, feeding and resting places are less clearly definable. However, if a place serves as a goal we may presume that the pattern of approach will be distinct from the usual manner of travel. We tested this hypothesis using the following definitions. Goal areas are places where the baboons stopped for at least 15 min. Fast marching is travelling with a mean speed of more than 2 km/h, i.e. more than double the average speed. We compared the sequential occurrence of fast travelling and resting. Fast marching occurred more often before resting than had been expected from random assumptions. In 93% of these cases a marked acceleration of more than 1 km/h was noted. This took place more than 500 m before the goal area was reached – a distance from where the condition of a feeding place could be estimated neither by eye nor by odour.

We found that the baboons often followed exactly the same pathway as taken on earlier marches. The most striking paths could be described as street segments of 500 m in length and 150 m in width. The fact that such street segments were repeatedly used excludes the hypothesis of random wandering. The connections between street segments vary greatly and this contradicts the assumption that the baboons simply follow topographical features.

It seems reasonable to conclude that baboons orientate themselves to intermediate goals during the march, using street segments for comfortable walking, while keeping to the general direction of the midday waterhole, or – in the afternoon – of the sleeping cliff. In other words, the mental map includes main goals, intermediate goals and route segments for comfortable travelling.

If the mental map has a major influence on determining the route taken, we may expect that Band I and Band II which use parts of their overlapping home range with different intensity will exhibit different route patterns, because of a difference in the accuracy of the mental map. On the other hand, if the environment is the most important factor determining the route patterns, similar patterns between the bands should be expected.

To test these hypotheses, we split up the routes into eight equal sectors corresponding to the midday waterholes. Both bands took shorter routes in their less-familiar sectors. In each sector, the 'foreign' band, i.e. the band who visited the sector the least, took shorter routes than the resident band. The 'foreign' band thus reached the waterhole by a more direct route than the resident one, probably because the resident band included more intermediate goals in its route.

Discussion

Now, two questions arise: Why do baboons visit the less-familiar regions? Why do they not explore the foreign region in more detail, i.e. taking longer routes?

There is absolutely no evidence that baboons change foraging regions because of a shortage of food or water. We also could not find any resource which was not present in both regions. In answer to the first question we must suppose that there is an exploratory tendency in the baboon's motivational system. Since there are seasonal changes in the food content of the main feeding areas as well as differences in the availability of and access to waterholes, the baboons' regular survey of alternative places where necessary resources can be found might be most important for their safety in case of a predator-induced change of the foraging region. However, this tendency to explore alternative places only makes sense if baboons are capable of keeping the geographical information in their memory.

The fact that the dayranges in the less familiar area are shorter may indicate that the exploratory tendency is limited, either because of a restricted learning or memory capacity or, more likely, because of motivational constraints. We suspect that exploration increases general arousal, while visiting well-known places decreases it. The postulation of an arousal homeostasis regulating the exploratory tendency coincides with other patterns which we have observed: (1) the difference in the length of route, as already mentioned; and (2) the fact that the preferred type of habitat, the wadi, is preferred more the further away it is from the cliff. These two patterns may also be influenced by a less-accurate mental map in less-visited areas which influences the baboons to follow prominent topographical structures to a greater extent. This is not the case in the following observations: on days after rainfall, when water is more abundant and therefore the risk of thirst is minimal, the baboons choose unusual directions more often than on other days. On the other hand, the baboons regularly choose directions leading to the most preferred places when the decision process in the morning has lasted for more than 1.5 h, indicating heavy disagreement between the different group members.

Therefore, we may assume that baboons not only have an idea of where to go, but also estimate the relative risk of failing to find the necessary resources on the basis of memorized information.

Acknowledgments
This study has been supported by the Swiss National Science Foundation (Prj.No.3.924.72). I thank Alex Stolba for collecting the main part of the data and many hours of discussion, and Hans Kummer for initiating the project and his help throughout the study. Helen Everett gave her assistance in writing the English manuscript.

References

Altmann, S. & Altmann, J. (1970) *Baboon Ecology: African Field Research.* Bibliotheca primatologica, No. 12. Basel: Karger

Boesch, C. & Boesch, H. (1984) Mental map in wild chimpanzees: an analysis of hammer transports for nut cracking. *Primates*, **25**, 160–70

Menzel, E. (1973) Chimpanzee spatial memory organisation. *Science*, **182**, 943–5

Menzel, E. (1978) Cognitive mapping in chimpanzees. In *Cognitive Processes in Animal Behavior*, ed. S. Hulse, H. Fowler & W. Honig, pp. 375–422. Hillsdale, NJ: Lawrence Erlbaum Associates

Sigg, H. & Stolba, A. (1981) Home range and daily march in a hamadryas baboon troop. *Folia Primatol.*, **36**, 40–75

Stolba, A. (1979) Entscheidungsfindung in Verbänden von *Papio hamadryas*. Phil.II dissertation, Zürich

Tolmann, E. D. (1948) Cognitive maps in rats and men. *Psychol. Rev.*, **55**, 189–208

III.3

Cognition, brain size and the extraction of embedded food resources

K. R. GIBSON

Introduction

Mammals in general, and primates in particular, have long been known for their enlarged brains (Jerison, 1973). As yet, however, there is little agreement about what, if any, survival advantage increased brain size might confer. Recent hypotheses suggest that brain size may be selectively neutral (Radinsky, 1982), an adaptation to particular feeding strategies (Eisenberg & Wilson, 1978; Clutton-Brock & Harvey, 1980), or an adaptation conferring intelligence and information processing capacity (Jerison, 1979, 1982; Parker & Gibson, 1979; Gibson & Parker, 1982). The adaptive significance of intelligence, however, is itself open to debate.

Nevertheless, physiological data indicate that increased brain size must confer some major survival advantage, for the brain is such a metabolically expensive organ that it would otherwise be under strong selective pressure to reduce in size, especially during periods of food shortage. The human brain, for example, utilizes fully 20% of the body's metabolic energy, while the macaque brain utilizes 9% (Armstrong, 1983). This enormous metabolic drain occurs 24 hours a day, 365 days per year.

Viewed from the standpoint of these metabolic demands, it is not surprising that recent findings indicate a strong positive correlation between brain size and body metabolic rate (Armstrong, 1982, 1983; Martin, 1982). In particular, metabolic rate appears to set an upper limit on brain size. Since body metabolism is fueled by food intake, it is likely that only animals who can sustain high caloric intake relative to energy expenditure on a year-round basis can afford large brains. Possible exceptions would be species who seasonally consume excess calories which are deposited in the form of fat.

These considerations suggest a relation between brain size and feeding strategies. Such relations have, in fact, been found. In mammalian evolution, for instance, brain size has tended to increase more rapidly among carnivores than among herbivores (Jerison, 1973), while among both bats and primates, frugivores possess larger relative brain sizes than insectivores (Eisenberg & Wilson, 1978; Clutton-Brock & Harvey, 1980).

Neocortical functions

Among mammals, increased brain size usually reflects increased size of the neocortex (Jerison, 1973, 1979). Examination of neocortical functions suggests that these functions may provide significant advantages for the location, recognition, procurement, and manipulation of foods. Hence, cortical enlargement may have arisen as an adaptation to feeding on varied foods which are difficult both to procure and to process.

First, the motor cortex provides mammals with the capacity for fine differentiated movements (Kuypers, 1962). Secondly, human data indicate that the cortex helps plan and execute coordinated sequential and simultaneous movement of various anatomical manipulators, as in the use of two hands simultaneously in the manipulation of an object or in the sequential use of several anatomical manipulators to meet a single end (Luria, 1964).

Thirdly, the neocortex perceives tactile, visual and auditory images by breaking sensory images into fine component parts and then reassembling them to construct complex perceptual wholes (Gibson, 1977, 1981; Luria, 1966; Mountcastle, 1978). These perceptual constructive processes can be either simultaneous, as in the construction of an image of a single object, or sequential, as in the perception of an action sequence or of a string of words (Luria, 1964). The mental anticipation of future sensory events may be an outgrowth of these sequential perceptual constructional processes.

Fourthly, the cortex, particularly the frontal cortex, provides mental flexibility which enables an organism to give diverse responses to the same stimulus, to construct a wide variety of perceptual images and thereby to recognize a wide variety of objects, and to construct an almost infinite variety of coordinated sequential and simultaneous motor actions (Warren & Akert, 1964).

Mammalian *versus* lower vertebrate feeding strategies

The potential feeding advantages of these cortically based capacities are readily illustrated by contrasting feeding techniques in lower vertebrates, who possess little in the way of a neocortex, with those of the larger-brained mammals. Frogs, for example, sit motionless in wait for flying insects. When one is spotted, the tongue lashes out, captures the insect and is withdrawn into the mouth. Prey are recognized by conformance to a rather stereotyped visual image of size and movement (Lettvin et al., 1959). No evidence exists for the ability to construct visual images of prey of highly varied shapes, sizes, and habits. If these animals can anticipate or follow prey movements, this is not evident from their feeding behavior. Escaping insects are not pursued.

From a manipulative standpoint, prey are captured through a single motor act involving one anatomical organ. Food is swallowed without prior breakdown and without chewing. This feeding technique is clearly efficient. It is also limiting. Frogs can feed only on insects that will identify themselves by movement and can be swallowed whole.

While not all lower vertebrates feed in as stereotyped a fashion as the frog, only a few extant amphibians and reptiles possess the ability to break food into pieces prior to swallowing. Most rely on visual movement or olfaction as a means of prey identification. Stalking or the pursuit of escaping prey, when present at all, is accomplished by the following of an olfactory trail rather than by visual or auditory anticipation of prey movement. Exceptions are the monitor lizards, whose feeding techniques parallel those of mammals and who are the largest brained of the extant reptiles (Auffenberg, 1978; Northcutt, 1978; Regal, 1978).

By contrast, mammalian feeding techniques commonly manifest simultaneously and sequentially coordinated movements of the lips, tongue, teeth and hands. As a consequence, mammals possess major feeding advantages over most reptiles. Specifically, they break their food into component parts prior to ingestion and, hence, can feed on foods too large to be swallowed whole or containing indigestible elements such as shells, bones, or attached sticks. When combined with masticatory skills and salivation, this ability to subdivide food prior to ingestion also results in a relatively speedier digestive system which can, in turn, fuel a higher metabolic rate and a larger brain size.

As a result, mammals possess the energy for active food search and expanded sensory capacities to facilitate this search, particularly in tactile (e.g. rhinarium and vibrissae) and auditory spheres. These capacities not only permit the procurement of food items not readily recognized by olfactory or visual senses but, when accompanied by the ability to construct highly varied and differentiated object images, may permit the recognition and utilization of stationary, cryptic and hidden foods of varied shapes and sizes. To the extent that auditory or visual anticipation of prey movement is present, anticipatory stalking and prey pursuit may also occur.

Extant primitive mammals exhibit many of these behaviors (Herter, 1972). Insectivores, for instance, routinely feed on large food items which are severed into component parts by sequential and simultaneous actions of varied anatomical manipulators. They also tend to eat relatively varied diets and often search in the leaf litter of the forest floor or in underground burrows in search of hidden and stationary prey. This suggests that expanded feeding capacities may have accompanied the initial placental radiation and initial neocortical expansion.

Primate feeding strategies

By comparison with insectivores, primates, particularly Anthropoidea, exhibit expanded visual and tactile capacities and increased fine motor control over the digits and facial musculature. In addition, even very simple primate feeding strategies, such as leaf eating, manifest both simultaneously and sequentially coordinated movements of the hands, lips, tongue and teeth. As a result, primates commonly utilize stationary or hidden foods which require considerable manipulation prior to ingestion. These expanded sensorimotor and feeding capacities are accompanied by concomitant neocortical expansion, particularly in cortical motor, tactile and visual areas (Le Gros Clark, 1971).

In certain primate taxa, the habit of utilizing foods which are difficult to find and to process has culminated in a particular type of feeding strategy: omnivorous extractive foraging (Parker & Gibson, 1977). Extractive foraging means feeding on foods that must first be removed from other matrices in which they are embedded or encased. Extractive foods include nut-meat, shellfish, snails, eggs, brains or bone marrow which must be removed from hard outer coverings, seeds and beans that must be removed from pods, tubers or roots that must be dug from the ground, ants and termites that must be removed from mounds or hills, insect larvae or pith that must be extracted from bark, and meat that is extracted from its hide prior to ingestion.

Extractive foods tend to be premium foods, high in both energy and protein. Many are readily available during dry seasons. Consequently, animals who can utilize these foods may gain access to the year-round high-calorie diet essential for the support of a large brain.

Extractive foods, however, often require complex processing skills.

As a consequence, many mammalian and avian species have developed specialized anatomical manipulators to handle such chores as cracking shells, nuts, seeds or bones, digging in the ground, gnawing or drilling bark, or extracting insects or nectar from tight crevices.

Animals who lack specialized anatomical manipulators must forego extractive foods or they must use tools (Alcock, 1972). In fact, nearly all mammals and birds who use tools in feeding endeavors do so in extractive contexts (Parker & Gibson, 1977, 1979; Beck, 1980). Examples include the use of stones to crack ostrich and emu egg shells by Egyptian vultures and Australian black buzzards, respectively (Chisholm, 1954; Van Lawick-Goodall & Van Lawick, 1966), the use of sticks to probe insects from bark by the Galapagos woodpecker finch and other avian species (Lack, 1947; Eibl-Eibesfeldt, 1961; Millikan & Bowman, 1967; Morse, 1968), the use of stones to crack mollusk shells by California sea otters (Kenyon, 1959; Hall & Schaller, 1964), and, among chimpanzees, the use of hammerstones to crack nuts, of termiting and anting sticks to extract insects from mounds or hills, and of leaf sponges to extract water from tree holes (Van Lawick-Goodall, 1968; Suzuki, 1969; Sabater-Pi & Jones, 1973; Teleki, 1975; McGrew, 1976; Sugiyama & Koman, 1979; Boesch & Boesch, 1981; McBeath & McGrew, 1982; Anderson, Williamson & Carter, 1983). Finally, the most accomplished tool users of all, human beings, crack nuts, shells, and bones with hammerstones, dig tubers from the ground, cut thick animal hides in order to extract meat, and, in technologically complex societies, use tools to extract milk from cows, grain from hulls, and sugar, alcohol and drugs from plants.

Previous works questioned whether intelligence, like tool use, correlates with extractive foraging and concluded that levels 5 and 6 of Piaget's sensorimotor intelligence series characterize neither all extractors nor all tool users (Piaget, 1952, 1954; Parker & Gibson, 1977, 1979). High sensorimotor intelligence is found, however, in those primate extractive foragers who are highly omnivorous and who extract varied foods requiring varied manipulative and identification skills, specifically *Cebus*, chimpanzees and humans. Gorillas and orangutans also exhibit high sensorimotor intelligence. Although not omnivorous, they do extract pith, tubers and/or hard-shelled fruits and may well have shared a more omnivorous common ancestry with chimpanzees and humans (Ciochen, 1983; Wolpoff, 1983).

Primate encephalization and neocortical progression indices also demonstrate a correlation between an enlarged brain relative to body size and omnivorous extractive foraging habits (Tables 1 and 2). These indices refer to the extent to which brain or neocortical size in a given species exceeds that predicted for an insectivore of similar body size (Stephan, 1972). In general, primates with relatively large brain sizes also have large neocortices (Jerison, 1973).

Among prosimians, New World Monkeys, Old World Monkeys, and Hominoidea, the largest relative brain sizes occur in omnivores (*Daubentonia, Cebus, Cercopithecus talapoin, Papio, Pan,* and *Homo*). Each of these omnivorous forms engages in at least some extractive foraging in the wild, such as unfolding leaves to obtain encased insects, opening pods to obtain seeds, and digging grass corms from the ground.

Four of these taxa possess unusually large relative brain sizes for their phylogenetic level and engage in extractive tasks requiring complex, cortically mediated, sensorimotor coordinations for both the recognition and processing of foods (*Homo, Pan, Cebus,* and *Daubentonia*). In *Pan*, for instance, like *Homo*, tools are required for extractive tasks.

Table 1. *Encephalization indices. These data are taken from Stephan (1972). The encephalization indices refer to the extent to which brain size in a given species exceeds that predicted for a basal insectivore of similar body size*

Encephalization indices	Genus			
28.8	*Homo*			
10–12	*Pan*	*Miopithecus*	*Cebus* *Lagothrix*	
8–10	*Pongo*	*Papio* *Macaca* *Cercopithecus*	*Ateles* *Saimiri*	
6–8	*Gorilla*	*Colobus*		*Daubentonia*
4–6			*Aotes* Callitrichidae *Alouatta*	Lemurini Tarsiidae Lorisidae Galagidae
2–4				Indriidae Cheirogaleinae Tupaiidae Lepilemurinae

In both *Daubentonia* and *Cebus* simultaneous and sequential coordinations of three or more sensorimotor tasks, tapping, probing, looking, and listening, are utilized in the location and recognition of bark-embedded insects, ripe palm nuts and, in *Cebus*, frogs and grasshoppers hidden within tree cavities (Petter & Petter-Rousseau, 1967; Izawa & Mizano, 1977; Izawa, 1978, 1979; Terborgh, 1983; Fragaszy, in press). Processing of these foods involves equally complex sensorimotor coordinations which in *Cebus* may involve all of the above plus the use of the tail, trunk, hands and feet. *Cebus* also extracts tiny seeds with fine finger movements (Chevalier-Skolnikoff, 1984), opens nuts and termite nests by banging them against tree trunks (Izawa & Mizano, 1977) and may anticipate the ripening of fruits in order to return to appropriate feeding spots at opportune times (Terborgh, 1983).

Moderately large brains tend to be found in primate omnivores or frugivores who engage in extractive tasks requiring somewhat less complex sensorimotor coordinations than that of the above, such as cracking nuts with the teeth, peeling fruits, opening bean pods and

Table 2. *Neocortical progression indices. These data are taken from Stephan (1972). The Neocortical progression indices refer to the extent to which neocortical size exceeds the size predicted for a basal insectivore of similar body size*

Neocortical progression indices		Genus		
156		*Homo*		
60		*Pan*	*Miopithecus*	{ *Cebus* *Lagothrix*
45			*Macaca* *Cercopithecus*	*Ateles*
30		*Gorilla*	*Colobus*		*Daubentonia*
20				*Aotes* Callitrichidae *Alouatta*	Lemurini Tarsiidae Lorisidae Galagidae
10					Indriidae Cheirogaleinae Lepilemurinae Tupaiidae

unrolling leaves to obtain hidden insects. (*Papio, Macaca, Ateles, Lagothrix, Pongo, Saimiri*) (Table 1). Small brains are found among insectivores, frugivores, omnivores and foliovores who eat exposed rather than hidden foods and who manipulate foods using only typical primate eye–hand–mouth coordinations.

Extractors who concentrate on one extractive food, for which they possess a rather specialized anatomy, also tend to have brain sizes which are small relative to body size. These include the gorilla, who pulls tubers from the ground with his massive strength, and marmosets, whose specialized dental apparatus permits the extraction of sap from bark (Sussman & Kinsey, 1984).

Consequently, relatively large brain and neocortical sizes, like tool use and sensorimotor intelligence, correlate with omnivorous extractive foraging in primates. Unlike tool use, however, extraction, *per se*, does not appear to be the critical variable with respect to brain size. Rather, neocortical size correlates with the complexity and variety of the sensorimotor coordinations needed for the finding and processing of foods. Omnivorous extractive foraging on a wide variety of extractive foods represents the feeding style which places the greatest premium on these neocortical capacities.

Omnivorous extractive foraging demands a metabolically expensive brain and time-consuming searching and manipulative activities. It may require metabolically expensive muscle structure and physical exertion. Why then utilize extractive foods? As stated above, these foods tend to be high in nutritive content and available all year round. As a result, omnivorous extractive foragers such as *Cebus, Pan, Papio*, and *Homo* can utilize seasonally lean environments not readily accessible to others (Parker & Gibson, 1977, 1979).

Summary

Evidence indicates that in mammals, as compared to lower vertebrates, an enlarged neocortex correlates with the presence of complex sensorimotor processes permitting the exploitation of varied stationary or hidden foods which are both difficult to find and process.

Among some primates, including *Daubentonia, Cebus, Pan*, and *Homo*, this trend has culminated in a feeding process, omnivorous extractive foraging, which demands both high intelligence and an enlarged brain but permits the exploitation of niches not available to other species.

References

Alcock, J. (1972) The evolution of tools by feeding animals. *Evolution*, **26**, 464–73

Anderson, J. R., Williamson, E. A. & Carter, J. (1983) Chimpanzees of Sapo Forest, Liberia: density, nests, tools and meat eating. *Primates*, **24**, 594–601

Armstrong, E. (1982) A look at relative brain size in mammals. *Neurosci. Lett.*, **34**, 101–4

Armstrong, E. (1983) Relative brain size and metabolism in mammals. *Science*, **220**, 1302–4

Auffenberg, W. (1978) Social and feeding behavior in *Varanus komodoensis*. In *Behavior and Neurology of Lizards*, ed. N. Greenberg & P. MacLean, pp. 301–31. Rockville, MD: National Institutes of Mental Health

Beck, B. (1980) *Animal Tool Behavior*. New York: Garland Press

Boesch, C. & Boesch, H. (1981) Sex differences in the use of natural hammers by wild chimpanzees: a preliminary report. *J. of Hum. Evol.*, **10**, 585–93

Chevalier-Skolnikoff, S. (1984) Abstract of paper presented at 10th Congress – Int. Primatological Society. *Int. J. Primatol.*, **5**, 307

Chisholm, A. H. (1954) The use by birds of 'tools' as instruments. *Ibis*, **96**, 380–3

Ciochen, R. L. (1983) Hominoid cladistics and the ancestry of modern apes and humans; a summary statement. In *New Interpretations of Ape and Human Ancestry*, ed. R. L. Ciochen & R. S. Corruccini. New York: Plenum

Clutton-Brock, T. H. & Harvey, P. H. (1980) Primates, brains and ecology. *J. Zool. (London)*, **190**, 309–23

Eibl-Eibesfeldt, I. (1961) Uber Werkzeugebrauch des Sprechtfinken. *Z. Tierpsychol.*, **18**, 343–6

Eisenberg, J. F. & Wilson, D. (1978) Relative brain size and feeding strategies in the Chiroptera. *Evolution*, **32**, 740–51

Fragaszy, D. (In press) Time budgets and foraging behavior in wedge-capped capuchins (*Cebus olivaceus*): age and sex differences. In *Proceedings of the IXth Congress of the International Primatological Society, Vol. 1. Current Perspectives in Primate Social Dynamics*, ed. D. Taub & F. King. New York: Van Nostrand Reinhold

Gibson, K. R. (1977) Brain structure and intelligence in macaques and human infants from a Piagetian perspective. In *Primate Biosocial Development*, ed. S. Chevalier-Skolnikoff & F. Poirer, pp. 113–57. New York: Garland

Gibson, K. R. (1981) Comparative neuroontogeny, its implications for the development of human intelligence. In *Infancy and Epistemology*, ed. G. Butterworth, pp. 52–82. Brighton: Harvester Press

Gibson, K. R. & Parker, S. T. (1982) Brain structure, Piaget, and adaptation. "No I think, therefore I eat." *Behav. Brain Sci.*, 28–293

Hall, K. R. L. & Schaller, G. B. (1964) Tool using behavior in the California sea otter. *J. Mammalol.*, **45**, 287–98

Hernandez-Camacho, J. & Cooper, R. (1976) Non-human primates in Columbia. In *Neotropical Primates: Field Studies and Conservation*, ed. R. J. Thorington & P. C. Heine. New York: National Academy of Sciences

Herter, K. (1972) The Insectivores. In *Grzimek's Animal Life Encyclopedia*, ed. H. C. B. Grzimek. New York: Van Nostrand, Reinhold Company

Izawa, K. (1978) Frog-eating behavior of wild black-capped capuchin. *Primates*, **19**, 633–42

Izawa, K. (1979) Foods and feeding behavior of wild black-capped capuchin. *Primates*, **20**, 57–76

Izawa, K. & Mizano, A. (1977) Palm fruit cracking behavior of wild black-capped capuchin (*Cebus apella*). *Primates*, **18**, 773–92

Jerison, H. J. (1973) *Evolution of the Brain and Intelligence*. New York: Academic Press

Jerison, H. J. (1979) The evolution of diversity in brain size. In *Development and Evolution of Brain Size: Behavioral Implications*, ed. M. F. Hahn, C. Jensen & B. C. Dudek. New York: Academic Press

Kenyon, K. W. (1959) The sea otter. *Ann. Rep. Smithsonian Inst.*, 1958, 399–407

Kuypers, H. G. J. M. (1962) Corticospinal connections: postnatal development in the rhesus monkey. *Science*, **138**, 678–80

Lack, D. (1947) *Darwin's Finches*. Cambridge: Cambridge University Press

Le Gros Clark, W. E. (1971) *The Antecedents of Man*, 3rd edn. Edinburgh: University Press

Lettvin, J. Y., Maturana, H. R., McCulloch, W. S. & Pitts, W. H. (1959) What the frog's eye tells the frog's brain. *Proc. Inst. Radio Engineers*, **47**, 1940–51

Luria, A. (1966) *Higher Cortical Functions in Man*. New York: Basic Books

Martin, R. P. (1982) Allometric approaches to the evolution of the primate nervous system. In *Primate Brain Evolution*, ed. E. Armstrong & D. Falk. New York: Plenum

McBeath, N. M. & McGrew, W. C. (1982) Tools used by wild chimpanzees to obtain termites at Mt. Assirik, Senegal: the influence of habitat. *J. Hum. Evol.*, **11**, 65–72

McGrew, W. C. (1976) Socialization and object manipulation in wild chimpanzees. In *Primate Biosocial Development*, ed. S. Chevalier-Skolnikoff & F. Poirer. New York: Garland Press

Millikan, G. C. & Bowman, R. I. (1967) Observations of Galapagos tool-using finches in captivity. *Living Bird*, **6**, 23–42

Morse, D. H. (1968) Use of tools by brown headed nuthatches. *Wilson Bull.*, **80**, 220–4

Mountcastle, V. B. (1978) An organizing principle for cerebral function: the unit module and the distributed system. In *The Mindful Brain*, ed. G. M. Edelmann & V. B. Mountcastle. Cambridge, Mass.: MIT Press

Northcutt, R. G. (1978) Forebrain and midbrain organization in lizards and its phylogenetic significance. In *Behavior and Neurology of Lizards*, ed. N. Greenberg & P. D. Maclean, pp. 11–64. Rockville, MD: National Institute of Mental Health

Parker, S. T. & Gibson, K. R. (1977) Object manipulation, tool use and sensorimotor intelligence as feeding adaptations in great apes and cebus monkeys. *J. Hum. Evol.*, **6**, 623–41

Parker, S. T. & Gibson, K. R. (1979) A developmental model for the evolution of language and intelligence in early hominids. *Behav. Brain Sci.*, **2**, 367–407

Petter, J. J. & Petter-Rousseau, A. (1967) The aye-aye of Madagascar. In *Social Communication Among Primates*, ed. S. Altmann, pp. 195–205. Chicago: University of Chicago Press

Piaget, J. (1952) *The Origins of Intelligence in Children*. New York: Norton

Piaget, J. (1954) *The Construction of Reality in the Child*. New York: Ballantine Books

Radinsky, L. (1982) Inferences about relative brain size. In *Primate Brain Evolution*, ed. E. Armstrong & D. Falk. New York: Plenum

Regal, P. J. (1978) Behavioral differences between reptiles and mammals: an overview of mental capabilities. In *Behavior and Neurology of Lizards*, ed. N. Greenberg & P. D. MacLean. Rockville, MD: National Institute of Mental Health

Sabater-Pi & Jones, C. (1973) Comparative ecology of *Gorilla gorilla* and *Pan troglodytes* in Rio Muni, West Africa. Basel: Karger

Stephan, H. (1972) Evolution of the primate brain: a comparative anatomical investigation. In *The Functional and Evolutionary Biology of Primates*, ed. R. Tuttle. Chicago: Aldine

Sugiyama, Y. & Koman, J. (1979) Tool-using and -making behavior in wild chimpanzees at Bossou, Guinea. *Primates*, **20**, 513–24

Sussman, R. W. & Kinsey, W. G. (1984) The ecological role of the Callitrichidae: a review. *Am. J. Phys. Anthropol.*, **64**, 419–49

Suzuki, A. (1969) An ecological study of chimpanzees in savannah woodland. *Primates*, **10**, 103–48

Teleki, G. (1975) Chimpanzee subsistence technology: material and skills. *J. Hum. Evol.*, **3**, 575–94

Terborgh, J. (1983) *Five New World Primates: a Study in Comparative Ecology.* Princeton, NJ: Princeton University Press

Van Lawick-Goodall, J. (1968) The behavior of free-ranging chimpanzees in the Gombe Stream Reserve. *Anim. Behav. Monographs*, **1**, 161–311

Van Lawick-Goodall, J. & Van Lawick, H. (1966) Use of tools by Egyptian vultures. *Nature*, **12**, 1468–9

Warren, J. M. & Akert, K. (1964) *The Frontal Granular Cortex and Behavior.* New York: McGraw Hill

Wolpoff, M. H. (1983) Ramapithecus and human origins. The anthropologist's perspective of changing interpretations. In *New Interpretations of Ape and Human Ancestry*, ed. R. L. Ciochen & R. S. Corrucini. New York: Plenum

III.4

Development of feeding selectivity in mantled howling monkeys, *Alouatta palliata*

J. M. WHITEHEAD

Introduction

Herbivorous primates are faced with the formidable task of detecting nourishing food from amongst a bewildering array of tree species and their chemical constituents, both toxic and nutritious. How does a primate select its diet?

This study of the mechanisms of diet selection is based on the work of Glander (1975, 1981) on the normative feeding patterns and the chemical determinants of feeding selectivity in Costa Rican mantled howling monkeys, *Alouatta palliata*. Glander reported that his group of howlers fed predominantly on 15 of 62 tree species in a forest composed of over 100 species of trees. Over 80% of feeding time was spent in these 15 species. The abundance of a species of tree in his study area was negatively correlated with the feeding frequency on that species ($r = -0.287$); high selection ratios (percentage of total feeding time on a species/percentage of total trees present) show also that the animals fed on rare trees. Most striking is Glander's observation that the howlers fed only on some of the individual trees within a species (e.g. *Gliricidia sepium* or *Bursera simaruba*) and neglected accessible conspecific trees. Chemical analysis of the consumed and rejected parts of *Gliricidia* and *Bursera* showed that the howlers were, by and large, avoiding high concentrations of secondary compounds. This feeding pattern could imply a detailed knowledge of the location and identity of resources. Glander (1981) concluded that the diet of the howlers results from optimization of net amino acid intake through selection of plant parts high in protein and low in secondary compounds.

How do mantled howling monkeys select their diet? As *Alouatta palliata* is distributed widely in Central America and northern South America (Napier & Napier, 1967), and as the distribution of plant species in their diet varies with locality (cf. the diets of mantled howlers from southern Mexico (Estrada, 1984), western Costa Rica (Glander, 1975, 1981) and Panama (Hladik & Hladik, 1969; Milton, 1980)), learning presumably plays an important role in diet selection. The question posed in this investigation is: which learning mechanisms play a role in diet selection?

The work of ethologists (e.g. Thorpe, 1956), as well as more recent research by comparative and biological psychologists (e.g. Rozin, 1976; Garcia & Hankins, 1977; studies in Kamil & Sargent, 1981 and in Barker, Best & Domjan, 1977; Whitehead, unpublished), has identified a wide range of learning mechanisms which might affect diet selection in animals. I divide the types of mechanisms into two categories: ones which require a social environment, such as observational learning (see discussion in Clayton, 1978); and others which can operate when an animal is solitary, such as trial-and-error learning or one-trial taste aversion learning. One can therefore seek evidence for these mechanisms by examining the social context of an infant's experiences with novel plant parts in the natural habitat. For example, animals that attend to social partners while eating plant parts for the first time show evidence for a socially dependent mechanism; animals that do not attend to social partners while eating a plant part for the first time reveal the action of a learning process independent of social influences.

Because of sedentary feeding habits in the howlers and because of the unusually good visibility within the forests of western Costa Rica, I was able to document the infant feeding sequence: the sequence in which novel plant parts were eaten or tasted by infants. In addition, previous work on phenological changes in the forest (Frankie, Baker & Opler, 1974) and the presence of identified trees in the study area (Glander, 1975) made it possible to note the seasonal changes in the availability of plant parts and to detect early experiences infants had with certain foods. For example, the tree *Byrsonima crassifolia* (Malpighiaceae) fruits in the early wet season, normally between May and July. During late April one can be confident that an infant less than 1 year old observed eating a *Byrsonima* fruit has had no prior experience with this food.

These two factors, the infant feeding sequence and phenological changes in a seasonal forest, permit discrete predictions about the behavioral manifestations of socially dependent or socially independent learning mechanisms; these predictions are shown in Table 1. The term social partner includes parents, alloparents and other adults, for in principle, any animal that is a competent forager will serve as an adequate model to a naive monkey.

The first requirement for a socially dependent learning mechanism is that the learning model be near the observing infant. Proximity to the model permits the infant to see clearly what part of a plant the social partner selects. The second requirement is that the infant look at the model, especially when the infant lacks prior experience with a novel plant part. The third requirement is the restriction of feeding by infants to periods when the social partner feeds. The fourth requirement is that parents will benefit directly from preventing their offspring from ingesting a toxic plant part, as has been reported for chacma baboons (Fletemeyer, 1978). This prediction is supported by kin-selection theory (Hamilton, 1964); such intervention will achieve large benefits to close relatives while incurring minor costs. The cost of intervention is based on the deleterious effects that secondary compounds could have on an infant's detoxification system; this sytem requires induction by exposure to certain chemicals which an adult's detoxification system, already induced by prior exposure, can handle (see Freeland & Janzen, 1974). The fifth requirement is that monkeys failing to attend to a feeding model might eat some unpalatable or toxic plant parts. This requirement is based on the assumption, largely

Table 1. *Consequences resulting from the action of two classes of learning mechanisms*

Predictions about characteristics of early feeding bouts	Type of learning mechanism	
	Socially dependent	Socially independent
1. Nearest monkey to infant feeds first	Yes	No
2. Infant looks at parent before feeding	Yes	No
3. Infant feeds only when parent feeds	Yes	No
4. Parent intervenes if infant samples a non-preferred plant part	Yes	No
5. Digestive illness	Unlikely	Possible
6. Infant eats only what group members eat	Yes	No

upheld by Glander's work, that the monkeys are minimizing amounts of aversive chemicals in the diet. Thus an inattentive monkey is likely to show signs of gastrointestinal illness. Finally, an infant that attends to what social partners are eating will not eat items outside of the model's diet. Likewise, the socially independent monkey will feed on whatever is nearby, not necessarily what other group members are eating.

Once the adult diet has been learned and feeding competence attained, later feeding bouts need not show these sequential relations because the rate of learning, an unknown in this investigation, could be rapid (as it is in one-trial taste aversion learning; see Garcia & Hankins (1977)).

Methods

These predictions were tested during a 3-month study in 1979 on a group of howling monkeys located in lowland tropical dry forest near Glander's site in Guanacaste Province, Costa Rica, Central America. The study site contains two types of forest: evergreen riverine forest and upland deciduous forest whose trees are leafless for a major portion of the 4- to 5-month dry season. The study area offers excellent conditions for clear observations, which permitted Glander (1975) to keep his focal animals in view 96.4% of the time and made possible the detailed observations required for this study.

After 2 months spent learning the local tree species and observing feeding bouts in the study group, I focused intensively on the feeding behavior of two mother–infant pairs during 164 hours of contact. The infants observed were between 4 and 12 months old and were classified as infant-2 (Carpenter, 1934) on the basis of their coat color (brown-black), and their position when being carried (dorsosacral). Because of their young age, I was able to observe their initial exposures to many seasonal plant parts. When group members entered the crown of a tree, I focused on one mother–infant pair (GR) and scored who ate first, attention of the infant towards the female, the plant part eaten by the mother, the plant part eaten by the infant, and the general activities of other group members. I obtained the same data, where possible, on one other mother–infant dyad (Whitefoot). At other times I observed the infants of the focal pairs to detect feeding or tasting when other group members were not feeding. Tasting is defined as placing a plant part in the mouth once with little mastication; feeding is defined as multiple placements of plant parts in the mouth followed by mastication and ingestion.

Results

Infants ate or tasted 24 distinct parts of 14 identified species during the 164 hours of observation. Of these plant parts, 88.9% were fruits and leaves. Fruits comprised 6 of the 24 plant parts from identified species in the infant feeding sequence, a lower proportion than would be consumed earlier in the wet season when fruits are more common (Frankie et al., 1974; Glander, 1975). At the beginning of the feeding observations, the canopy in the upland deciduous forest had been fully foliose for about 3 months. There is broad overlap between the focal pairs' diet and the diet of the group studied by Glander (1975) for the comparable period of the wet season: during the 3 months of observations, the study group fed on 16 of 18 tree species available to and used by Glander's group in a contiguous area of forest. One of the two species not eaten by Glander's group (*Cassia grandis*) was eaten only by an infant.

The analysis of all observations of feeding and tasting leaves (Table 2) reveals the action of a socially dependent form of learning.

Table 2. *Results for mature and new leaves*

Predictions about characteristics of early feeding bouts	Type of learning mechanism	
	Socially dependent	Socially independent
1. Nearest monkey to infant feeds first	Yes[a]	No
2. Infant looks at parent before feeding	Yes[b]	No
3. Infant feeds only when parent feeds	Yes[b]	No
4. Parent intervenes if infant samples a non-preferred plant part	Yes	No[b]
5. Digestive illness	No[b]	Yes
6. Infant eats only what group members eat	Yes[c]	No

[a] Of 45 observations of infants feeding on leaves, infants followed the nearest adult neighbor 41 times (binomial test, $p < 0.001$).
[b] Supported by the majority of observations.
[c] Tasting the leaves of *Hymenea courbaril*, *Cassia grandis* and the new leaves of *Astronium graveolens* was observed outside of the context of feeding, largely during play bouts.

Significantly more feeding sequences involved the mother eating the leaf before the infant than the infant before the mother (binomial test, $p < 0.001$ (Siegel, 1956)). The four exceptions to this rule (the leaves of *Ficus* spp., *Anacardium excelsum*, *Terminalia chiriquensis*, and *Lysiloma seemannii*) were probably already part of the group diet. These species are not seasonally deciduous and hence the likelihood of the infants' having prior experience with them is quite high. Adults were observed feeding on mature leaves of all four species later in the sampling period.

Females and infants maintained significantly greater proximity to each other than did any other pair of group members during feeding bouts (mean distance infants–females = 1.28 m; Whitehead, unpublished). When the mother was clearly visible, infants were observed attending to her as she began to feed. For example, during feeding bouts on leaves of *Astronium graveolens* (Anacardiaceae), mothers and infants were simultaneously observable 11 of 12 times the feeding bouts began, and infants clearly attended to the mother 9 of 11 times observed ($p < 0.05$, binomial test, one-tailed (Siegel, 1956)). In general, infants were not seen eating when most of the group did not. Even so, no instances of an adult removing a plant part from an infant's mouth were observed. Nor was there any evidence of digestive illness. Finally, infants ate leaves from only those species that other group members fed on.

It is difficult to quantify the striking parallel between the mother's and infant's behavior. Often a mother entered the crown of a tree and waited until the infant had caught up with her. Then she moved to another part of the crown and adopted a feeding posture, often hanging by her tail, with the infant not far behind. Frequently the infant would then adopt the same posture. As the mother drew a branch to her mouth, the infant grasped leaves from the same branch and, following the mother, ate them. These observations provide general support for the influence of social learning on the ingestion of leaves.

In contrast to leaves, the results for fruits (Table 3) show the effects of socially independent feeding processes and are less influenced by socially dependent mechanisms. For example, during eight fully observed feeding bouts on fruits of *Ficus* spp. (Moraceae), the infant watched the female only 50% of the time before it started to feed. Although mothers fed first on *Ficus* fruits on six of eight occasions, infants fed first on fruits of *Muntingia calabura* (Elaeocarpaceae) and on the seed pods of *Cassia grandis* (Mimosaceae). Whereas infants ate leaves only after observing a social partner eating them, infants alone ate the seed pods of *C. grandis*. In addition, infants fed vigorously on fruits and tried to feed on fruit-like objects normally during the context of group feeding. Prior to the onset of systematic observations, infants attempted to eat galls, found commonly on the terminal twigs of *Licania arborea* (Rosaceae), and woody seed cases of *Luehea candida* (Tiliaceae), which are extremely hard and proved unbreakable under pressure from the infants' jaws. Nevertheless infants, along with juveniles, persisted in attempting to feed on them; in contrast adults did not attempt to feed on these plant parts. Thus, although infant feeding on fruits was still largely confined to periods when other group members fed, infant feeding on fruits was generally less coordinated with adult feeding: infants were less likely to attend to adult feeders, to wait for adults to feed first or to feed on the same plant part that adults ate.

Table 3. *Results for fruits or fruit-like objects[a]*

Predictions about characteristics of early feeding bouts	Type of learning mechanism	
	Socially dependent	Socially independent
1. Nearest monkey to infant feeds first	Yes	No[b]
2. Infant looks at parent before feeding	Yes	No[b]
3. Infant feeds only when parent feeds	Yes[b]	No
4. Parent intervenes if infant samples a non-preferred plant part	Yes	No[b]
5. Digestive illness	No[b]	Yes
6. Infant eats only what group members eat	Yes	No[b]

[a] A fruit-like object is defined as any plant part with columnoid or spheroid shape. Young animals mouthed or chewed on, but did not ingest, galls on branches of *Licania arborea* and woody seed pods of *Luehea candida*. They also chewed and ingested parts of green seed pods of *Cassia grandis*. None of these plant parts is in the adult diet.
[b] Supported by the majority of observations.

Table 4 presents the infant feeding sequence and thus reflects the first time I observed an infant eating a fruit or a leaf. In the majority of feeding sequences adults preceded infants. When infants were observed feeding first, it is possible that I had missed the infant's first encounter with these plant parts. There are two reasons for thinking this: first, because each plant part in this category was later identified as a part of the group diet, and also because each part had been commonly available since the beginning of the wet season (for about 3 months). In the only instance of its kind, an infant tasted but did not eat a leaf (*C. grandis*) outside the group's diet. Adults associated with infants were not seen tasting novel plant parts. Adults away from infants do taste them (Glander, 1981 and personal observation). This could support Glander's (1981) concept of sampling and suggests a relationship to the development of dietary preferences in young monkeys. An adult that samples a new plant part away from the group reduces its chances of being a model to an infant.

Discussion and conclusions

Two different mechanisms appear to control incorporation of plant parts into the diet of mantled howlers. The identity of selected

Table 4. *The sequence of actions by adults and infants when observed eating or tasting a novel leaf or fruit for the first time*

Action	Adult acts first	Infant acts first
Feed	15	4^a
Taste	0^b	3^c

[a] Two of the four incidents involved fruits (*Muntingia calabura*) or seed pods (*Cassia grandis*); the remainder involved new leaves of *Anacardium excelsum*, a species in the group diet, and leaves of *Terminalia chiriquensis*, a plant part later identified as a constituent of the group diet.
[b] No adults close to infants were observed tasting. Since tasting is the prelude to feeding or to the hypothesized addition of a new species to the feeding repertoire, it seldom occurs near infants that are in process of learning the group diet.
[c] Leaves of *Cassia grandis*, *Ficus* and *Guazuma ulmifolia*. The last two species are part of the group diet. Thus the only novel plant parts tasted by infants were the leaves of *Cassia grandis*.

leaves appears to be transmitted through a learning mechanism, such as observational learning, which requires a social context. This mechanism leads to conservative selection of the leafy portion of a herbivore's diet within a slowly changing environment and to minimal hazards to an infant's detoxification system resulting from ingestion of a toxic food (Freeland & Janzen, 1974). On the other hand, the ingestion of fruit-like objects, often containing sugary inducements to some seed dispersers, is not constrained by social learning. Thus howling monkeys tend to ingest fruits, to disperse seeds of a manageable size (Howe, 1980; Milton, 1980), and to enhance germination rates of some seeds (Estrada & Coates-Estrada, 1984). These two classes of foods have served as selective forces favoring the evolution of two separate feeding mechanisms. The evolutionary response to leaves, chemically well protected against herbivores (Freeland & Janzen, 1974; Feeny, 1976; Rhoades & Cates, 1976; cf. Janzen, 1978), has been a conservative, socially dependent mode of transmission of known resources, plus an individuated habit of sampling. Fruits, on the other hand, have stimulated the evolution of a permissive form of learning which ensures at least a loose coevolutionary link (Wheelwright & Orians, 1982) with tree species with fruits of an appropriate size and palatability that serve as food resources and whose seeds howlers disperse (Estrada & Coates-Estrada, 1984; for more general discussions of seed dispersal by vertebrates, see Morton, 1973; McKey, 1975; Janzen, 1983). The tendency of young monkeys to eat frugiform objects implies that the howlers may become dispersal agents for any pioneer species of tree whose fruits possess the appropriate characteristics.

A simplified behavioral model of feeding selectivity and innovation that incorporates the two-fold learning process is presented in Table 5. The contrasting approaches to fruit and leaves could be elicited by visual, gustatory and olfactory cues (as suggested by Hladik *et al.*, 1971) provided by potential food objects. Thus in eating fruits, infants attempt to taste and eat anything with primarily a fruit-like shape and secondarily a sweet taste; prior experience, in contrast, constrains adults to select fruits only, though not always ripe ones (e.g. *Ficus*). The ontogeny of leaf selection is more complex. Infants will taste leaves that adults do not, but they ingest only what they see adults eating. Adults track the changing chemical composition of the trees; upon identifying a new nutritious species, they can introduce other social partners to it through social learning. Thus members of the group can change their diet so that it satisfies variable needs (e.g. during pregnancy, during infestation with gut parasites) and responds to changes in the chemical composition of the forest (Glander, 1981).

The selection of an adequate diet, optimal or not, is one of the ecological tasks (cf. Johnston, 1981) an animal must complete. Natural selection affects the behavioral design of animals to produce competent foragers. A good principle of design (*sensu* McFarland & Houston, 1981; McFarland, 1982), undoubtedly arrived at through natural selection, is redundancy. One expects to find that the design of a biological system responsible for a task as important as diet selection is redundant; numerous mechanisms have overlapping functions which, in the face of the failure of one, can insure completing the task under most conditions. Feeding selectivity relies in part on gustatory and olfactory cues, on perception of gastrointestinal events and, in the case of leaves, on attention paid to the actions of conspecifics. Although a socially dependent learning mechanism can explain the maintenance of feeding selectivity, a socially independent mechanism is more likely to account for the origination of a particular diet. In all likelihood, both types of mechanisms are present in the monkeys and are capable of working in a complementary fashion to ensure foraging competence.

Future research should document the infant feeding sequence through a full annual cycle and address the following questions. Are any plant phenophases sufficiently toxic (e.g. *Hura crepitans*, Euphorbiaceae) to elicit parental intervention? What role does the color (or

Table 5. *Behavioral maintenance of feeding selectivity*

Infant behavior	Adult behavior
Fruits	
Taste anything with a fruit-like shape and, if not harmful, eat	Same; greater experience leads to greater selectivity
Leaves	
Ingest after adult does	Eat available plant parts in the group diet; serve as a model to other group members
When alone, taste and do not ingest	Taste and assess consequences; if leaf is not harmful, ingest it away from group and assess consequences; if unharmed, eat in presence of other group members

other features) of a plant part play in promoting or deterring its selection by herbivores (e.g. Janson, 1983; A. Gautier-Hion et al., 1985)? Are new leaves different in color from mature leaves eaten ad lib. or only under social constraints? What mechanisms appear to regulate the ingestion of flowers, a major constituent of the diet during the dry season (i.e. are they regarded as fruit-like objects because of their color and the sugar from floral nectaries)? Does adult sampling result in socially dependent learning about changing leaf quality? Finally, the roles of socially dependent or socially independent learning mechanisms in the development of other behaviors remain to be investigated. Likely candidates for such behaviors are predator-specific vocalizations, patterns of parental care, selection of arboreal pathways, and age- and context-specific emission of long-distance vocalizations.

Acknowledgements

For discussions of the original plans for this study I thank Drs Doyle McKey, Kent Redford, John Robinson, and James K. Russell. Dr William Haber helped with plant identification. Both the Organization for Tropical Studies and the Hagnauer family helped in numerous ways to make field work productive. Particular thanks are given to Dr Ken Glander whose prior work laid a solid foundation for this pilot project. Grants from the Organization for Tropical Studies and the National Science Foundation funded the study. Participants in the symposium, 'Primate cognition and behavior in nature' at the Xth Congress of the International Primatological Society, gave helpful suggestions after my talk in Nairobi. Drs Peter Klopfer, Mark Rausher, and Linda Taylor provided stimulating discussion of some ideas presented in this paper. Drs Dorothy Fragaszy, Helmut Mueller, John Oates, R. Haven Wiley and Mr Paul Bauer made constructive comments on the manuscript.

References

Barker, L. M., Best, M. R. & Domjan, M. (Eds) (1977) *Learning Mechanisms in Food Selection*. Austin, Texas: Baylor University Press

Carpenter, C. R. (1934) A field study of the behavior and social relations of howling monkeys (*Alouatta palliata*). *Comp. Psychol. Monographs*, **10**, 1–168

Clayton, D. A. (1978) Socially facilitated behavior. *Q. Rev. Biol.*, **53**, 373–92

Estrada, A. (1984) Resource use by howler monkeys (*Alouatta palliata*) in the rain forest of Los Tuxtlas, Veracruz, Mexico. *Int. J. Primatol.*, **5**, 105–31

Estrada, A. & Coates-Estrada, R. (1984) Fruit eating and seed dispersal by howling monkeys (*Alouatta palliata*) in the tropical rain forest of Los Tuxtlas, Mexico. *Am. J. Primatol.*, **6**, 77–91

Feeny, P. P. (1976) Plant apparency and chemical defense. *Recent Adv. Phytochemistry*, **10**, 1–40

Fletemeyer, J. R. (1978) Communication about potentially harmful foods in free-ranging chacma baboons. *Papio ursinus*. *Primates*, **19**, 223–6

Frankie, G. W., Baker, H. G. & Opler, P. A. (1974) Comparative studies of trees in tropical wet and dry forests in the lowlands of Costa Rica. *J. Ecol.*, **62**, 881–913

Freeland, W. J. & Janzen, D. H. (1974) Strategies in herbivory by mammals: the role of plant secondary compounds. *Am. Naturalist*, **108**, 269–89

Garcia, J. & Hankins, W. G. (1977) On the origin of food aversion paradigms. In *Learning Mechanisms in Food Selection*, ed. L. M. Barker, M. R. Best & M. Domjan, pp. 3–19. Austin, Texas: Baylor University Press

Gautier-Hion, A., Duplantier, J.-M., Quris, R., Feer, F., Sourd, C., Decoux, J.-P., Dubost, G., Emmons, L., Erard, C., Hecketsweiler, P., Moungazi, A., Roussilhon, C. & Thiollay, J.-M. (1985) Fruit characters as a basis of fruit choice and seed dispersal in a tropical forest vertebrate community, *Oecologia*, **65**, 324–37

Glander, K. (1975) Habitat and resource utilization: an ecological view of social organization in mantled howling monkeys. Ph.D. dissertation, University of Chicago

Glander, K. (1981) Feeding patterns in mantled howling monkeys. In *Foraging Behavior: Ecological, Ethological and Psychological Approaches*, ed. A. C. Kamil & T. D. Sargent, pp. 231–58. New York: Garland

Hamilton, W. D. (1964) The genetical theory of social behaviour. I, II. *J. Theoret. Biol.*, **7**, 1–52

Hladik, A. & Hladik, C. M. (1969) Rapports trophiques entre végétation et primates dans la forêt de Barro Colorado (Panama). *Terre Vie*, **1**, 25–117

Hladik, C. M., Hladik, A., Bousset, T., Valdebouze, P., Viroben, G. & Delort-Laval, J. (1971) Le régime alimentaire des Primates de l'Ile de Barro Colorado (Panama). Resultats des analyses quantitatives. *Folia Primatol.*, **16**, 85–122

Howe, H. F. (1980) Monkey dispersal and waste of a neotropical fruit. *Ecology*, **6**, 944–59

Janson, C. H. (1983) Adaptation of fruit morphology to dispersal agents in a Neotropical forest. *Science*, **219**, 187–9

Janzen, D. (1978) Complications in interpreting the chemical defenses of trees against tropical arboreal plant-eating vertebrates. In *The Ecology of Arboreal Folivores*, ed. G. G. Montgomery, pp. 73–84. Washington, D.C.: Smithsonian Institution Press

Janzen, D. (1983) Dispersal of seeds by vertebrate guts. In *Coevolution*, ed. D. J. Futuyma & M. Slatkin, pp. 232–62. Sunderland, Mass.: Sinauer

Janzen, D. & Liesner, R. (1980) Annotated check-list of plants of lowland Guanacaste province, Costa Rica, exclusive of grasses and non-vascular cryptograms. *Brenesia*, **18**, 15–90

Johnston, T. D. (1981) Contrasting approaches to a theory of learning. *Behav. Brain Sci.*, **4**, 125–73

Kamil, A. C. & Sargent, T. D. (Eds) (1981) *Foraging Behavior: Ecological, Ethological and Psychological Approaches*. New York: Garland

McFarland, D. (1982) *Functional Ontogeny*. Marshfield, Mass.: Pitman

McFarland, D. & Houston, A. (1981) *Quantitative Ethology*. Marshfield, Mass.: Pitman

McKey, D. (1975) The ecology of co-evolving seed dispersal systems. In *Coevolution of Animals and Plants*, ed. L. E. Gilbert & P. Raven, pp. 159–91. Austin, Texas: University of Texas Press

Milton, K. (1980) *The Foraging Strategy of Howler Monkeys. A Study in Primate Economics*. New York: Columbia University Press

Morton, E. S. (1973) On the evolutionary advantages and disadvantages of fruit-eating in tropical birds. *Am. Naturalist*, **107**, 8–22

Napier, J. R. & Napier, P. H. (1967) *A Handbook of Living Primates*. New York: Academic Press

Rhoades, D. F. & Cates, R. G. (1976) Toward a general theory of plant anti-herbivore chemistry. *Recent Adv. Phytochem.*, **10**, 168–213

Rozin, P. (1976) The selection of foods by rats, humans and other animals. *Adv. Stud. Behav.*, **6**, 21–76

Siegel, S. (1956) *Nonparametric Statistics for the Behavioral Sciences*. New York: McGraw-Hill

Thorpe, W. H. (1956) *Learning and Instinct in Animals*. London: Methuen

Wheelwright, N. T. & Orians, G. H. (1982) Seed dispersal by animals: contrasts with pollen dispersal, problems of terminology, and constraints on coevolution. *Am. Naturalist*, **119**, 402–13

III.5

The development of social behaviour and cognitive abilities

H. O. BOX AND D. M. FRAGASZY

As a final contribution to this section we would like to consider questions which concern the interfacing between the development of social behaviour and that of cognitive abilities. There are many gaps in our knowledge here, and it is appropriate to take this opportunity to raise some relevant issues, and make one or two suggestions for future investigation. Although the main area of interest of these papers is behaviour in nature, studies in captivity have been particularly instructive in the area so far and will provide the bulk of examples in our paper. We focus on the influence of varying social and physical factors upon the development and expression of cognitive performance. Findings drawn from captive studies are our most useful source of hypotheses concerning the role of these same factors on natural settings.

One issue concerns the physical and social conditions under which skilled performance develops. There are many potential experiments which we might consider. Some of Galdikas' (1982) work, for example, has shown that orangutans, which do not use tools extensively in nature, may readily do so if they are raised in captivity in the context of close social bonds with individuals, albeit of another species, namely our own, by whom those skills are frequently used. By contrast with wild orangs, captive-reared and subsequently rehabilitated orangs used tools in many different circumstances; they dug holes, probed, pried objects loose, hit conspecifics and other species, and so on.

Galdikas' observations draw attention to a very important potential facilitation effect of early experience (potential in the sense that individuals of whatever species, for whatever reasons, and to whatever extents, are capable of showing quantitatively different behaviour than that which they demonstrate in nature). This indicates a considerable latitude for individual experience to affect cognitive mechanisms such as attention span, response to novel objects, reactions to frustration, and so forth that underlie many aspects of normal behaviour in natural settings.

There are several points for general consideration here. First, social influences may achieve their *principal* importance by providing the appropriate context for the development of very general species-typical cognitive abilities, as with the practice of attentional skills. Individuals may develop their own skills by attending to others, and the behaviour of the other individual during this period will influence the development of such skill in the young. Thus, there is a transgenerational influence on basic cognitive mechanisms. This is certainly not a new idea, but from a comparative point of view, there are all kinds of fascinating comparisons and predictions to be made, as for instance, between the rapidly shifting attention of squirrel monkeys (*Saimiri sciureus*) and that of the more persistent attention of capuchins (*Cebus* spp.). Developmentally, it seems probable that young capuchins could practice extended concentration in the process of watching conspecifics perform routine activities, whereas young squirrel monkeys would not, at least in part, because they see individuals moving quickly from one activity to another.

Further, we know that among infants of anthropoid species the development of visuo-motor abilities is strongly influenced by interacting influences which include the attributes of the physical environment, the infant's own body and its social interactions with its mother. In human children, for example, the powerful influence of social interaction in cognitive development is evidenced by a good deal of mutual eye contact; that visual attention of the infant soon begins to focus on physical objects and is encouraged by manipulations and language. Intense mutual eye contact does not seem to be typical of the interactions among mothers and new offspring of other primate species, including chimpanzees (*Pan troglodytes*) in their first 3 months. Again, comparative data are required. In this context, it is interesting that Vauclair (1982) reported that he has studies in progress with Savage-Rumbaugh on the development of object manipulation in humans and the two species of chimpanzee, within the context of mother–infant communication.

We have few data on the interactive influences of social relationships of varying qualities with varying complexities of the physical environment among species. Jensen, Bobbitt & Gordon (1968) showed that mother–infant relationships in pigtail macaques (*Macaca nemestrina*) are affected by the richness of the physical environment. In general, however, much remains to be learned about the specificity of influence of mother–infant interactions and their influences upon the interactions with the physical environment and cognitive development generally.

We have so little information from non-human primates on *any* of these issues and especially in nature, that it is certainly worth mentioning that research on humans has reported interesting data on the influence of the quality of social relationships and their social attentional propensities within which young children develop practical skills. For example, Cohler & Grunbau (1977) found that children of depressed mothers showed significantly less-than-normal spans of attention. Again, Belsky, Goode & Most (1980) showed that verbal and physical attention-focusing strategies by normal mothers towards 18-month-old children were positively related to more developed cognitive functioning in these infants. The interest of some workers is now concentrating on the fact that 'while children show normal development amid a wide variety of upbringings and environmental conditions, it is possible to establish specific association between characteristics of maternal style, taken singly or together, and developmental outcome as evidenced by child cognitive emotional and social functioning' (Maggie Mills, unpublished).

These kinds of concerns will perhaps encourage us to acquire more information on the analysis of social relationships, on patterns of interaction, with reference to a variety of behaviour involving social-cognitive skills in nonhuman primates. We do know, for example, that relatively high or low social status within a particular social group may influence individual performance in various species in formal learning tasks, as in squirrel monkeys for example (Clark & Gay, 1978). However, dimensions of comparison have been relatively crude; we really do not know what the social influences are, whether there is a fearfulness by low status animals in avoiding new situations, or whatever. In any case, we need developmental studies, irrespective of our dimensions of comparison. Stevenson-Hinde & Simpson (1981) have made a start on this task by attempting to correlate maternal and infant characteristics.

A way in which we could improve our knowledge here is to draw more heavily upon experimental studies of responsiveness to new situations (e.g. Menzel, 1966, 1969, 1973; Fragaszy & Mason, 1978, 1983). In short, the techniques may be useful for comparing a range of comparative abilities with reference to the detection of cumulative change within the physical environment, as well as a powerful and realistic technique for assessing individual responsiveness to change under various social conditions. The situations do not have to be novel; any aspect of the natural environment, for instance, may be modelled, and there is a good example here from Andrews' work on spatially distributed resources (see Chapter III.1). Novel environments or stimuli may be particularly useful, however, in obtaining first-time measures of latency to respond, persistence of responses, diversity of strategies of responding, social interactions, and so on.

The work of one of us (HOB) on the behavioural development of twin common marmosets (*Callithrix jacchus*) involves techniques using 'novelty'. We are recording a wide range of behaviours and comparing the relative proportions and relations among behavioural activities within twin groups, compared with those of each individual twin with regard to all other members of their natal family groups (usually four individuals). We emphasise individual behavioural ontogeny in captive animals without intervention. Occasionally, the physical environment is also changed somewhat, as for example by the addition of new objects.

Experiments of this kind provide ways of looking at who does what with whom, the persistence of the developing relationships, the emancipation of the young, and so on. It can at least give clues to the 'social value' of different individuals for any one individual within a social group, and how the social attention of a developing animal changes over time. Such techniques are also obvious candidates for comparative studies. As we know, differences among species in their manipulative skills, curiosity, emotionality, etc., may be substantial, and there has been a history of experimental work going back to the mid-1960s by Jolly (1964), Glickman & Sroges (1966) and later by Mason and Fragaszy (Mason, 1968, 1974; Fragaszy & Mason, 1978, 1983), showing that different psychological attributes of behaviour are related to the behavioural ecology of the various species. What we need now is much more information on the social developmental domain of ecological variability. There is obviously a lot of scope.

Now let us develop some of these themes in a rather more natural context. The social contexts of learned behaviour, for example, provide their emotional and motivational climate, as well as providing models of behaviour. Social influences can both facilitate and inhibit behaviour among group members. The whole emotional climate of a social unit, together with the prevailing ecological conditions, influence behaviour. Kawamura (1965), for instance, found that individual Japanese macaques (*Macaca fuscata*) affected the rate at which a new food habit occurred within different troops. In some troops the leaders were socially relaxed and individuals accepted food easily. In a troop in which the leader was aggressive, however, it took longer for monkeys to accept the food. In fact, differences among troops in the acceptance of food when the same methods of provisioning were used ranged from 1 week to 1.5 years.

Another good example of the influence of individual differences on the acquisition of a new feeding habit is given by Strum (1981) in her studies of hunting and amicable sharing of meat within a troop of olive baboons (*Papio anubis*) over a period of 3 years. Among other things, her observations suggested that the behaviour may have been influenced by particular behavioural characteristics of individuals. In one case, an adult male displaced in status another male who had previously been the most successful at both predatory behaviour and holding on to food. The 'new' male engaged in a good deal more predatory behaviour and ate much more meat thereafter. Interestingly, he was independent in this behaviour, ranging further and further away on his own in his predatory behaviour. At the same time he maintained friendly and integrated relationships with all members of his troop. Strum suggested that the idiosyncratic behaviour of this animal may have facilitated increased predatory behaviour among other males.

As a general point, we need to develop techniques to make quantitative assessments of individual profiles of behaviour in the field, where conditions are appropriate. There are certainly valid techniques for assessing personality among individuals in captivity (e.g. Stevenson-Hinde & Simpson, 1981).

Questions related to behavioural *innovators* are important ones. Kummer (1971) has emphasised that inventive individuals increase the repertoire of behaviour of a group, and the range of adaptations which it makes to the environment. Field observations are obviously often very difficult to make, however, unless one has long-term studies of accessible individuals, but captive studies may be useful here. Moreover, *captivity* may also provide opportunities to study situations in which the innovation of an idiosyncratic behaviour may or may not be taken up by other members of a social unit. There are many questions we can ask. For example, various species of baboons may show tool-using by individuals in captivity but the skill does not spread (Beck, 1980). Why not? We may suspect some kind of social influence (as in some troops of Japanese macaques that we mentioned) – but what? We need to look closely at patterns of social interactions with reference to skilled performance, as we have also emphasised earlier.

Innovations are often initiated by young animals and we want to know more about their individual characteristics and social developmental histories. Again, individuals which are innovative in one kind of behaviour are often innovative in others. The young female Japanese macaque which invented potato washing for instance, also started to wash wheat. Moreover, the development of a practical skill, such as washing food in water, gave rise to the use of water in other ways, such as in swimming. In fact, it is generally assumed that an important difference between higher primates and other taxonomic groups is the extent to which socially mediated learning may be cumulative and the kinds of environmental conditions in which such learning may take place. It is important to obtain comparative data across mammalian orders, as well as within the primates to confirm and refine this assumption. One may expect substantial differences in socially mediated learning among species that possess differential behavioural adaptability. We need techniques with which to make comparisons. In this context we may note that Parker & Gibson (1977) advocate the use of a Piagetian model as a useful framework within which to compare aspects of social learning as in imitation and observational learning, i.e. by correlating imitative ability with various degrees of sensorimotor intellectual development.

We also need to know more about the natural contexts in which individuals learn from other individuals. There is plenty of circumstantial evidence that particular social influences are important, as with young chimpanzees learning to fish for termites in the presence of their skilled mothers (McGrew, 1977). Moreover, as we have already mentioned, social factors influence the extent to which individuals attend to one another, and to their behaviour. Again, studies of Japanese monkeys are important here. They are very long term and represent cases of behavioural innovation which were observed from the start, and subsequently become social traditions, so that the routes of social transmission can be studied also. These are classic examples, but we need much more comparative data. An experimental field study by Cambefort (1981) is a particularly interesting and rare extension to the more familiar work. In one experiment chacma baboons (*Papio ursinus*) and vervet monkeys (*Cercopithecus aethiops*) were compared in situations where palatable bait was clued by various markers. Cambefort found that members of both species learned, but that the two species differed considerably in the social patterns by which a new food habit was propagated. It is obvious that we need to have more such data and to correlate species differences in this domain, with various aspects of behavioural ecology.

Finally we should emphasise a point which we very briefly alluded to earlier, namely that innovation and the influence of innovators are influenced by current ecological conditions. In Strum's study, for example, wide-ranging predatory behaviour was much less hazardous for these baboons because large carnivores within the area had been removed. Further, Kawai (1965) noted that similar inventive behaviour was observed among some few individuals in a number of troops of Japanese macaques, but that it did not become part of the traditional behaviour of all those troops. He pointed out that on Koshima Island (of potato washing and other fames), a combination of different types of vegetation, sea and sand, gave more variety of feeding grounds, and that these helped to facilitate both the acquisition and the transmission of new behaviour.

In this general context also, we should mention the question of time budgets of behaviour under different ecological conditions as providing varying degrees of social and practical experience of the environment. Two examples will suffice. Baldwin & Baldwin (1972) showed that the amount of play shown by different troops of squirrel monkeys varied from zero to 3 hours per day, depending on the ecology of the area and the size of the troop. Hamilton, Buskirk & Buskirk (1978) emphasised the 'economics' of behaviours such as tool-using by comparing the behaviour of different troops of the same species of baboon living in a desert ravine area and a swampy area. For example, with regard to feeding strategies, sandy conditions and regular annual flooding in the desert ravine did not facilitate a dense enough population of termites to make it profitable for the baboons living in this area to move quantities of debris in searching for them. Similarly, it may not have been profitable for the baboons living in the swampy area to mine seeds and figs in the relatively hard soil of the swamp, and in which buried food was less abundant anyway.

Some of the above comments relate to material covered in this section, but not to all of it. We have mentioned issues which are of particular interest to our own thinking in this area and which will hopefully provoke discussion. The contributions to this section represent a good range by discussing four main areas, namely: resource exploitation, observational learning, cognitive mapping and instrumental sensory-motor skills. In so many cases in the literature cognitive abilities are discussed in very limited contexts, as in formal laboratory tasks, or in tool-using situations in nature. We are now beginning to appreciate, however, that cognitive abilities are especially important in influencing social relationships and vice versa among higher primates, compared with their influence in any other animal taxa. We need to follow up this appreciation by developing appropriate methods for the study of cognitive aspects of social life (Mason, 1982). There is also a growing appreciation that cognitive attributes are expressed in everyday non-social activities such as moving in space, feeding, resting and sheltering from rain, as well as in social activities (Chevalier-Skolnikoff, Galdikas & Skolnikoff, 1982; Fragaszy, 1985). It happens that with limited space here we have chosen to emphasise particular examples. However, in more general terms it is a healthy methodological and theoretical state of affairs that we are now beginning to know what kinds of questions to ask about primate cognition in nature; that we are beginning to find out what different species know about their physical environments and about each other (Seyfarth & Cheney, 1983). On the other hand, the fact that there is no comprehensive theory which embraces the interfacing of the cognitive social nexus in any context represents a serious gap in behavioural biology. There is a need, as Alison Jolly has emphasised, to develop an integrated description of perceptual, motor, cognitive and social aspects of behaviour.

References

Baldwin, J. D. & Baldwin, J. I. (1972) The ecology and behaviour of squirrel monkeys (*Saimiri oerstedi*) in a natural forest in Western Panama. *Folia Primatol.*, **18**, 161–84

Beck, B. B. (1980) *Animal Tool Use*. New York: Garland STPM Press

Belsky, J., Goode, E. & Most, R. (1980) Maternal stimulation and infant-exploratory competence: cross sectional, correlational and experimental analysis. *Child Dev.*, **51**, 1168–78

Cambefort, J. P. (1981) A comparative study of culturally transmitted patterns of feeding habits in the chacma baboon *Papio ursinus* and the vervet monkey (*Cercopithecus aethiops*). *Folia Primatol.*, **36**, 243–63

Clark, D. L. & Gay, P. E. (1978) Behavioural correlates of social dominance. *Biol. Psychiatry*, **13**, 445–54

Chevalier-Skolnikoff, S., Galdikas, B. M. F. & Skolnikoff, A. Z. (1982) The adaptive significance of higher intelligence in wild orang-utang: a preliminary report. *J. Hum. Evol.*, **11**(7), 639–52

Cohler, E. A. & Grunbau, F. (1977) Disturbance of attention among schizophrenic, depressed, and well mothers and children. *J. Child Psychol. Psych.*, **18**, 115–35

Fragaszy, D. M. (1979) Squirrel and titi monkeys in a novel environment. In

Captivity and Behavior, ed. J. Erwin, T. Maple & G. Mitchell. New York: Holt, Rinehart and Winston

Fragaszy, D. M. (1985) Cognition in squirrel monkeys. A contemporary perspective. In *Handbook of Squirrel Monkey Research*, ed. L. A. Rosenblum & C. Coe. New York: Plenum

Fragaszy, D. M. & Mason, W. A. (1978) Response to novelty in *Saimiri* and *Callicebus*: influence of social context. *Primates*, **19**, 311–31

Fragaszy, D. M. & Mason, W. (1983) Comparative studies of feeding in captive squirrel and titi monkeys. *J. Comp. Psychol.*, **97**, 310–26

Galdikas, B. M. F. (1982) Orangutan tool-use at Tanjung Puting Reserve, Central Indonesian Borneo (Kalimantan Tengah). *J. Hum. Evol.*, **11**, 19–33

Glickman, S. E. & Sroges, R. W. (1966) Curiosity in zoo animals. *Behaviour*, **26**, 151–88

Hamilton, W. J., III, Buskirk, R. E. & Buskirk, W. H. (1978) Environmental determinants of object manipulation by chacma baboons (*Papio ursinus*) in two Southern African environments. *J. Hum. Evol.*, **7**, 205–16

Jensen, G. D., Bobbitt, R. A. & Gordon, B. N. (1968) Effects of environment on the relationship between mother and infant pigtailed monkeys (*Macaca nemestrina*). *J. Comp. Physiol. Psychol.*, **66**, 259–63

Jolly, A. (1964) Prosimian manipulation of simple object problems. *Animal Behaviour*, **12**, 560–70

Jolly, A. (1974) The study of primate infancy. In *The Early Growth of Competence*, ed. K. Conolly & J. Bruner. New York: CIBA Foundation & Academic Press

Kawai, M. (1965) Newly acquired pre-cultural behaviour of a natural troop of Japanese monkeys. *Seibutsu Shinka*, **2**(1)

Kawamura, S. (1965) Sub-culture among Japanese macaques. In *Monkeys and Apes – Sociological Studies*, ed. S. Kawamura & J. Itani. Tokyo: Chuokoronsha

Kummer, H. (1971) *Primate Societies: Group Techniques of Ecological Adaptation*. Chicago: University of Chicago Press

Mason, W. A. (1968) Naturalistic and experimental investigations of the social behaviour of monkeys and apes. In *Primates: Studies in Adaptation and Variability*, ed. P. Jay. New York: Holt, Rinehart and Winston

Mason, W. A. (1974) Comparative studies of social behaviour in *Callicebus* and *Saimiri*: behaviour of male–female pairs. *Folia Primatol*, **22**, 1–8

Mason, W. A. (1982) Primate social intelligence: contributions from the laboratory. In *Animal Mind – Human Mind*, ed. D. R. Griffin, Dahlem Konferenzen. Berlin: Springer-Verlag

McGrew, W. C. (1977) Socialization and object manipulation of wild chimpanzees. In *Primate Biosocial Development*, ed. F. E. Poirier & S. Chevalier-Skolnikoff, pp. 216–88. New York: Garland

Menzel, E. W., Jr (1966) Responsiveness to objects in free-ranging Japanese monkeys. *Behaviour*, **26**, 130–50

Menzel, E. W., Jr (1969) Naturalistic and experimental approaches to primate behaviour. In *Naturalistic Viewpoints in Psychological Research*, ed. E. P. Willems & H. L. Raush. New York: Holt, Rinehart and Winston

Menzel, E. W., Jr (1973) Chimpanzee spatial memory. *Science*, **182**, 943–5

Parker, S. T. & Gibson K. R. (1977) Object manipulation, tool use and sensory motor intelligence as feeding adaptations in cebus monkeys and great apes. *J. Hum. Evol.*, **6**, 623–41

Seyfarth, R. M. & Cheney, D. L. (1983) How monkeys see the world: a review of recent research on East African vervet monkeys. In *Primate Communication*, ed. C. Snowdon, C. Brown & M. Peterson, pp. 239–52. Cambridge: Cambridge University Press

Stevenson-Hinde, J. & Simpson, M. J. A. (1981) Mother's characteristics, interactions and infants characteristics. *Child Dev.*, **52**, 1246–54

Strum, S. C. (1981) Processes and products of change: baboon predatory behaviour at Gilgil, Kenya. In *Omnivorous Primates: Gathering and Hunting in Human Evolution*, ed. R. S. O. Harding & G. Teleki. New York: Columbia University Press

Vauclair, J. (1982) Sensorimotor intelligence in human and non-human primates. *J. Hum. Evol.*, **11**, 257–64

Part IV

Perception and Performance: Growth, manipulation and communication

Editors' introduction

The previous two sections have emphasised the new and exciting research underway on primate cognition. They have concentrated on how to define cognition in primates, and how to describe it in natural environments. In this section, the issues of how cognitive skills arise in development are explored in several papers on early brain and physical development. The development of the brain and body are then related to the locomotory and cognitive abilities of different species of primates. Most of these papers have in common a comparative approach where the brain development, motor skills and communication of monkeys, apes and humans are compared and contrasted. The issues of how monkeys perform in and perceive their world are used to clarify our understanding of human growth, cognitive development, and communication.

In Chapter IV.1, DeVito *et al.* discuss the relation between foetal brain size and body weight among rhesus monkeys. They find allometrically increasing volumes of different brain structures with increasing foetal age, and differential rates of growth for various brain subdivisions. While their study has particular applications for understanding developmental abnormalities related to prematurity, they provide normative data on both foetal and infant brains. Sharma & Lal (Chapter IV.2) investigate body growth, weight gain, and dentition among rhesus monkeys up to one year old. They find the highest rates of growth in the first 3 months and little sexual dimorphism in monkeys less than 1 year old. Both of these studies allow us to make generalisations about the normal patterns of development in body and brain size.

Dienske (Chapter IV.3) relates rates of foetal and postnatal brain growth and behaviour, such as sleep patterns, contact with mother and social stimuli, to differences between monkeys, apes and humans in the timing of the appearance of different developmental 'milestones' and motor capacities. He suggests that maternal interference in the activities of infants is more pronounced in humans and leads to an increase in the rate at which behavioural items are added to the human repertoire over that of monkeys or apes.

Spinozzi & Natale (Chapter IV.4) then use a similar comparative approach to assess the interaction between rates and timing of cognitive development. They examine independent locomotion in particular for its influence on the development of prehension activities and on the exploration of external objects.

The final two papers deal with aspects of primate communication. In Chapter IV.5, Peters suggests that the classification of primate vocalisations by observers into separate calls may overlook the importance of a graded sequence of calls. In these types of vocalisations, one call can occupy a variable position along one or more gradients. She suggests that in a graded sequence, an immense number of calls can be generated by a small number of contributing variables in terms of form and frequency. Liska (Chapter IV.6) examines the communication of primates in order to provide clues to the origins and evolution of human symbolic behaviour. Her approach, like so many of the papers in this section is comparative, examining the natural communication of different species, including humans.

IV.1

Morphometry of the developing brain in *Macaca nemestrina*

J. L. DeVITO, J. GRAHAM, G. SCHULTZ,
J. W. SUNDSTEN AND J. W. PROTHERO

Introduction

Among mature specimens of Mammalia, brain weight has been found to scale allometrically with body weight, with slopes ranging from 0.67 to 0.76 for the regression line (Holt *et al.*, 1975; Armstrong, 1983). Brain weight is also related to life span and metabolic rate (Sacher, 1959; Armstrong, 1983; Hofman, 1983); for example, animals with small brains usually have high metabolic rates and short life spans. During ontogenetic growth, brain weight appears to be directly proportional to body weight, the regression coefficient for the two variables being close to unity. This relation has been shown for the human fetus and for fetal *Macaca mulatta* (Holt *et al.*, 1975), and we have confirmed it in a series of fetal *M. nemestrina*.

Expansion of the cerebral cortex is the major factor in the growth of the primate brain (Jenkins, 1921; Stephan & Andy, 1964), and the greatest rate of expansion generally occurs before birth (Cheek, 1975). The cerebellum starts its growth spurt later than the cerebral cortex, and its rate of growth is higher than that of the cerebral cortex. These periods of rapid growth in the fetal and infant brain have been termed 'critical periods', and it is thought that developing structures are most vulnerable to external influences at these times (Dobbing, 1970; Winick, Rosso & Waterlow, 1970; Chase, 1973; Dobbing & Sands, 1973; Cheek, 1975). Growth rates have been determined for the major dissectable subdivisions of the brain (Jenkins, 1921; Portman, Alexander & Illingworth, 1972; Dobbing & Sands, 1973; Cheek, 1975; Holt *et al.*, 1975). There is little documentation, however, for the growth rates of neural structures that require microscopic identification.

We are preparing serial histological sections of the brains of age-dated fetal and infant *M. nemestrina*. We have identified and outlined various brain subregions, and calculated their areas and volumes, in order to determine the absolute and relative growth rates of a large number of neural entities. This paper presents some initial findings for specific brain regions. Preliminary data have been published in abstracts (DeVito et al., 1984; Sundsten et al., 1984).

As well as providing important information on neural development, the analysis of this series of fetal and infant brains will serve as a normative base for assessing abnormalities related to prematurity (Sackett, 1981).

Methods

To test the normalcy of our collection, a regression line was fitted to the log of brain weight plotted against the log of body weight for 24 age-dated *M. nemestrina* fetuses. The specimens ranged in age from 60 to 167 days post-conception; brain weights ranged from 1.34 to 61.1 g and body weights from 8.1 to 585 g.

Histological sections were prepared of brains from six *M. nemestrina* of post-conception ages 105–311 days (Table 1). The 105-day fetus was perfused with 4% paraformaldehyde in 0.1M phosphate buffer and the brain was postfixed in 10% neutral buffered formalin for 8 months. The other five brains were fixed by immersion in 10% neutral buffered formalin for 2–17 months. Brains were embedded in celloidin and cut coronally at 40 μm. A 1-in-20 series through the telencephalon and a 1-in-10 series through the rest of the brain were stained with cresyl violet for Nissl substance; a 1-in-20 series was stained with Weil's method for myelin. For the 105-day specimen an additional series of

Table 1. *Developmental data on animals used for morphometry*

Animal number	Age (days)			Sex	Weight (g)	
	Gestational	Post-natal	Post-conception		Body	Brain
T83044	105	0	105	M	231.3	25.30
F83254	137	0	137	F	277.0	39.29
T83002	141	1	142	F	328.0	45.99
T82456	163	0	163	M	485.0	60.52
T82459	190	22	212	M	638.5	66.33
F83033	164	147	311	F	755.0	67.53

All body weights were taken on the day of death, except in the case of animal T83002, in which the body weight was taken one day before death. All brain weights were taken just before histological processing. M, male; F, female.

1-in-20 through the telencephalon and 1-in-10 through the rest of the brain was stained with hematoxylin and eosin.

Projection drawings of the major subdivisions of the brain (1-in-40 series of sections stained with cresyl violet) and various nuclei and fiber tracts (1-in-10 series of sections stained with cresyl violet) were made with the aid of a Bausch and Lomb microprojector or an Omega enlarger. A computer-based morphometric system (Sundsten & Prothero, 1983) was used to digitalize the profiles and to calculate the areas and volumes of specific neural structures and the volume of the whole brain. Volumes of neural structures are combined values for both sides of the brain, except where indicated.

To determine the shrinkage factor, we divided the weight of the unprocessed brain by the volume of the processed brain as calculated with the computer. To obtain corrected volumes, we multiplied the computer-calculated volumes of neural structures by the shrinkage factor. The specific gravity of the brains averaged 1.045 (1.040–1.051). Because of the small range of specific gravities and the values being close to unity, no correction was made for this factor.

Results

For the sample of 24 fetal *M. nemestrina*, the slope of the regression line of log brain weight plotted against log body weight was 0.976 ($r = 0.995$).

Fig. 1A shows the absolute values for the volumes of the major brain subdivisions plotted against post-conception age. For all structures, the greatest rate of growth occurred in the last 30 days of gestation. With the exception of the cerebellum, most of the volumetric growth also occurred before birth. In the case of the cerebellum, its size nearly doubled in the five post-natal months. The prosencephalon and telencephalon increased slightly in volume post-natally, and then plateaued. The diencephalon, mesencephalon, and medulla/pons either decreased slightly or remained fairly constant in the post-natal period.

In Fig. 1B the volumes of the major components of the prosencephalon are expressed as percentages of prosencephalic volume and plotted against post-conception age. The contribution of gray matter to the total prosencephalic volume increased dramatically up to 212 days and then stabilized. There was a corresponding decrease and subsequent leveling off in the volume per cent of prosencephalic white matter. The relative contributions of the diencephalon and telencephalic nuclei remained almost constant throughout the sample period.

Fig. 1. A. Volumes of the major subdivisions of the brain plotted against post-conception age. Note change in Y-axis scale. B. Components of the prosencephalon plotted as per cent of prosencephalon (volume of component divided by volume of prosencephalon × 100) against post-conception age. Right half-brains only. Arrows indicate average length of gestation (170 days) in *Macaca nemestrina* in this study.

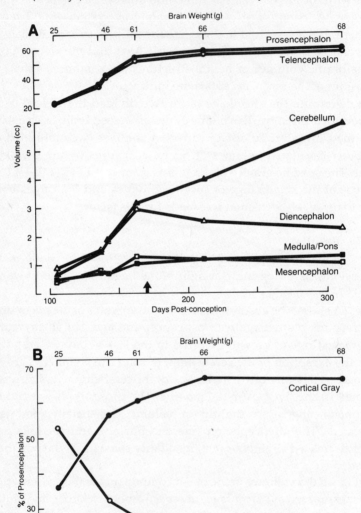

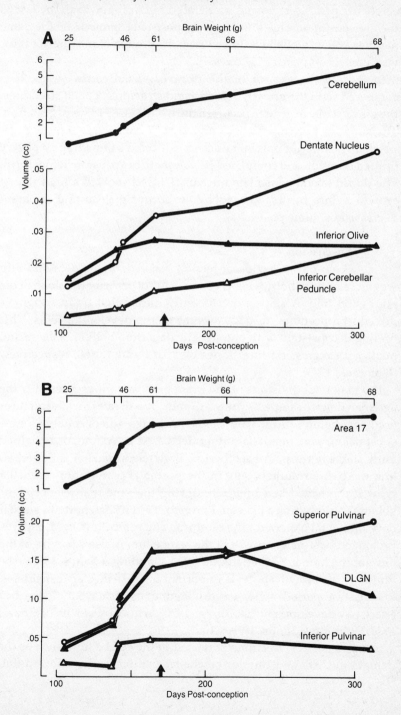

Fig. 2. A. Volumes of the cerebellum and associated structures plotted against post-conception age. B. Volumes of area 17 and several subcortical visual structures plotted against post-conception age. Note changes in Y-axis scales. Arrows indicate average length of gestation (170 days) in this study.

In Fig. 2A the growth of the cerebellum is compared with that of some associated structures. The dentate nucleus and the inferior cerebellar peduncle showed accelerated rates of growth in the last 30 days of gestation and a continuing post-natal growth in the same manner as the cerebellum. In contrast, the growth of the inferior olive was almost complete at 137 days.

In Fig. 2B the growth of the primary visual cortex (area 17) is compared with the growth of associated structures. For all structures, the greatest rate of growth occurred during the last 30 days of gestation. The superior pulvinar nucleus and area 17 continued to grow post-natally. The inferior pulvinar nucleus exhibited a growth spurt of very limited duration and completed its volumetric expansion before birth. The dorsal lateral geniculate nucleus (DLGN) showed a high rate of growth before birth, followed by an abrupt plateau and apparent decrease in volume post-natally.

Discussion

From our collection of fetal *M. nemestrina*, the data on brain weight relative to body weight gave a value of 0.976 for the slope of the regression line. A slope close to unity means that brain weight is directly proportional to body weight in the developing fetus. This finding is consistent with ontogenetic data from both *M. mulatta* and man, with regression line slopes of 1.0164 and 1.0160, respectively (Holt *et al.*, 1975).

The major subdivisions of the brain grew at various rates over the age span that we sampled. The maximum growth rates for all structures occurred before birth. The stem structures (diencephalon, mesencephalon, pons/medulla) either decreased in volume or remained fairly stable in the post-natal period. Both telencephalon and cerebellum increased volumetrically in the postnatal period, but cerebellar growth proceeded at a high rate during the time that telencephalic volume was reaching a plateau. Portman *et al.* (1972) reported a similar developmental pattern for the cerebrum and cerebellum in *M. mulatta*. Prolonged post-natal growth of the cerebellum is also a feature of the developing human brain (Jenkins, 1921; Dobbing & Sands, 1973). The timing of the growth spurts is important because they are considered to be critical periods when environmental insults have their greatest effects on development (Dobbing, 1970; Winick *et al.*, 1970; Chase, 1973; Dobbing & Sands, 1973; Cheek, 1975).

Some structures functionally related to the cerebellum, such as the dentate nucleus and inferior cerebellar peduncle, also showed volumetric increases during the post-natal period. However, the inferior olivary complex, which has reciprocal connections with the cerebellum and projects through the inferior cerebellar peduncle, had its peak growth rate early in gestation (prior to 137 days). This finding is consistent with the report by Robertson & Stotler (1974) that in *M. mulatta* the inferior olive shows its greatest cellular differentiation and fiber development between 60 and 130 days of gestation. In the cerebellum, Purkinje-cell development is delayed until after 100 days, when climbing fibers from the inferior olive first appear and begin to induce dendritic formation (Kornguth & Scott, 1972).

Prosencephalic growth was largely attributable to an increase in cortical gray matter. The volume per cent of cortical gray matter increased dramatically during the second half of gestation and continued to rise for about 40 postnatal days. As the percentage of cortical gray matter increased, white matter showed a proportionate decrease. After 212 days, the volume per cent of white matter stabilized. Examination of our Weil-stained sections revealed that between 163 and 212 days, myelination of fiber tracts becomes well established, and this may account for the plateau in volume per cent of white matter. The contribution of the telencephalic nuclei and diencephalon to prosencephalic volume remained relatively stable over all gestational ages.

In the visual system, the volume of area 17 (primate visual cortex) continued to increase post-natally, whereas the DLGN appeared to stop growing at birth and even to decrease in size post-natally. A similar overshoot in DLGN growth has been reported for the tree shrew (Kip, 1978). In man, Preobraschenskaja (1966) found a reduction in the rate of growth of DLGN shortly after birth, but growth continued at a slower rate for several years. We do not yet have enough post-natal data to determine whether there is a true difference between man and *M. nemestrina* in DLGN development. The difference in the growth curves of the superior and inferior divisions of the pulvinar is rather surprising, and not easily explained by differences in connectivity (Campos-Ortega & Hayhow, 1972; Mathers, 1972; Benevento & Fallon, 1975; DeVito, 1978). The inferior pulvinar appears to reach peak volume well before birth, whereas the superior pulvinar grows post-natally with the same time course as area 17.

These are preliminary results of a project designed to establish growth curves for a wide range of developing neural structures. More specimens will be added at each developmental horizon, and the age range will be extended both pre- and post-natally. We have shown that the growth pattern of a major subdivision such as the diencephalon cannot predict the growth patterns of specific substructures. Some components are decreasing in volume at the same time that others are growing rapidly and still others have reached a plateau. By comparing rates of growth and the timing of growth spurts among various neural structures, we hope to provide new information on brain development. In addition, these experiments will serve as a normative base for experimental manipulations.

Acknowledgments

We wish to thank Kathleen D. Schmitt for editing and typing the manuscript, Thomas A. Stebbins and Kathleen E. Sweeney for the illustrations, and Anne C. Chambers, Karen E. Robbins and Carolyn A. Thostenson for the histology.

References

Armstrong, E. (1983) Relative brain size and metabolism in mammals. *Science*, **220**, 1302–4

Benevento, L. A. & Fallon, J. H. (1975) The ascending projections of the superior colliculus in the rhesus monkey (*Macaca mulatta*). *J. Comp. Neurol.*, **160**, 339–62

Campos-Ortega, J. A. & Hayhow, W. R. (1972) On the organization of the visual cortical projection to the pulvinar in *Macaca mulatta*. *Brain, Behav. Evol.*, **6**, 394–423

Chase, H. P. (1973) The effects of intrauterine and postnatal under-nutrition on normal brain development. *Ann. N.Y. Acad. Sci.*, **205**, 231–44

Cheek, D. B. (1975) The fetus. In *Fetal and Postnatal Cellular Growth: Hormones and Nutrition*, ed. D. B. Cheek, pp. 3–22. New York: John Wiley & Sons

DeVito, J. L. (1978) A horseradish peroxidase-autoradiographic study of parietopulvinar connections in the *Saimiri sciureus*. *Exp. Brain Res.*, **32**, 581–90

DeVito, J. L., Graham, J., Schultz, G., Sundsten, J. W. & Prothero, J. W. (1984) Morphometry of the developing brain in *Macaca nemestrina*. *Int. J. Primatol.*, **5**, 331

Dobbing, J. (1970) Undernutrition and the developing brain. The relevance of animal models to the human problem. *Am. J. Dis. Child.*, **120**, 411–15

Dobbing, J. & Sands, J. (1973) Quantitative growth and development of human brain. *Arch. Dis. Child.*, **48**, 757–67

Hofman, M. A. (1983) Energy metabolism, brain size and longevity in mammals. *Q. Rev. Biol.*, **58**, 495–512

Holt, A. B., Cheek, D. B., Mellits, E. D. & Hill, D. E. (1975) Brain size and the relation of the primate to the nonprimate. In *Fetal and Postnatal Cellular Growth: Hormones and Nutrition*, ed. D. B. Cheek, pp. 23–44. New York: John Wiley & Sons

Jenkins, G. B. (1921) Relative weight and volume of the component parts of the brain of the human embryo at different stages of development. *Contrib. Embryol.*, **13**, 41–60

Kip, G. (1978) Qualitative und quantitative Untersuchungen des Corpus Geniculatum Laterale an einer ontogenetischen Reihe von männlichen *Tupaia belangeri*. *J. Hirnforsch.*, **19**, 345–70

Kornguth, S. E. & Scott, G. (1972) The role of the climbing fibers in the formation of Purkinje cell dendrites. *J. Comp. Neurol.*, **146**, 61–82

Mathers, L. H. (1972) The synaptic organization of the cortical projection to the pulvinar of the squirrel monkey. *J. Comp. Neurol.*, **146**, 43–60

Portman, O. W., Alexander, M. & Illingworth, D. R. (1972) Changes in brain and sciatic nerve composition with development of rhesus monkey (*Macaca mulatta*). *Brain Res.*, **43**, 197–213

Preobraschenskaja, N. S. (1966) Die zytoarchitektonischen Besonderheiten der Rinde des Okzipitalgebietes und einiger subkortikaler Bildungen des Gehirns während der Entwicklung. *J. Hirnforsch.*, **8**, 269–81

Robertson, L. T. & Stotler, W. A. (1974) The structure and connections of the developing inferior olivary nucleus of the rhesus monkey (*Macaca mulatta*). *J. Comp. Neurol.*, **158**, 167–90

Sacher, G. A. (1959) Relation of life span to brain weight and body weight in mammals. In *Ciba Foundation Colloquia on Ageing*, Vol. 5, *The Life Span of Animals*, ed. G. E. W. Wolstenholme & M. O'Connor, pp. 115–41. London: Churchill

Sackett, G. P. (1981) A nonhuman primate model for studying causes and effects of poor pregnancy outcomes. In *Preterm Birth and Psychological Development*, ed. S. L. Friedman & M. Sigman, pp. 41–63. New York: Academic Press

Stephan, H. & Andy, O. J. (1964) Quantitative comparisons of brain structures from insectivores to primates. *Am. Zool.*, **4**, 59–74

Sundsten, J. W. & Prothero, J. W. (1983) Three-dimensional reconstruction from serial sections: II. A microcomputer-based facility for rapid data collection. *Anat. Rec.*, **207**, 665–71

Sundsten, J. W., Schultz, G., Prothero, J. W. & DeVito, J. L. (1984) Quantification of the development of forebrain subdivisions in *Macaca nemestrina*. *Anat. Rec.*, **208**, 176A

Winick, M., Rosso, P. & Waterlow, J. (1970) Cellular growth of cerebrum, cerebellum, and brain stem in normal and marasmic children. *Exp. Neurol.*, **26**, 393–400

IV.2

Age-related growth patterns of colony-born rhesus monkeys

D. N. SHARMA AND K. C. LAL

Introduction

The rhesus monkey is the most extensively used non-human primate for research on problems related to human health and welfare. A captive breeding colony is maintained at the All-India Institute of Medical Sciences for their use in fertility control research.

Very few reports are available on the eruption of deciduous and permanent teeth, on weight gain and other body measurements (Schultz, 1933; Eckstein, 1948; Hurme & Van Wagenen, 1953, 1961; Van Wagenen & Catchpole, 1956; James, 1960). The present investigation was undertaken to assess the general relations between weight gain, body measurement and tooth eruption of colony-born rhesus monkeys.

Materials and methods

The study began in 1981 with 6 male and 6 female rhesus born in the colony. Subsequently, 9 males and 9 females were added to the study, making a total of 15 animals of each sex. Infants were kept with their mothers until 4–6 months of age, after which they were weaned. After weaning, infants were paired. Environmental conditions (temperature, humidity, and light) and nutrition were held constant for all animals.

The aim of the study was to relate the age of the monkeys to the following body parameters: (1) weight gain from birth to adulthood; (2) body measurements – (a) circumference of head, chest and pelvic girth; (b) length of trunk, forelimb and hindlimb; and (3) eruption of deciduous teeth and the timing of their replacement by permanent teeth.

Recording of parameters

The different parameters were recorded within 72 hours of birth for all newborns. Weight (in grams) was recorded by placing the infant on a single pan balance. The circumferences of the head, chest and pelvic girth were recorded in centimetres using a measuring tape. The length of the trunk was measured as the distance from the base of the head to the base of the tail, and length of forelimbs was measured from the shoulder joint to the tip of the middle finger. The length of the hind limb was measured from the hip joint to the tip of the second toe. These lengths were measured in centimetres. During a physical examination, the oral cavity was examined for the eruption of deciduous teeth. From birth onwards the parameters were measured once per month and eruption of deciduous teeth examined on a weekly basis. After 60 days of age the animals required sedation in order to obtain the measurements. This was carried out using ketamine hydrochloride (Vetlar) at a standard dose of 5–10 mg/kg body weight.

Results

Birth weights

The mean birth weights for males and females were 513 ± 43.9 g and 446 ± 39.8 g, respectively (Table 1). A linear increase in body weight was found for both sexes. Unlike data for adult rhesus (personal observation) this difference between the sexes in weight was not significant during the first year.

The gain in body weight was highest during the first 3 months of life for both sexes. From birth to 3 months, males gained 439 g as compared to 401 g for females. Beginning in the fourth month, the rate of weight gain tended to slow down for both sexes. The increases in weight beween 3 and 6 months were 259 g for males and 214 g for females. This tendency for the rate of weight gain to become slower with increasing age was observed throughout the first year of life (Table 2).

Body measurements

At the time of birth the head circumference was greater than those of the chest and pelvic girth. Chest circumference increased more rapidly than did that of the head or pelvic girth. There were no significant differences between the sexes in these measurements.

As was found for body weight, the highest rates of growth in circumferences of head, chest and pelvic girth were observed during the first three months of life. A similar pattern was observed for growth in length of the trunk, forelimb and hind limb (Tables 1 and 2).

Table 1. Weight gain and growth during the first year of life in colony-born rhesus monkeys (Mean ± SD)

	Males (age, in months)					Females (age, in months)				
	0[a]	3	6	9	12	0[a]	3	6	9	12
Body weight (g)	513.0 ± 43.9	952.0 ± 108.3	1211.0 ± 103.3	1422.0 ± 143.3	1604.0 ± 183.5	446.0 ± 39.8	847.0 ± 110.6	1061.0 ± 113.5	1261.0 ± 147.3	1435.0 ± 212.8
Circumference (cm)										
Head	19.0 ± 0.9	22.2 ± 1.0	22.8 ± 1.0	23.2 ± 0.7	24.6 ± 0.6	18.7 ± 1.0	21.5 ± 0.9	22.2 ± 0.7	22.7 ± 0.7	23.3 ± 0.6
Chest	14.8 ± 1.1	18.7 ± 1.2	20.5 ± 0.8	22.1 ± 0.7	23.3 ± 4.8	14.5 ± 1.5	17.9 ± 1.4	19.8 ± 0.9	21.2 ± 0.9	22.4 ± 0.6
Pelvic girth	11.2 ± 1.0	14.2 ± 1.4	15.6 ± 0.1	17.0 ± 1.1	17.4 ± 1.4	11.4 ± 1.1	13.4 ± 1.1	14.7 ± 0.9	16.2 ± 0.9	17.1 ± 1.1
Length (cm)										
Trunk	13.3 ± 1.4	17.4 ± 1.2	19.0 ± 1.0	20.5 ± 1.2	22.5 ± 0.5	13.7 ± 1.1	17.1 ± 1.2	19.1 ± 1.0	20.5 ± 0.9	22.2 ± 1.1
Forelimb	15.5 ± 1.2	21.2 ± 2.0	23.2 ± 1.5	24.8 ± 1.3	26.6 ± 0.8	15.5 ± 1.5	20.2 ± 1.7	23.1 ± 1.2	24.8 ± 1.2	26.5 ± 1.7
Hindlimb	19.2 ± 2.5	26.0 ± 2.9	29.9 ± 1.3	31.6 ± 1.8	33.0 ± 1.3	18.6 ± 1.4	25.1 ± 3.1	28.9 ± 1.4	30.5 ± 1.7	31.9 ± 1.5

[a] Birth weight

Emergence of deciduous teeth

The eruption of teeth begins just after root formation is initiated and is accompanied by an increase in root length. This increase in the root length together with the vertical increase in bone deposition at the apex of the tooth can be used to assess the eruption rate.

Observations of tooth eruption indicated that all the monkeys in this sample had at least one deciduous incisor visible in the oral cavity by 45 days. By 1 year, all the deciduous teeth were found to have erupted. As in man, there are 20 deciduous teeth (I 2/2, C 1/1, M 2/2). Mean values and standard deviations for the eruption of deciduous teeth are given in Table 3. All incisors had erupted within 45 days, canines after 95 days, and molars after 216 days from birth.

Discussion

In the present study, the data presented on the rate of growth for body weight and size have shown that there is no significant difference(s) between the male and female between birth and 1 year old. However, in both the sexes, there is a rapid increase in body weight and measurement during the first 3 months. The rate of increase in body weight gradually decreases and there is about a 50% fall in rate of increase in both the sexes from the third to the sixth month. There is also an abrupt fall in rate of increase in the parameters

Table 2. *Quarterly increase in body weights and measurements (mean) of colony-born rhesus monkeys*

	Males (age, in months)				Females (age, in months)			
	3	6	9	12	3	6	9	12
Body weight (g)	439	239	211	182	401	214	200	174
Circumference (cm)								
Head	3.2	0.6	0.4	1.4	2.5	0.7	0.5	0.6
Chest	3.9	1.8	2.4	1.2	3.4	1.9	1.4	1.4
Pelvic girth	3.0	1.5	1.4	0.4	2.0	1.3	1.4	0.9
Length (cm)								
Trunk	4.0	1.7	1.5	1.9	3.4	2.0	1.4	1.7
Forelimb	5.7	1.9	1.6	1.8	4.7	2.9	1.7	1.7
Hindlimb	6.8	3.9	1.9	1.1	6.6	3.6	1.6	1.4

studied for the body size from the third to the sixth month; from the sixth month onwards the increase in size is very gradual. One of the factors that may contribute to the abrupt decrease in the growth rate observed from the third to the sixth month is the weaning of the young ones from their mothers to where they depend exclusively on the diet provided for their nourishment.

By 1 year of age, all the deciduous teeth (total 20) were found to have erupted. The rate of growth of erupting incisors in the rhesus monkeys up to 40 days is 0.04 mm per day (Hurme & Van Wagenen, 1953). According to Kenney (1975), the pattern of development and structure of teeth in the rhesus monkey closely resemble those of man.

The present observation that all the incisors have erupted by 45 days, the canines by 95 days and the molars by 216 days is in accordance with the earlier reports of Hurme & Van Wagenen (1953).

The study is being continued in these monkeys to determine the age at which significant differences between the two sexes manifest in the growth pattern, as well as to determine the reproductive profile at the onset of sexual maturity.

Acknowledgements

The Primate Research Facility under which the breeding of rhesus monkeys is undertaken is supported by the World Health Organization. We are grateful to Dr G. F. X. David, Officer-in-Charge of the facility, for his guidance.

Table 3. *Days after birth at which deciduous teeth were found erupted in male and female colony-born rhesus monkeys. The data represent the mean ($\pm$SD) values for 15 monkeys of each sex*

	Maxilla		Mandible	
	Males	Females	Males	Females
Incisors				
Central	20.6 ± 6.4	20.6 ± 9.7	23.0 ± 6.4	20.6 ± 9.7
Lateral	41.3 ± 8.1	40.3 ± 15.1	33.5 ± 8.1	36.5 ± 18.1
Canines	95.8 ± 5.2	66.6 ± 14.1	94.0 ± 29.1	77.1 ± 26.0
Molars				
1	93.0 ± 37.0	114.8 ± 40.8	90.5 ± 37.1	114.8 ± 40.8
2	216.0 ± 35.1	189.5 ± 10.7	169.0 ± 37.3	195.7 ± 10.7

References

Eckstein, P. (1948) *Growth and Development of Rhesus Monkeys*. Proceedings of the Annual Meeting of the British Medical Association. London: Butterworth

Hurme, V. O. & Van Wagenen, G. (1953) Basic data on the emergence of deciduous teeth in monkey (*Macaca mulatta*). *Proc. Am. Phil. Soc.*, **97**, 291–315

Hurme, V. O. & Van Wagenen, G. (1961) Basic data on the emergence of the permanent teeth in the rhesus monkeys (*Macaca mulatta*). *Proc. Am. Phil. Soc.*, **105**, 105–40

James, W. W. (1960) *The Jaws and Teeth of Primates*. London: Pitman Medical

Kenney, E. B. (1975) Development and eruption of teeth in rhesus monkey. In *The Rhesus Monkey*, Vol. I, ed. G. H. Bourne, pp. 145–66. New York: Academic Press

Schultz, A. H. (1933) Growth and development. In *The Anatomy of the Rhesus Monkey*, ed. C. G. Hartman & W. L. Strans, Ch. II. New York: Hafner Publishing Co.

Van Wagenen, G. & Catchpole, H. R. (1956) Physical growth of rhesus monkeys (*Macaca mulatta*). *Am. J. Phys. Anthropol.*, **14**, 245–70

IV.3

A comparative approach to the question of why human infants develop so slowly

H. DIENSKE

Introduction

A typical precocious mammal gives birth to a large, single offspring that has fur, is not cached in a nest and is capable of locomotion within a day. Primates are precocious according to the first four characteristics. Locomotion, however, develops later. Unlike, for instance, guinea-pigs or sheep, primate infants do not follow the mother soon after birth but must be carried during the substantial period of time that the infant is not capable of competent locomotion. This period lasts for several days in monkeys, a few months in apes and no less than half a year in man. The development of various other motor capacities is likewise considerably delayed (Hines, 1942; Riesen & Kinder, 1952; Dienske, 1984). Gould (1977) regarded human development as neoteny. The question asked in this paper is why this slow motor development has evolved.

Gestation and growth

A reason for the longer helplessness of human babies might be that a larger, more advanced neonate could not pass the pelvic passage, which is just wide enough. It seems possible that this constraint has resulted in a shorter pregnancy. This turns out to be unlikely however. Gestation time is primarily a function of female body weight (and thus metabolic efficiency). Human gestation time is in keeping with that for precocious mammals in general (Martin & MacLarnon, 1985).

The explanation that prolonged helplessness resulted from a limiting pelvic passage size is also weakened by other data. In rhesus monkeys, the pelvic passage closely matches neonatal head dimensions; in apes, however, the passage is much larger than the newborn's head, which permits easy delivery (Leutenegger, 1982). Nevertheless, the development of apes is considerably delayed in comparison with monkeys. It is not parsimonious to accept pelvic limitations as an explanation for immaturity in man but not in apes.

It could be that the motor behaviour of the human neonate is relatively immature because the brain is disproportionately small in size. Again, this is not supported by the data: human neonatal brain weight is comparable to that of other primates (Leutenegger, 1982) (Fig. 1a), namely about 12% of the body weight. This applies not only to the newborn but is also true for the entire period of fetal growth (Holt *et al.*, 1975; Sacher, 1982; Hofman, 1983). Qualitative observations suggest that the human baby has more subcutaneous fat than nonhuman primates. If this extra fat is disregarded, human neonatal brain weight is about 14% of body weight. This suggests only a minor increase of the fetal brain-to-body ratio during human evolution.

In man, the adult brain is three times as large as expected for a nonhuman primate of that body weight. It is possible that the long human helplessness can be understood from the way such a large brain develops. Prenatal brain growth seems to contribute little, but there are two other mechanisms.

First, rapid fetal brain growth rate can be extended beyond birth (Fig. 2). This seems to be the case in rhesus monkeys; data of Holt *et al.* (1975) suggest deceleration in brain growth between birth and 2 months of age in different individuals. The deceleration in chimpanzees seems to occur appreciably later. Data given by Count (1947) by extrapolation suggest a change when the infant weighs around 3 kg; this weight is attained at about 6 months of age. Budd, Smith & Shelley (1943) also found a change in skull size growth at 6 months in one chimpanzee. In man, the rapid fetal brain growth is continued for more than a year after birth (see Fig. 2).

Fig. 1. Relations among neonatal brain weight, neonatal body weight and maternal weight in primates. After Leutenegger (1973, 1982).

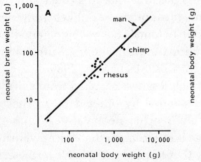
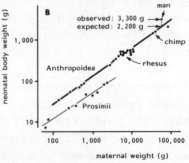

Secondly, fetal growth rate can be accelerated. This manner of achieving a larger adult brain took place when higher primates evolved from prosimian ancestry: newborn higher primates have relatively heavier bodies (Leutenegger, 1973) (Fig. 1b) as well as brains (Passingham, 1982; Sacher, 1982) than prosimians. During human evolution, another 'jump' has taken place. On a double log scale, maternal and neonatal body weights of monkeys and apes fit the same straight line (Fig. 1b). This graph predicts a human neonatal weight of 2200 g (Gould, 1977; Leutenegger, 1982). In Western countries, the actual value is more than 1 kg heavier. Fig. 2 shows a hypothetical human brain growth based on a birth weight of 2200 g of which one-tenth is brain weight. The gain due to acceleration appears to be considerable.

Prolongation beyond birth (mainly apes and man) and acceleration (man) of fetal brain growth rate result in a heavier adult brain. This implies that the neonatal brain is a smaller proportion of its final adult weight. In rhesus monkeys, chimpanzees and man, these proportions are about two-thirds, one-half and one-quarter (Passingham, 1982; Hofman, 1983). A lower value implies a greater immaturity if many parts of the neonatal brain are still in a vestigial stage of functioning.

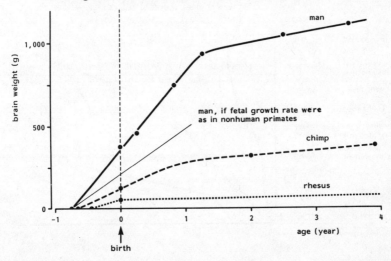

Fig. 2. Brain growth as a function of age. After data given by Holt *et al.* (1975), Passingham (1982) and Hofman (1983). As the scale is considerably enlarged in comparison with the original data, accuracy is limited. The hypothetical human brain growth is based on a neonatal body weight of 2200 g predicted from nonhuman primates (see Fig. 1b), of which one-tenth is brain weight.

This would explain why the early motor helplessness increases in the order of rhesus monkey, chimpanzee and man. Is there a connection between the first appearances of behavioural capacities and the percentage of the adult brain weight that has developed?

Development of behaviour repertoires

A schematic, preliminary representation of the ages at which various characteristics appear is given in Fig. 3 (sources are given in the legend). For motor capacities and smiling, the ages at which half of the individuals studied had developed a trait is given. For off-mother and social play, the mean ages at which a rapid increase starts and terminates are illustrated. The accuracy of the given data points is very limited. However, as most differences among the species are large the global, qualitative conclusions given below are unequivocal.

Developments in the rhesus monkey start earlier and are faster (indicated by a steeper line) than in the other two species. Chimpanzees are 3–7 months earlier than human infants, depending on the

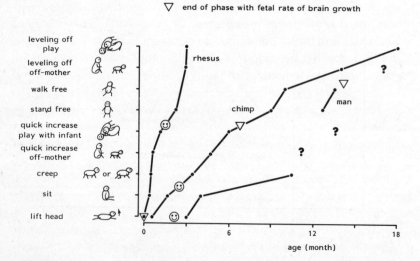

Fig. 3. Ages at which certain behaviour patterns are first shown. Data for rhesus monkeys are based on Hines (1942), Hinde & Spencer-Booth (1967) and de Jonge *et al.* (1981). Chimpanzee data are after Riesen & Kinder (1952), van Lawick-Goodall (1967), Plooij (1979, 1984) and Horvat, Coe & Levine (1980). Those for man are based on Riesen & Kinder (1952), Touwen (1976) and Illingworth (1980). The final decision on the given mean age, however, is the author's responsibility. This applies especially to the question marks.

trait. Chimpanzees seem to develop initially faster and subsequently slower than human infants, as is especially suggested by more extensive data of Riesen & Kinder (1952), but comparability is limited (see Dienske, 1984).

The developmental lines in Fig. 3 can be used to address the question of whether the delays in the developments of apes and man are connected with the percentage of the adult brain that infants have attained at various ages, as given in Fig. 2. As mentioned earlier, the proportions of the adult brain that are already developed at birth are very different for rhesus monkeys, chimpanzees and man ($\frac{2}{3}, \frac{1}{2}, \frac{1}{4}$). From Fig. 3 it can be derived that the development of motor capacities, off-mother and play are roughly comparable for the three species at the ages of 1, 6 and 12 months. At these ages, the percentages of adult brain weights are not so different (about 80, 65 and 60), although the monkey is already closer to the 'ceiling'. In spite of being far from accurate, these values support the supposition that behavioural maturation is globally connected with the infantile brain weight expressed as a percentage of adult brain weight. This supposition makes it understandable that in the species considered a larger adult brain is accompanied by a slower development. An explanation based on brain morphology and functions, however, would be much more accurate.

Another question is whether the age at which the change from a rapid 'fetal' to a slower 'infantile' brain growth occurs is marked by similar behavioural developments in the three species. The approximate ages suggested in Fig. 2 are indicated by triangles in Fig. 3. It appears that the rather sudden decrease in brain growth rate has taken place at a progressively later stage of motor maturation in the phylogenetic sequence of the species. At this point, the development of brain size and motor behaviour do not seem to match each other.

In Fig. 3, another behaviour is given that does not correspond with the developmental sequences of motor capacities, off-mother and social play. The play-face (cf. smile) in the rhesus monkey appears at about the same time as social play. It is shown mainly during the wrestling-type play between infants and only occasionally if a peerless infant plays with the mother. In chimpanzees, however, smiling develops at a much earlier stage, long before playful interactions with other infants are common. Unlike in the rhesus monkey, the chimp smile is first shown during manual and oral interactions with the mother (Plooij, 1979). In man the smile develops at about the same (chronological) age as in chimpanzees, i.e. again at an earlier stage of development. So the slower the development, the earlier the smile appears.

Moreover, the richness of the behavioural repertoire of acts that are shown specifically between mothers and infants in the species considered increases with the duration of helplessness (see Dienske, 1984). Rhesus mothers carry, reject, retrieve and groom their infants. Chimpanzees do the same, but tickling, smiling, gazing and kissing are common maternal acts not present in monkeys. Human parent–infant interactions are enriched with talking, offering objects and special time devoted to teaching. A slower development apparently is accompanied by a larger repertoire of mother–infant interactions.

Discussion

It can be concluded that the long motor helplessness of the human infant cannot be explained by a comparatively short gestation period, a narrow pelvic passage or a small neonatal brain. Some connection with the heavy human neonatal body weight is possible, but it is not clear how to account for that. The facts available at present suggest that the helplessness corresponds with the low percentage of the adult brain that has developed. So it seems that a low percentage indicates a small degree of functional maturation of many parts of the brain.

This applies primarily, however, to motor capacities, off-mother and play. As walking is required for leaving the mother, and as off-mother is necessary for agile social play, it is understandable that these acts appear in a similar order in the three species. Interactions between mother and infant follow a different pattern. The slower a species' motor capacities develop, the earlier the appearance of the play-face or smile and concomitant mother–infant interactions in the sequence of motor developments. If this were not the case, the early postnatal life would be spent in an inertia that may be regarded as an unavoidable 'by-product' of a larger adult brain. In man at least, inertia and lack of stimulation due to hospitalization or parental neglect is disadvantageous for behavioural development. During evolution, the selection pressure favouring a larger adult brain might have been counteracted by negative effects of inertia. Consequently, selection favoured infants which were earlier capable of giving positive communicatory feedback. This would explain the shift of smiling to an earlier phase of development.

On the one hand, we found a tentative answer to the question of why the human motor development is so slow: at birth and shortly thereafter, the brain is only a small proportion of the adult brain, which may be presumed to imply a low degree of functional maturity.

On the other hand, it became clear that this slowness does not concern all behaviour. During phylogeny, smiling and concomitant interactions between infant and mother developed at an increasingly earlier ontogenetic phase. Simultaneously, the number of repertoire items and skills expanded, which also is a mode of greater developmental speed. These data show, besides neotony, apparent heterochrony in the sense of Gould (1977). Although a human infant takes time to develop, the outcome is amazingly rich and complex.

Acknowledgements

I am grateful to Drs C. Goosen, M. A. Hofman, F. X. Plooij, H. F. R. Prechtl and A. C. Ford for their comments on an earlier draft, to Mrs D. de Keizer for typing the manuscript, and to Mr E. J. van der Reyden for the preparation of the figures.

References

Budd, A., Smith, L. G. & Shelley, F. W. (1943) On the birth and upbringing of the female chimpanzee 'Jacqueline'. *Proc. Zool. Soc. London, Series A*, **113**, 1–20

Count, E. W. (1947) Brain and body weight in man: their antecedents in growth and evolution. *Ann. N. Y. Acad. Sci.*, **46**, 993–1122

Dienske, H. (1984) Early development of motor abilities, daytime sleep and social interactions in the rhesus monkey, chimpanzee and man. In *Continuity of Neural Functions from Prenatal to Postnatal Life*, ed. H. F. R. Prechtl, pp. 126–43. Clinics in Developmental Medicine, No. 94. London: Spastics International Medical Publications

Gould, S. J. (1977) *Ontogeny and Phylogeny*. Cambridge, Mass.: Belknap

Hinde, R. A. & Spencer-Booth, Y. (1967) The behaviour of socially living rhesus monkeys in their first two and a half years. *Anim. Behav.*, **15**, 169–96

Hines, M. (1942) The development and regression of reflexes, postures and progression in the young macaque. *Contrib. Embryol.*, **196**, 155–209

Hofman, M. A. (1983) Evolution of brain size in neonatal and adult placental mammals: a theoretical approach. *J. Theoret. Biol.*, **105**, 317–32

Holt, A. B., Cheek, D. B., Mellits, E. D. & Hill, D. E. (1975) Brain size and the relation of the primate to the nonprimate. In *Fetal and Postnatal Cellular Growth and Nutrition*, ed. D. B. Cheek, pp. 23–44. New York: Wiley

Horvat, J. R., Coe, C. L. & Levine, S. (1980) Infant development and maternal behavior in captive chimpanzees. In *Maternal Influences and Early Behavior*, ed. R. W. Bell & W. P. Smotherman, pp. 285–309. Lancaster, UK: MTP Press

Illingworth, R. S. (1980) *The Development of the Infant and the Young Child, Abnormal and Normal*. 7th edn. Edinburgh: Churchill Livingstone

Jonge, G. de, Dienske, H., Luxemburg, E.-A. van & Ribbens, L. (1981) How rhesus monkey infants budget their time between mothers and peers. *Anim. Behav.*, **29**, 598–609

Lawick-Goodall, J. van (1967) Mother–offspring relationships in free-ranging chimpanzees. In *Primate Ethology*, ed. D. Morris, pp. 287–346. London: Weidenfeld & Nicolson

Leutenegger, W. (1973) Maternal–fetal weight relationships in primates. *Folia Primatol.*, **20**, 280–93

Leutenegger, W. (1982) Encephalization and obstetrics in primates with particular reference to human evolution. In *Primate Brain Evolution: Methods and Concepts*, ed. E. Armstrong & D. Falk, pp. 85–95. New York & London: Plenum Press

Martin, R. D. & MacLarnon, A. M. (1985) Gestation period, neonatal size and maternal investment in placental mammals. *Nature*, **313**, 220–3

Passingham, R. E. (1982) *The Human Primate*. Oxford: Freeman

Plooij, F. X. (1979) How wild chimpanzee babies trigger the onset of mother–infant play and what the mother makes of it. In *Before Speech*, ed. M. Bullowa, pp. 223–44. Cambridge: Cambridge University Press

Plooij, F. X. (1984) *The Behavioral Development of Free-living Chimpanzee Babies and Infants*. Monographs on Infancy, vol. 3. Norwood, NJ: Ablex Publishing Co.

Riesen, A. H. & Kinder, E. F. (1952) *Postural Development of Infant Chimpanzees*. New Haven: Yale University Press

Sacher, G. A. (1982) The role of brain maturation in the evolution of primates. In *Primate Brain Evolution: Methods and Concepts*, ed. E. Armstrong & D. Falk, pp. 97–112. New York & London: Plenum Press

Touwen, B. (1976) *Neurological Development in Infancy*. Clinics in Developmental Medicine, no. 58. Spastics International Medical Publications. London: Heinemann

IV.4

The interaction between prehension and locomotion in macaque, gorilla and child cognitive development

G. SPINOZZI AND F. NATALE

Introduction

Our interest in the development of prehension activities in gorilla and macaque arose in the context of investigating their general cognitive development and comparing it with that of the human child. According to the Piagetian framework, within which our comparative study is carried out, the development of prehension schemata plays a fundamental role in the organization of the infant's cognition. In fact, it is the progressive intercoordination of prehension schemata with those involving suction, hearing and, especially, vision, that makes possible the unification of the hitherto separate sensorimotor spaces and the beginning of the cognitive construction of the external reality as such (see Piaget, 1952, 1954).

We investigated the development of prehension activities in so far as they are relevant to and in relation with the development of cognition. Our study is thus different from previous studies on prehension development in nonhuman primates, which were concerned with the anatomical and motor development of hand-functioning in prehension (Halverson, 1943; Connolly & Elliot, 1972).

We centered instead on the functional aspects of prehension schemata and their intercoordinations. This perspective led us to take into account not only the role of the hand but also the role of the mouth in prehension. In fact, beside being used as an effective prehensile organ, the mouth represents, during the first stages of cognitive development, one of the fundamental knowledge-sources for the infant: most of the cognitive exploration of the external world is carried out through the mouth.

The present study analyzes, within this perspective, the development of prehension schemata in macaque, gorilla and human infants.

Subjects and methodology

The macaque subject was an infant male of *Macaca fascicularis* born in captivity, separated from its mother since birth and hand-reared. The experimental sessions took place three times a week and consisted of 1-hour-long observations of spontaneous behaviors of the infant while interacting with a variety of objects offered by the experimenter.

The gorilla subject was a female infant born and housed in the Rome Zoo; she was followed with the same methodology as the macaque. Experimental sessions consisted of 1-hour-long weekly videotaped observations of the spontaneous behavior of the infant.

Data on the human children were collected for critical periods of prehension development; five infants, four male and one female, covering a total age span from 1 to 12 months, were examined and videotaped.

Results

Data analysis makes reference to Piaget's identification of five stages of prehension development in the human infant (see Piaget, 1952).

Stage I is characterized by impulsive movement and pure reflex in grasping.

Stage II is characterized by the first acquired adaptations, like circular reactions relative to hand movements, without any schemata intercoordination.

Stage III is characterized by the appearance of the first intercoordination: that of prehension and sucking.

Stage IV is characterized by the first level of coordination between vision and prehension: the child succeeds in grasping an object that he sees but only if both the hand and the object are in his visual field.

Stage V achieves complete reciprocal coordination between prehension and vision: the child can grasp anything he sees.

These five stages span across the first two stages of sensorimotor intelligence development: with the achievement of stage V of prehension development the infant passes into stage 3 of sensorimotor intelligence.

Fig. 1 shows the sequence and time of attainment of these five stages of prehension; furthermore, the figure also shows the time of onset of independent locomotion for each subject (end of the dashed line). Age is given in weeks for macaque, and in months for gorilla and child.

As can be seen in Fig. 1, the macaque attains the second stage of prehension development in its first week of life, and the fifth stage of prehension in its fourth week of life.

On the other hand, both gorilla and child pass into the second stage of prehension development at the age of 1 month, and attain the fifth stage at 5 and 6 months of age, respectively.

With regard to the development of locomotion, the macaque moves independently in its first week of life, the gorilla at the age of 3 months, and the child at the age of 8 months.

Two major trends clearly emerge: the first is the much faster general development in the macaque as compared to the gorilla and child; the second is the progressive delay, going from the macaque to the gorilla and from the gorilla to the child, of the appearance of autonomous locomotion. It must be noted that the onset of locomotion is delayed not only in absolute but also in *relative* terms: the macaque begins to move independently in space when still in the second stage of prehen-

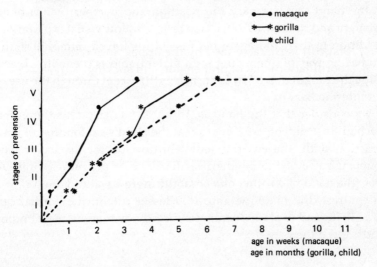

Fig. 1. The sequence and the time of the attainment of the five stages of prehension development for macaque, gorilla and human infants. Age is given in weeks for macaque and in months for gorilla and child.

sion development, the gorilla begins to move while in stage III of prehension development, and the child can move independently in space only long after the completion of the whole sequence of prehension development. Therefore, both macaque and gorilla are capable of moving themselves in space before the completion of hand/sight coordination: they can and do reach external objects directly with their mouth before they are capable of grasping them at sight. The human child cannot reach objects directly with his mouth.

Macaque and gorilla therefore develop a prehension pattern based on mouth use which is never seen in children. The macaque directly uses its mouth from the very beginning of its interaction with the external objects, and this pattern of prehension remains frequent even after the full achievement of hand/sight coordination. The gorilla is not capable of moving itself in space when interest in external objects arises, and thus it begins to try to contact objects with its hands. When it starts to walk, the hand/sight coordination is not fully developed and, as a consequence, we find an explosion of mouth use in prehension. Contrary to the macaque's case, however, this pattern progressively decreases as hand/sight coordination develops.

These differences have important consequences in determining the patterns of interaction with external objects. Throughout a long period of his cognitive development the child can interact with objects only through the use of his own hands and, eventually, through the use of intermediary objects. Both his oral and his visual exploration of objects depend on his hand-grasping. Since neither his body nor his head can move, it is the hand that must carry objects to the mouth and it is the hand that can vary the position and perspective (through movement and rotation) of objects in order to allow visual exploration. In addition to that, and again because of his forced immobility, the child has no way of interacting with distant objects (i.e. with objects outside the range covered by his arm length) except through the use of other intermediary objects.

If we consider that the hand is, in an elementary sense, the first 'detached' instrument, we can see that the child's exploration of, and interaction with, the environment will develop for the major part through the use of intermediaries: the hand at the more primitive levels, (the hand plus) other objects at the more advanced ones.

In contrast, the macaque and, to a lesser extent, the gorilla, can interact with and explore the objects directly by moving their bodies to and around them.

Discussion

The difference in the relative time of onset of locomotion in a developmental course which for other relevant aspects is similar in the three species, very soon channels the cognitive interaction with the external world into two different modes: action through intermediaries in the human child and direct action through body movements in the macaque, with the gorilla lying somewhere in between.

It should be noted that the first mode carries with it a concentration on the effects produced by actions on objects, and on the causal and spatial relations of objects with each other. In other words it bears directly on the cognitive construction of physical causality and object-to-object relations.

In fact, in later development we find a clear divergence in these cognitive domains both between child and gorilla and between gorilla and macaque. On the one hand, both macaque and gorilla, as opposed to the human child, do not appear to develop schemata involving object–object relations, such as hitting an object against another, putting an object into a container or on the top of another, or in general combining one object with another (Redshaw, 1978; Antinucci & Spinozzi, 1983; Poti', unpublished, cited with permission). On the other hand, there appears to be a clear differentiation between the macaque and the gorilla in the development of physical causality, as seen, for example, in the mastering of the spatial and dynamic constraints governing the use of detached tools (Natale, Poti' & Spinozzi, 1984).

References

Antinucci, F. & Spinozzi, G. (1983) Lo sviluppo cognitivo di un gorilla nel suo primo anno di vita. *Eta' Evolutiva*, **15**, 46–55

Connolly, K. & Elliot, J. (1972) The evolution and ontogeny of hand function. In *Ethological Studies of Child Behavior*, ed. N. Blurton Jones, pp. 329–71. London: Cambridge University Press

Halverson, H. M. (1943) The development of prehension in infants. In *Child Development and Behavior*, ed. R. G. Barker, J. S. Kounin & H. F. Wright, pp. 49–65. London: McGraw-Hill

Natale, F., Poti', P. & Spinozzi, G. (1984) Development of tool use in macaque and gorilla. Paper presented at the *Xth Congress of the International Primatological Society*, Nairobi, Kenya. *Int. J. Primatol.*, **5**, 365

Piaget, J. (1952) *The Origins of Intelligence in Children*. New York: W. W. Norton and Company Inc.

Piaget, J. (1954) *The Construction of Reality in the Child*. New York: Ballantine Books

Redshaw, M. (1978) Cognitive development in Human and Gorilla infants. *J. Hum. Evol.*, **7**, 133–41

IV.5

Grading in the vocal repertoire of Silver Springs rhesus monkeys

E. H. PETERS

Introduction

The purpose of this chapter is to re-examine 'grading' as a characteristic of primate vocalizations and to suggest a new way of thinking about the organization of calls which may clarify the generation of extreme variety in vocal repertoire. The graded nature of primate calls was first reported by Rowell & Hinde (1962), following their pioneering application of sound spectrography to rhesus monkey calls. Working with sound samples from a captive troop of animals, these investigators defined a finite number of 'basic noises' but prefaced this list with the observation that there exists 'an almost infinite series of intermediates between the main sounds' and the suggestion that calls may 'intergrade with one another so that the variation is immense and it is impossible to give the full vocabulary written equivalents'. In a follow-up analysis of agonistic vocalizations, Rowell (1962) suggested that calls in this family could be arranged in a continuous graded series, with the particular points along the continuum identified and labeled in a somewhat arbitrary fashion.

Much of the difficulty which these and other investigators have encountered in the description of primate vocal repertoire may be related to overly strict adherence to the typological paradigm which emerges as an underlying feature of all such descriptive attempts. The standard procedure for describing primate sound signals continues to be that of defining a finite number of physically distinctive, relatively stereotyped acoustic units (i.e. the vocalization or call), each with its own meaning or message. This procedure can be taken to breathtakingly subtle levels of analysis, as Green's (1975) fine-grained description of Japanese macaque vocalizations has shown. The danger with an exclusively typological approach is that characteristics which cross-link different calls may be obscured, making it more difficult to discern how pattern and meaning are actually generated and discerned by the animals themselves. The current practice of referring to the existence of 'intermediates' between the types or 'variants' as departures from the type may reify categories which have no such primacy of position in the mind of the animal.

Typological constraint in the description of primate sound signals may be related to the precedent set by the highly successful ornithologists, a group whose leadership in the analysis of animal sound is undoubtedly related to the salience of aural communication in birds. However, it is not always appreciated that the success of typological sound analysis for birds and many other animals is related to the prevalence of a high degree of stereotypy and contrast in their signals. These characteristics, as Morris (1957) has pointed out, enhance signal clarity and decrease ambiguity. Given the perceptual-cognitive limitations shown to characterize many species of animals, such stereotypy and high contrast might be necessary for reliable signal processing. Suppose, however, evolutionary processes produced an animal with the perceptual acuity to process incoming sound signals even when the important cues are subtle in nature and embedded in a miscellany of extraneous stimuli. Suppose this animal had the cognitive complexity to synthesize simultaneously the information in multiple cues. Suppose it was sensitive to small changes in meaning wrought by small changes in signal form. In short, given an appropriate level of perceptual-cognitive complexity, neither stereotypy nor high contrast may be necessary for reliable processing of conspecific sound signals. For any species, the nature of its sound signal system is likely to be correlated with its perceptual-cognitive capacity. As this capacity shows evidence of a change in evolutionary grade, it is reasonable to look for emergent properties in the co-evolving signal system. It is also reasonable to suggest that the same system of signal description which works well with birds may not work quite so well with monkeys.

Methods

The data on which my own analysis of rhesus monkey vocalizations is based were collected from a troop of feral rhesus monkeys which inhabit the lowland floodplain adjacent to the Silver River near Silver Springs, Florida. This troop is part of a colony of several hundred animals descended from monkeys released into the wild in the 1940s as an exotic and seemingly natural wildlife exhibit. Recordings of vocalizations and associated behavior were collected between October 1978 and May 1979 from a single troop of approximately 30 individually identifiable animals. Spectrograms of vocalizations were made on a Kay Sonagraph model 6061A using a 150 Hz filter to scan the 80–80 000 Hz range.

Results

It was while attempting to sort vocalization samples from the above monkeys using the standard typological paradigm that difficulties with this procedure became apparent. These included problems both in defining what Altmann (1967) has called 'criteria of sequential demarcation' (given a temporal continuum of sound, where does a specific call begin and where does it end?) and in defining 'criteria of class membership' (when are two sounds sufficiently similar so that they should be considered members of the same class?).

For many animal sound signals the beginning and the end of a call can be easily and reliably determined. When a single unit of sound (unit, as defined by Struhsaker (1967), refers to a continuous tracing along the temporal axis of the sonogram) is surrounded by substantial boundaries of silence, then that single unit can be reliably understood to be the call. Rhesus tonal coos and many barks and grunts (noisy, one-unit sounds with either a high pitch or a low pitch, respectively) fall into this category. Other species-typical sound signals are made up of multiple units, but if the component units are similar, the time interval between units fairly regular, and the time interval between component units shorter than the time interval between groupings of units, then it is again relatively easy to distinguish the beginning and the end of the call. The geckering (Rowell & Hinde, 1962) of immature rhesus and the chortling (Lindburg, 1971) of adult and juvenile females are examples of multi-unit calls for which sequential demarcation between individual calls is still fairly obvious.

Problems in defining criteria of sequential demarcation exist for calls composed of multiple units with irregularly shaped, highly variable components and irregular spacing between units. The family of calls which Rowell (1962) called 'agonistic' often includes examples which fall into this category. It includes high-pitched screams (Gouzoules, Gouzoules & Marler (1984) have called these 'aid enlistment' calls) and low-pitched threat calls. Many of these fear- and threat-expressing calls are composed of multiple units, with component units often of dissimilar form, number and/or spacing. The response to a single stimulus situation might be a train of irregular sound units which are several seconds long. How much of this train should be considered a single call? Given the highly variable nature of the component units and the resultant whole, how can any two sampled calls be considered members of the same class or type? The difficulty and the importance of this last issue has been underscored by the recent discovery of Cheney & Seyfarth (1982) that even spectrographically similar-looking vervet monkey grunts can function in distinctly different ways.

Recently Gouzoules et al. (1984) have made some progress in this issue by defining repeatable features found in 'aid enlistment' screams. For example, screams which contain a wide bandwidth of noise ('noisy screams') were classified separately from those with energy restricted to a narrow range ('tonal screams'), regardless of the number of units, the shape of the units or the spacing between them. As useful as this approach is as an essential first step, it is only the beginning of an explanation. Fully 25% of the uninterrupted vocal units which Gouzoules et al. sampled either could not be classified or consisted of characteristics representative of more than one of their defined scream classes. Their analysis also gives no indication of the

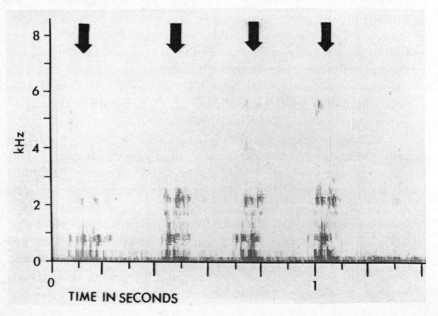

Fig. 1. A multi-unit threat call composed of four distinctive units each showing evidence of vertical striation. Arrows clarify the cadence discernible to the human ear. Note that energy emphasis is below 3000 Hz in this particular call.

fact that units of different classes may be combined into long sequences of response to a single stimulus situation. All of this seems to indicate that rhesus vocal skill encompasses something more than a simple typological repertoire.

My own consciousness about rhesus vocal behavior was raised when I began to think about a missing element in Rowell & Hinde's (1962) cataloguing of rhesus threat calls. Specifically, these investigators distinguished between a one-unit threat call with vertical striation (the 'growl') and a one-unit threat call without vertical striation (the 'bark'). These in turn were categorically separated from all multi-unit threat calls which they called 'pant threats'. Now multi-unit threat calls can vary enormously in the number of units and also vary somewhat in the size, shape and spacing of the units, in pitch and in volume. They can also show vertical striation as assuredly as the one-unit 'growl' (see Fig. 1). Consistency in the use of cataloguing criteria would therefore demand that striated multi-unit threat calls be categorically distinguished from unstriated multi-unit calls just as Rowell & Hinde (1962) did for one-unit threat calls. On the other hand, if one-unit threat calls were separable from multi-unit calls, why not distinguish two-unit calls from three-unit calls, etc? Each of these in turn would have to be split into striated (Fig. 2) and unstriated

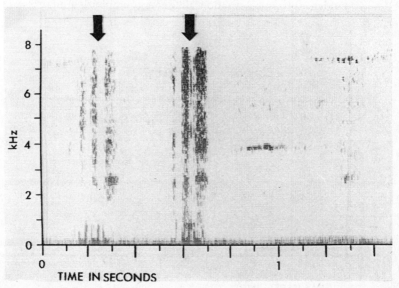

Fig. 2. A two-unit threat call with well-defined vertical striation (ignore background noise after 800 ms). Note that energy is broadly distributed to 8000 Hz and beyond.

versions (Fig. 3). The potential for splitting generated by the logical extension of a few simple cataloguing criteria makes it desirable to seek a more parsimonious way of conceptualizing this immense variety.

In threat situations, the two parameters under examination, number of units and vertical striation, seemed to be varied independently by the monkeys. Furthermore, each parameter showed evidence of existing as a continuum. Number of units could typically range from one to seven (and sometimes more). Vertical striation could be well defined (Fig. 2), partially visible (Fig. 1) or completely absent (Fig. 3). Volume, cadence, and the length of and distribution of energy in the constituent units also showed evidence of being both variable along a continuum and at least somewhat independently manipulable (note, for example, differences in the frequency distribution of energy in Figs 1, 2 and 3). Given the observation that the form of any specific threat call could vary enormously depending on the specific, separable contribution of each of the component-form gradients, then a new conceptualization of the physical basis of call variety becomes possible. Stereotyped calls can be conceived of as calls for which selection has produced a typical and relatively invariant position along each of the several contributing form gradients (e.g. number of units, degree of striation, volume, pitch, cadence, length of constituent units, etc.). On the other hand,

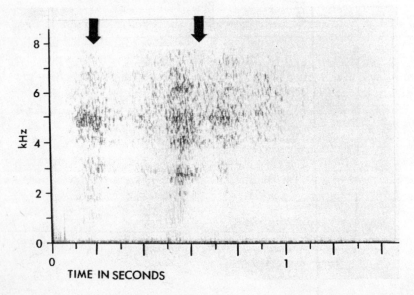

Fig. 3. A multi-unit threat call which the human ear perceives as composed of one short and one longer unit. This two-unit call lacks evidence of vertical striation. Note energy emphasis around 5000 Hz.

'graded' calls and extreme variety of vocal repertoire result when *more than one* position along each of the contributing form gradients is possible.

Discussion

The existence of component-form gradients and the possibility that position along each gradient makes a separate contribution to the emergent form of any given call leads one to ask whether semantic gradients accompany the form gradients (i.e. whether component meaning gradients, not just categorical shifts in meaning, characterize rhesus monkey calls). Conceivably, the meaning of a specific vocalization may be the result of the convergence of effects wrought by position along all component form-meaning gradients. Thus each call might be customized both in form and meaning and any attempt to set up typological call categories would only weakly display the richness of the system. In practice, some typologizing has proven to be a useful way to understand information transfer in non-human primates. Just as assuredly, however, the immense variability of the individual sound signals produced by certain species of non-human primates has continued to thwart comprehensive typologizing and to suggest a more flexible, multi-faceted system of signal generation.

The typologizing habit seems to be an ancient thread in Western civilization (witness Plato's concept of ideal forms) and it may even be a standard cognitive tool of the human mind. Despite its proven usefulness, however, it is not necessarily the only way to conceptualize the organization of animal sound signals and it may not always be the best way to understand what is going on. A better understanding of both graded and non-graded signals (and of the evolutionary relationship between them) may emerge if we loosen this strictly typological approach and try to include some understanding of component gradients and their potential for generating enormous variety in signal structure and meaning. Perhaps the future study of non-human primate vocalizations will not just be a search for the meaning of a finite number of categorical call types but will also be a search for contributing gradient constituents of large call families and for any change in meaning wrought by change in position along each contributing gradient.

References

Altmann, S. A. (1967) The structure of primate social communication. In *Social Communication among Primates*, ed. S. Altmann, pp. 325–62. Chicago: University of Chicago Press

Cheney, D. L. & Seyfarth, R. M. (1982) How vervet monkeys perceive their grunts: field playback experiments. *Anim. Behav.*, **30**(3), 739–51

Gouzoules, S., Gouzoules, H. & Marler, P. (1984) Rhesus monkey (*Macaca mulatta*) screams: representational signalling in the recruitment of agonistic aid. *Anim. Behav.*, **32**, 182–93

Green, S. (1975) Communication by a graded system in Japanese monkeys. In *Primate Behavior*, Vol. 4, ed. L. A. Rosenblum, pp. 1–101. New York: Academic Press

Lindburg, D. G. (1971) The rhesus monkey in north India: an ecological and behavioral study. In *Primate Behavior: Developments in Field and Laboratory Research*, ed. L. A. Rosenblum, pp. 1–106. New York: Academic Press

Morris, D. (1957) 'Typical intensity' and its relation to the problem of ritualization. *Behaviour*, **11**, 1–12

Rowell, T. E. (1962) Agonistic noises of the rhesus monkey (*Macaca mulatta*). *Symp. Zool. Soc. London*, **8**, 91–6

Rowell, T. E. & Hinde, R. A. (1962) Vocal communication by the rhesus monkey (*Macaca mulatta*). Proc. Zool. Soc. London, **138**, 279–94

Struhsaker, T. T. (1967) Auditory communication among vervet monkeys (*Cercopithecus aethiops*). In *Social Communication among Primates*, ed. S. A. Altmann, pp. 281–324. Chicago: University of Chicago Press

IV.6

Symbols: the missing link?

J. LISKA

Introduction

This paper constitutes an attempt to apply some aspects of human communication theory to the study of nonhuman primate communication. The extent to which signs are arbitrary and thus symbolic, as well as the nature of what constitutes communication behavior, has been of central concern to those of us in human communication who are also interested in the evolution of symbolic communication behavior and the comparability of those communicative behaviors in extant species. Specifically, the thesis of this paper is that the communicative abilities of any species can be placed in a continuum of sign arbitrariness, with 'symptoms', the least arbitrary signs, defining one end of the continuum and 'symbols', the most arbitrary signs, defining the other end of the continuum. Marler (1980) has suggested that while the general view is that animal sign behavior 'tends to be associated with 'affective' processes, in contrast to the 'symbolic' processes that are typical of our own species, it may be fruitful to think of symbolic and affective signaling as representing extremes on the same continuum of specificity in these relationships' (p. 227). Marler goes on to illustrate the degree of specificity found among primate vocalizations and thereby argues 'that symbolic and affective signals may be viewed as differing in degree rather than kind' (1980, p. 229).

Arranging signals along a continuum of sign arbitrariness seems especially useful to the study of communication. First, one could describe a particular species in terms of its most characteristic communication behaviors. We might discover, for example, that some species' communication behaviors fall at the affective or functional/consummatory end of the continuum, whereas other species' communicative repertoire includes signs of varying degrees of arbitrariness, with some signs possibly approaching the symbolic or purely arbitrary end of the continuum. Additionally, the apex of the species' communication behaviors and abilities might be specified. That is, the maximum degree of sign sophistication or arbitrariness of that species might be specified. Secondly, cross-species comparisons can be made. Tanner & Zihlman (1976) point out two research areas they see as 'particularly promising for understanding the evolution of human communication'. One of those suggestions calls for 'detailed comparisons of the configuration of human, ape and monkey communication systems . . .' (p. 475). Research on the extent to which human and nonhuman primates differ along a number of dimensions including differences in anatomy, behavior, biochemistry and neural functioning clearly illuminates the close evolutionary relationship we share. Thus, nonhuman primates, especially the great apes, have been accepted as a valuable model or subject for 'hypothesizing evolutionary continuity with the behavior of the human primate' (Tanner & Zihlman, 1976, p. 468). Human primates engage in symbolic communication and appear to be 'uniquely' equipped for manipulating symbols. 'It is not that other animals are totally incapable of such activity, but that humans are uniquely equipped to engage in it more easily, more frequently, even compulsively, perhaps even necessarily' (Cronkhite, 1984, p. 195). Now, the extent to which symbolic communication is a species-specific activity seems to be a largely unanswered question since the literature does not yield a systematic analysis/comparison of the communication behaviors or signs of two or more species along a continuum of the type proposed here or by Marler. Thirdly, we can assess the extent to which species arrangement on this continuum of sign arbitrariness sheds light on the evolutionary history of symbolic communication within and across species. We might find, for example, that the degree of evolutionary relationship between and among species may be reflected or clarified in their placement on a continuum of sign arbitrariness. Seuren (1978, p. 910) poses a parallel question: 'Is there a linguistic continuum in nature, we may ask, a phylogenetic scale of language that has a position for every relevant species?' (If we substitute 'sign arbitrariness' for 'linguistics' and 'language', then the question Seuren raises is too similar to be left out of this discussion.) We might expect the signs used by nonhuman primates to show increasing arbitrariness whereas the signs used by reptiles ought to fall at the symptomatic end of this continuum. Thus, 'symbolicity', or the extent to which a sign is arbitrary, may provide another useful dimension for evaluating and comparing the evolutionary relatedness among species.

The bulk of this paper will be devoted to defining and explaining this continuum of 'symbolicity' or sign arbitrariness and illustrating its utility by providing examples from the repertoire of nonverbal signs used by human and nonhuman primates. A major portion of this endeavor will be devoted to issues of definition. The continuum that I am proposing consists of sign behaviors I shall describe as 'symptoms', 'rituals', or 'symbols' depending upon the degree of arbitrariness of the particular sign. A 'sign' will be 'viewed as that which, in some sense, stands for something else – its significate' (Cronkhite, 1984, p. 53). This definition is rooted in the definitions of signs offered by Peirce (1940) and Morris (1938). These terms, 'symptom', 'ritual', and 'symbol', provide the organizational scheme for the body of this paper. There is, however, a definitional distinction that must be resolved before turning to the task of defining a continuum of symbolicity. This distinction is between 'communication' and 'interpretation'.

Communication can be and has been defined in a number of different ways. Scientific definitions of communication generally refer to the process of information transmission between a source and a receiver who share signs in common. Generally, definitions of communication are very broad, sometimes so broad as to compel researchers to describe *all* human (or animal) behavior and activities as communication. Certainly communication is behavior, but not *all* behavior is communication. Consequently, I shall narrow the scope of what constitutes communication behavior and define communication as an organism's response to a stimulus that is at least to some extent symbolic.

Humans and other animals behave as though they assign meaning to, or 'interpret', nonsymbolic signs. Certainly, the transfer and interpretation of information via nonsymbolic signs provides useful, even vital information. Recognizing and interpreting the signs of estrus is vital to a male nonhuman primate's reproductive success but the redness and swelling of the sex skin are purely *symptoms* of sexual readiness and as such do not constitute communication by my definition.

Defining communication in the way presented here poses a problem for researchers interested in animal species that may not use symbols. Carpenter (1969) distinguishes between social and communicative behavior. He suggests that communicative behavior is more specific and limited than is social behavior. His definition of social behavior seems to refer to behaviors that are not arbitrary. 'Although nursing, keeping the infant warm, grooming, and restricting its exploratory movements have threads of communicative elements; these kinds of behavior are more social than communicative in quality and effect' (1969, p. 44). We might then refer to 'social communication' as an organism's response to non-symbolic signs and use the term 'symbolic communication' to refer to an organism's response to a symbol. Thus social communication occurs to the extent that an organism responds to nonarbitrary signs, which I shall refer to as 'symptoms.'*

Symptoms, rituals and symbols

I am proposing a continuum of symbolicity or sign arbitrariness along which nonverbal sign behaviours may be placed. Marler (1980) has proposed a continuum of 'referential specificity'. Marler writes that in distinguishing between affective and symbolic signs, 'it is common to appeal to the specificity of the multiple relationships between a signal, its external referents and associated physical states and behaviors. Referential specificity refers to the class of objects or events represented by the signal' (1980, p. 229). To the extent that a sign has a large number of referents Marler considers it to be less of a symbol whereas I have specified that the placement of any given sign along the continuum depends upon the extent to which that sign bears a functional/consummatory relation to a physical/physiological state of the organism. I propose using sign arbitrariness in addition to referential specificity since there are signs that are quite specific but not arbitrary. It is my understanding, as one example, that termite fishing is fairly specific, but not arbitrary. Consequently, by Marler's conception it would be 'symbolic'; by mine it would not.

Symptoms

A 'symptom' is a sign that bears a purely causal or functional relation to some physical or physiological state of an organism. Symptomatic signs directly achieve their effects or, put another way, directly satisfy biological needs. The category of signs that I shall refer to as 'symptomatic' includes signs that Marler terms 'affective', signs that

---

* Of course I do not intend to imply that symbolic communication is not also social. This is one alternative approach to distinguishing between types of communication. While this distinction is a convenient way to clarify the two processes it is not the approach I have selected.

Buck (1982, p. 36) refers to as 'spontaneous', that is, 'responses [that] have evolved as externally accessible signs of the occurrence of internal motivational-emotional states', plus other sign behaviors that are functional or consummatory and directly achieve their effect.

Purely symptomatic sign behaviors seem to be best exemplified by such nonhuman primate signs as sex-skin swelling during estrus, the silverback of male gorillas, which may be *interpreted* by other gorillas as indicating credibility or dominance but which appears to be a symptom of age, and the bristling hair of a threatened animal, which probably serves the function of making the animal look bigger and therefore more ominous.

In human primates, symptomatic sign behaviors are represented, for example, by an individual's stomach growling as a symptom of hunger, increase in eyeblink rate, perspiration and/or pupil dilation as symptoms of anxiety, and micromomentary facial expressions that are considered to be reflections or symptoms of an individual's underlying emotional state. These symptoms, while important to an individual trying to understand and evaluate the behavior of another, do not themselves constitute communication, but are essential to its interpretation.

Symbols

A pure symbol is a sign that bears no conceivable natural or functional relation to its significate. With the exception of onomatopoeic words, all words are symbols. Onomatopoeic words appear to be a mixture of symptom and symbol; that is, they are to some extent arbitrary but the choices are restricted to mimicry. Thus, in English, our cats 'meow', 'hiss' and 'yowl', our dogs 'bark' or 'yip', our stomachs 'grumble', horses 'whinney' and insects 'buzz'.

Many nonverbal behaviours are arbitrary and thus symbolic. Ekman & Friesen (1969*a*) describe a category of nonverbal signs that have relatively precise verbal definitions, are learned and culture-bound and are actually an extension of the language. They call these signs 'emblems'. Examples include the 'V' sign for victory or peace, the 'O.K.' sign, and the signs used by baseball managers and football coaches. These 'emblems' are all arbitrary and thus symbolic.

Many of the signs used in AMESLAN (American Sign Language) are also symbols while some of the signs used in AMESLAN are icons. The definition of ritual I will offer subsumes iconic signs. Icons bear a resemblance to their significates and, like onomatopoeic words, mimic their significates to varying degrees. The sign for hotdog in

AMESLAN, a fist, and brandishing a stick are all examples of what I mean by icons.

Rituals

I will use the term 'ritual' to refer to those sign behaviors that are neither totally arbitrary nor totally symptomatic. A sign that is a ritual bears a perceptual similarity or resemblance to its significate or to the symptomatic behavior on which it is based. Therefore, a sign behavior is a ritual to the extent that it resembles a symptom or its significate but is still to some extent arbitrary. A line drawing of a person is a ritual that resembles its significate, whereas many facial expressions such as a threat or fear grimace are rituals that resemble the symptom, in this case the physiological/emotional state of fear. Some rituals begin life as symptoms but become exaggerated or transplanted into other situations so their use is more than symptomatic. This process of exaggeration is frequently referred to as 'ritualization' or 'formalization' (Jolly, 1972; Smith, 1977).

The task at hand is to place some examples of ritual sign behaviors used by nonhuman and human primates along this continuum of symbolicity. Many sign behaviors are sometimes symptomatic and sometimes rituals. That is, while smiling and frowning, for example, are sometimes symptomatic of happiness or sadness, those signs are frequently exaggerated or used when an individual is not experiencing happiness or sadness. Furthermore, an ironic or wry smile is so exaggerated and bears so little relationship to a symptom or significate as to be called 'symbolic'.

Tongue-showing is a ritual used by both man and gorillas and may also be used by other nonhuman primate species. Smith claims that this sign 'apparently derives from acts of ejecting food or a nipple from the mouth when an infant is being fed. After infancy, the visible tongue showing display is done much more often in interactions that have nothing to do with feeding than those that do. It still indicates that interaction, or some aspect of interacting, is being shunned by the communicator, although objects against which the tongue can push need no longer be available to be rejected' (Smith, 1977, p. 318). This sign behavior is a ritual in that it resembles the symptom of rejecting food yet has been exaggerated to indicate rejection of a variety of unpleasant stimuli.

Tanner & Zihlman (1976) talk about nonverbal signs of 'reassurance' such as hugging and kissing as 'more elaborate, complex, and more specifically "inter-animal" in nature (i.e. truly communicatory) than would be expected if they were strictly expressions of somatic or motivational states such as fear, anxiety, anger or excitement' (p. 469). These signs of reassurance, not unlike facial expressions, may be used when the individual is not experiencing positive feelings toward another individual but instead wishes to manipulate or appease another. In any event, the behavior has some degree of arbitrariness and as such is a ritual.

Signs of greeting such as waving appear to be rituals. It may be that waving started as reaching behavior that has been modified and exaggerated. Possibly waving as a greeting is the result of reaching out to touch another and since touching others is a restricted activity in many circumstances and in many cultures, waving became the socially accepted alternative.

Ekman & Friesen (1969b) refer to behaviors such as footshuffling as 'vestigial'. That is, footshuffling is a symptom of leave-taking or escape but it is no longer functional and therefore is a ritual sign.

Social grooming is another example of ritual sign behavior. Jolly (1972, p. 153) writes that 'certainly, grooming helps to remove dirt and parasites, but it is much more than that – it is the social cement of primates from lemur to chimpanzee'. Social grooming, then, is functional to the extent that it is necessary to cleanliness but it is a ritual because it has been exaggerated and transplanted into other situations so its use is more than symptomatic and to some extent arbitrary.

An outstretched hand with upturned palm may be a request for food or even reassurance. It is not necessarily just a receptacle for donation; it is, to some extent, a *symbolic* request for food or attention.

There are, of course, numerous other ritual sign behaviors to discuss but space does not permit such a detailed analysis. I have selected a range of sign behaviors from the literature that I believe are useful in illustrating how the continuum I propose might be used.

Rituals: the missing link?

Kenneth Burke, a literary critic and philosopher of communication, has described humans in the following way: 'Man is a symbol-using (symbol-making, symbol-misusing) animal inventor of the negative (or moralized by the negative) . . .' (1968, p. 3). Since there is considerable evidence that apes can be taught to use symbols,* it may

---

* The question of *how* apes acquire language is not relevant here. That is, whether they acquire signs as a result of response reinforcement or some higher order cognitive processes is not at issue here. Neither am I concerned, at the moment, with insinuating myself into the debate over what is language and/or what constitutes language behavior.

be less chauvinistic and more accurate to characterize humans as the symbol-*creating* animal. It appears, for the moment at any rate, that the creation of a new symbol is uniquely human. Human children reach a point in their development at which they realize that anything can stand for anything and they begin making up their own symbols with rampant abandon. This is a point that the apes do not seem to reach. At least, it is difficult to find evidence in the literature of an ape creating a pure symbol.

The chimpanzee Washoe's invention of the term 'waterbird' has received considerable attention. Apparently she combined two signs that she knew, 'water' and 'bird' to name an animal for which she did not have a sign. Fouts & Rigby (1980) describe another instance of Washoe combining signs she knew to describe or name objects for which she did not have a sign. 'Washoe was observed making a new combination, "gimme rock berry". When the experimenter approached to question her about the seemingly incorrect sign, he found that Washoe was pointing to a box of brazil nuts on the other side of the room' (p. 280). This process of combining two signs to describe an otherwise unnamable object is essentially the same activity used to create 'rituals', i.e. taking a sign and using it in a context in which it has not been used. Certainly these signs are rituals in that they resemble their significates. It is my hypothesis, then, that the exaggeration of symptoms and/or the transplantation of those signs into other contexts so their use is more than symptomatic constitutes the intermediary process between symptomatic and symbolic communication.* Thus, rituals and the process by which rituals are created appear to be the missing link.

The process by which a symptom becomes increasingly more arbitrary is not the same activity as creating a purely arbitrary sign. It seems very likely that it was the concept of absence that elicited the first arbitrary sign. It seems reasonable to guess that coordinating game hunts and food-gathering, as well as rounding up lost or stray youngsters, placed considerable pressure upon humanoids to develop a way to indicate that an area had been searched and no game, food or

---

* Watzlawick, Beavin & Jackson (1967) suggest that 'ritual may be the intermediary process between analogic and digital communication, simulating the message material but in a repetitive and stylized manner that hangs between analog and symbol' (p. 104). For Watzlawick, *et al.*, analogic communication includes *all* signs that are not arbitrary and thus subsumes *both* symptoms and rituals. Since Watzlawick *et al.* do not distinguish between symptoms and rituals, their approach differs considerably from mine.

stray youngsters were found. I can imagine that they might have used a mark on a tree to indicate the negative of absence. This mark would certainly have been an early symbol.

The negative of absence is the most likely candidate for the first symbol because it would have been the first abstract – that is, non-referential – concept for which early humanoids would have needed a sign. I want to emphasize that I am referring to the negative of absence, not the negative of prohibition or refusal. The negative of prohibition has a referential symptomatic behavior; the negative of absence has no referent. Therefore, without symbols, the negative of absence and other abstract concepts are quite literally unspeakable. It is, then, this process of creating symbols that appears to distinguish *Homo sapiens'* from simian communicative abilities.

References

Buck, R. (1982) Spontaneous and symbolic nonverbal behavior and the ontogeny of communication. In *Development of Nonverbal Behavior in Children*, ed. R. S. Feldman, pp. 29–62. New York: Springer-Verlag

Burke, K. (1968) *Language as Symbolic Action*. Berkeley, Calif.: University of California Press

Carpenter, C. R. (1969) Approaches to the naturalistic communicative behavior in nonhuman primates. In *Approaches to Animal Communication*, ed. T. A. Sebeok & A. Ramsay, pp. 40–70. The Hague & Paris: Mouton

Cronkhite, G. (1984) Perception and meaning. In *Handbook of Rhetorical and Communication Theory*, ed. C. Arnold & J. W. Bowers, pp. 51–229. Boston, Mass.: Allyn & Bacon, Inc.

Ekman, P. & Friesen, W. (1969a) The repertoire of nonverbal behavior: categories, origins, usage, and coding. *Semiotica*, **1**, 49–98

Ekman, P. & Friesen, W. (1969b) Nonverbal leakage and clues to deception. *Psychiatry*, **32**, 88–106

Fouts, R. & Rigby, R. L. (1980) Man–chimpanzee communication. In *Speaking of Apes: A Critical Anthology of Two-Way Communication with Man*, ed. T. A. Sebeok & J. Umiker-Sebeok, pp. 261–85. New York: Plenum Press

Jolly, A. (1972) *The Evolution of Primate Behavior*. New York: Macmillan Publishing Co. Inc.

Marler, P. (1980) Primate vocalization: affective or symbolic? In *Speaking of Apes: A Critical Anthology of Two-Way Communication with Man*, ed. T. A. Sebeok & J. Umiker-Sebeok, pp. 221–9. New York: Plenum Press

Morris, C. W. (1938) Foundations of the theory of signs. In *International Encyclopedia of Unified Science*, Vol. 1, No. 2. Chicago: University of Chicago Press

Peirce, C. S. (1940) *The Philosophy of Peirce: Selected Writings*. (Ed. J. Buchler.) Harcourt, Brace, Jovanovich, Inc.

Smith, W. J. (1977) *The Behavior of Communicating*. Cambridge, Mass.: Harvard University Press

Seuren, P. A. M. (1978) Language and Communication. In *Recent Advances in Primatology*. Vol. 1. *Behavior*, ed. D. J. Chevers & J. Herbert, pp. 909–17. London: Academic Press

Tanner, N. & Zihlman, A. (1976) The evolution of human communication: what can primates tell us? *Ann. N. Y. Acad. Sci.*, 467–80

Watzlawick, P., Beavin, J. H. & Jackson, D. D. (1967) *Pragmatics of Human Communication: A Study of Interactional Patterns, Pathologies, and Paradoxes.* New York: W. W. Norton & Co., Inc.

Part V

Functional aspects of development

Introduction *P. C. Lee and P. Bateson*

Among many species of primates, the period of social and physical development is a relatively long proportion of their lifespan. The developmental period has been the focus of a great deal of research in primatology. However, young animals have commonly been viewed as imperfect adults with incomplete behavioural repertoires, rather than as individuals behaving in ways relevant to their immediate competence and survival. Much of the considerable body of research relevant to development has tended to overlook the direct consequences of the behaviour of immature animals and the selective pressures acting on individuals while young.

The omission is made good when we ask what the behaviour of young animals is for; what is its function? The papers in this section were drawn together with a view to considering immaturity as an adaptive phase in the life history of individuals and the adaptive significance of what takes place during development. We have approached this issue in three ways, highlighted by Bateson in Chapter V.1. First, we examine the behaviour of juveniles as specialisations necessary for surviving the immature phase. Secondly, we consider the role of behaviour used in development as part of the developmental assembly process. Thirdly, we distinguish those elements of juvenile behaviour that are retained or incorporated into the behaviour of adults, those that are necessary for adults as well as immatures. From an understanding of the current use of behaviour, we are able to focus on the rules that are important in the assembly process and the different tactics or options open to immatures during development. Ultimately, this leads us to ask how a particular juvenile behaviour may have arisen in the course of evolution.

The papers in this section take several different perspectives. Mother–infant relationships, paternal interactions, and relationships between siblings and peers, the classic subjects of studies of development, are discussed and integrated with issues of learning, communication, exploitation of the environment and social partners, and the stability of relationships using the approaches outlined above.

Hauser, in Chapter V.2, examines mother–infant relationships using a games theory analysis of the matching and mismatching of signals between mothers and infants. The exchange of information between the mother and offspring was measured using the vocalisations of vervet monkeys. He concludes that an analysis of the effectiveness of care-eliciting behaviour illustrates how and when infants convey information about their needs and how, in turn, mothers perceive the content of the information. He emphasises the sensitivity of infants to their mothers' rules for the allocation of care. In his paper, learning, communication, and mother–infant relationships are placed in the perspective of the immediate costs and benefits of interactions between individuals.

In Chapter V.3, Collins deals with the nature of infants' relationships with adult males in yellow baboons. He proposes that infant handling by adult males can be related to paternal protection in many cases, but also serves independent and important functions in the relationships between adult males. While infants with an adult male partner or 'friend' may be protected from other males, and more likely to survive if orphaned or separated, the adult males are able temporarily to enhance their relative rank and defuse potential aggression from other males. The young infant's relationship with an adult male has immediate and long-term functions that are different for each partner. The importance of a functional perspective on development is demonstrated here by the observed separation in the consequences of an interaction for each of the partners.

Datta (Chapter V.4) takes a further step towards unravelling the complexities of social interactions among young animals. She discusses the problem of the acquisition and maintenance of maternal dominance rank among rhesus monkeys. The support that an immature rhesus receives from others during fights is shown to influence that immature's ability to challenge others effectively and thus to acquire its maternal rank. Here, we see how the events during juvenile life have important consequences for adult success. However, Datta emphasises that such events may have different outcomes when they occur at different stages in development. Considering the needs of an immature, the immediate and long-term consequences of behaviour, and the interactions with other group members allows us to understand more fully the phenomenon of rank inheritance.

The final paper by Lee (Chapter V.5) deals with how the physical environment acts as a variable influencing the social development of immatures. Overall patterns and qualities of relationships between mothers and offspring and between peers can be affected by local ecological conditions. In understanding both the behaviour used during development and the nature of juvenile specialisations, it is important to know the range of possible behavioural options and the variety of constraints acting on those options. Furthermore, we can directly compare juvenile responses to ecological variation with those of adults, seeking for generalities in such responses. Ecological conditions can be as influential as social variables in determining the tempo and mode of development, and perhaps can give us a better perspective on the survivorship values of options for development.

The approach outlined here can be, we believe, usefully extended to other aspects of development. Studies on adolescence, such as those by Kraemer *et al.* (1982) and Pusey (1983) on chimpanzees, highlight the nature and timing of transitions between immature and adult strategies. Studies on physiology, growth and the effects of hormones on behaviour also help us understand further how behaviour is used during development and how it has been shaped during evolution.

References

Kraemer, H. C., Horvat, J. R., Doering, C. & McGinnis, P. R. (1982) Male chimpanzee development focusing on adolescence: integration of behavioural with physiological changes. *Primates*, **23**, 393–405

Pusey, A. (1983) Mother–offspring relationships in chimpanzees after weaning. *Anim. Behav.*, **31**, 363–77

V.1

Functional approaches to behavioural development

P. P. G. BATESON

Introduction

Over the years the people studying behavioural development have been ridding themselves of a confrontational style that so often consisted merely of demonstrating that somebody else had underestimated the importance of genes or, reciprocally, the environment. Admittedly, the emerging ideas have verged at times on the obscurantist. To propose, as sometimes has been done, that in development everything interacts with everything is hardly an encouraging piece of advice to anybody who seeks to understand what is going on. It really is not good enough to repeat piously that development is all a matter of interaction. We need to translate a general and valid point about ontogeny into particular examples of how the developmental processes actually work.

Behavioural development is complex, particularly in animals like primates with large and intricate nervous systems that take a long time to develop relative to body size. In order to make the subject concrete and tractable, there are some major advantages in approaching it from a functional standpoint. In this chapter I shall explore what those benefits might be. Although I use some primate examples, the points I seek to make are general ones.

The danger of muddling different issues is ever present, so it is necessary to be clear what exactly is being done when the question of function is brought to bear on development. Tinbergen (1963) recognised that function was one of the four problems raised by the study of behaviour, the others being control, development and evolution. He properly treated these problems as being logically distinct. So, for instance, when we ask what something is for, any answer we get is not going to tell us how that thing works. Nonetheless, a recurring refrain in ethology has been how helpful it is to consider the function of a behaviour pattern, even when the primary concern is with mechanism (see Hinde, 1982). Why should this be? To obtain an answer, it is necessary to observe what happens as functional explanations are derived and discarded.

Not every speculation about the function of a behaviour pattern is equally acceptable. Both logic and factual knowledge can be used to decide between competing claims. Of course, initially good-looking ideas often seem silly as soon as the animal is studied in its natural environment. The skilled and patient observer notices the circumstances in which an activity is performed and those in which it never occurs. The correlation provides a crucial clue as to what the behaviour might be for. Even at this stage, several hypotheses might be in contention, as was the case when Tinbergen *et al.* (1962) were trying to explain why ground-nesting black-headed gulls remove egg shells from the nest-site after a chick has hatched. They were able to exclude a number of the candidates because another gull, the kittiwake, nests safely on cliffs and does not remove egg shells from the nest. Finally, they were able to show by direct experiment that egg shells, conspicuous because of the white inside, reveal to predators the presence of cryptically marked eggs. This classic demonstration of the feasibility of functional analysis has not been often emulated. Nonetheless, it shows the power of studying behaviour in its natural context and also the value of the comparative approach.

Asking what a given developmental process might be for is an important question in its own right. What, though, is the benefit of the functional approach for someone primarily interested in the study of developmental mechanism? I believe the answer is, first, that we are provided with a clear way of distinguishing between independent mechanisms controlling behaviour. Secondly, we are led fruitfully to the important controlling variables of each one. This is important when we start to design experiments in which, inevitably, only a small number of independent variables are actually manipulated while the others are held constant or randomised. The experiment is a waste of time if important conditions that are not going to be varied are badly arranged. A functional approach can provide the knowledge that prevents us from making expensive mistakes.

In what follows I shall first examine how the functional approach helps us to distinguish between the various activities of a developing animal. Sometimes everything that happens between conception and adult life is treated as though it were all relevant to understanding the same kind of developmental problem. A functional approach quickly leads to a different type of conclusion. I will illustrate this point with two types of question that have been applied to developmental studies.

Behaviour of young animals
Specialisations

It is obvious that most animals have to support themselves long before they are fully mature. Furthermore, they are emphatically not miniature adults. This point raises the functional question: is a given behaviour pattern an adaptation of the young animal to the particular conditions in which it lives? The young animal may have to survive in an environment that is radically different from that of the adult and consequently may have specialisations for the conditions in which it lives.

The adaptations of larval forms with totally different phenotypes from those of adults provide the most striking examples. However, mammals also have their juvenile adaptations. Hall and his colleagues have shown that in rats the control system involved in suckling is distinct from that involved in adult eating (Hall & Williams, 1983). The suckling system cannot be satiated easily and time since the last meal does not affect the readiness of the pup to take a nipple and suck. Presumably the more milk the offspring can get, the better off it is. The end of a bout of feeding is provided by the mother's limitations in providing milk. Within a few days of birth and long before the weaning process starts, the pup can be induced to lap milk like an adult. When it feeds in this way, time since the last meal does affect its responsiveness. In other words, both juvenile and adult feeding systems are independent, work differently and are fully developed at the same time. Normally, of course, the rat's adult feeding system does not come into operation until at least 2 weeks after the juvenile one.

Developmental scaffolding

Consider now a different type of functional question: Does a given behaviour pattern function as part of the assembly process, falling away like scaffolding round a building when the animal matures? This is what I called 'behaviour used in development' (Bateson, 1976). The biological function of some of the behavioural mechanisms found in many developing animals seems to be the gathering of information. For instance, precocial birds actively work to present themselves with a form of input that will start off the process of imprinting (Bateson & Reese, 1969). Another possible example is mammalian play. In a number of good quantitative studies, the duration and frequency of play categories have been found first to increase with age and then to decline (see Fagen, 1981). Fig. 1 shows changes with age in captive rhesus macaques observed by Lilyan White (1977). Adults may not give up playing altogether, but unquestionably they do less of it than juveniles.

The precise functional significance of play still excites a great deal of discussion (Martin & Caro, 1985). A conventional but unproven view is that play enhances skills involved later on in life in serious fighting, eluding predators and other such activities fraught with danger to the individual. Even an ineptly caught prey animal can cause serious injury to a predator. The argument is that play has delayed benefits and is a preparation for the risks of adult life. If this view is correct, then play does, indeed, function as scaffolding that has little or no use once the edifice of skills has been assembled. Some of the benefits of play may, however, be immediate and, if so, play should be thought of as partly a juvenile specialisation (Martin & Caro, 1985).

The conclusion, so far, is that many of the things that young animals do are not incompetent or incomplete versions of adult behaviour. The growing awareness of those facts has been an important benefit of adopting a functional approach to development. It is possible to go much further when considering the development of behaviour as opposed to behaviour in development.

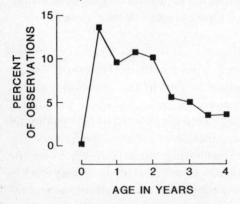

Fig. 1. Changes with age in the social play of rhesus monkeys. The medians of the percentage number of 30-s intervals in which social play was observed are shown. Males and females did not differ on this measure and so the data for both sexes are combined. The minimum sample size at any one point was 11. (Redrawn from White, 1977.)

Functional rules of assembly

When we are able to witness the actual assembly of adult behaviour, we can also ask functional questions about what we see. Why do they do it that way? When particular environmental conditions trigger a certain mode of development, why should that happen? What is the functional significance of alternative developmental routes? I shall focus here on the issue of alternative mating tactics in the sense used by Dominey (1984).

The study of individual differences has been revolutionised in recent years. More and more examples of striking differences in reproductive behaviour are being found between members of the same sex in the same species (see Davies, 1982). The appropriate measures of adult behaviour are sometimes distributed bimodally or multimodally. The functional hypotheses proposed for the existence of these different modes of behaviour have provided attractive alternatives to the blankly negative view that individual variation merely represents noise in the system. At first it was widely assumed that these instances represented evolutionarily stable mixes of genetically distinct individuals. However, another view, which is growing in popularity, is that each individual is capable of developing in more than one way. For instance, Dunbar (1984) has suggested that in gelada baboons all males can either become harem-holders or female-like followers that sneak their copulations.

You can think of the individual animal, which is capable of developing in a number of different ways, as a jukebox. It can play many tunes but, in the course of its life, may only play one. The particular tune it does play is triggered by the conditions in which it grows up (Caro & Bateson, in press). One developmental problem is: What are those conditions? Are we given any guidance by the functional approach? In the case of the male gelada baboons, Dunbar's (1984) functional suggestion was that in an environment populated by lots of harem-holding males, a young male is likely to have more offspring if he becomes a small-bodied follower than if he develops as a big-bodied harem-holder. With fewer harem-holders about, he does better if he becomes one himself. As always, it is easy to think of other functional explanations. However, Dunbar's suggestion does have clear implications for developmental analysis. The young male should delay committing himself to becoming a follower or a harem-holder until the point when further delay would adversely affect his chances of becoming a big harem-holder, if that turns out to be the better option. His decision needs to be based on the most up-to-date assessment of the troop in which he is likely to settle as an adult. At the decision point he should actively gather information about the number of harem-holding males relative to the number of available females. These are strictly issues that arise from design considerations, a topic which I shall turn to next.

Before leaving the topic of alternative tactics, it is as well to remember the massive capacity for behavioural change found in most adult primates. The long-term effects of the metaphorical jukebox may become irrelevant if the environmental conditions should change. We have to be alert to the possibility that animals developing in different ways may end up behaving in rather similar ways if circumstances alter.

Optimal design

The final issue I shall consider is the optimal design of developmental systems. If we were to design a machine to solve a given problem, what would be the best way to do it? A functional study seeks to answer the question of what a character is for. By contrast, an optimal design study assumes knowledge of the function, next proposes the best engineering solution to the problem faced by the animal, and finally examines whether use of the character conforms to the theoretical expectation (see Krebs & McCleery, 1984). Clearly one type of study can lead into the other. For instance, work in the field can generate some strong hunches about what a given pattern behaviour was for and this in turn can lead to thoughts about the best way to solve that problem if the animal were to be designed by a good engineer. I have already suggested how such an argument could be developed in the case of the gelada baboon.

Already the behavioural ecologists' studies of optimal foraging have intersected promisingly with the experimental psychologists' analysis of animal learning. As an animal gathers information about its fluctuating environment, what rules should it use in deciding where it should feed? Should it go to a place where the food is always available in small amounts or one in which it is periodically available in large amounts? Ideas about the best ways to sort out such conflicts between foraging in different places are providing sharp new insights into the nature of the mechanism (see Kamil & Yoerg, 1982; Marler & Terrace, 1984).

Another important question is when in development an animal should learn about its kin. Parenthetically, I should note that a variety of explanations has been offered for the evolution of kin-recognition. The first and most famous is that it evolved as a consequence of kin-selection (Hamilton, 1964). Once aid was given to collateral kin, natural selection would favour careful discrimination between kin of different degrees of relatedness. The second evolutionary explanation is that parental discrimination between own and others' offspring would be favoured by natural selection when parental care was costly – as it usually is (e.g. Burtt, 1977). Thirdly, offspring discrimination between parents and non-parents would be favoured by selection when parents were liable to attack unfamiliar young of their own species (e.g. Evans, 1970). Fourthly, familiarity with the parent and siblings of the opposite sex could be advantageous in mate choice when it was important to minimise the costs of inbreeding and those of too much outbreeding (e.g. Bateson, 1983). Use of close kin as standards when finding a mate of an optimal degree of relatedness may be highly beneficial.

When considering the best time to learn about kin, perhaps the most intriguing puzzle is the one posed when siblings of the opposite sex are used as standards in the choice of a slightly novel mate. Obviously, the visual appearance of siblings is liable to change a great deal as they develop. Good design considerations suggest that either the animal must learn selectively about cues (such as the siblings' smell) that are invariant across development, or it must delay learning about its siblings until they are sufficiently adult-like in appearance. In the second case the optimal design suggestion is that an animal should not tune its reference point for mating preferences too early in development lest it obtains information about the juvenile appearance of its siblings that could not be used effectively when the time comes to choose a mate. On the other side it must not tune its mating preferences too late in development after the family group has broken up and it is likely to be exposed to non-kin.

What has been gained by the design argument about timing? How does it help the study of developmental mechanism? It provides a way of thinking about the difference in timing between sexual imprinting and filial imprinting. Indeed, the evidence suggests that birds delay sexual imprinting until their siblings have moulted into adult plumage (see Bateson, 1979). If the two types of imprinting are treated as part of the same general process, the difference seems to be of no importance and is quickly forgotten about. However, with attention focussed on the problem, we can attempt to analyse the mechanisms responsible for the difference in timing. The point is, then, that the optimal design approach frames and stimulates research on the process of development.

Conclusion

The postulation of a function is not to tell a Kiplinesque Just-So story about how the leopard got her spots (cf. Gould & Lewontin, 1979). It is a story about what the spots are for, here and now. Confusion about the meaning of function has arisen because the word 'adaptation', often used as a synonym for function, has two widely different meanings in biology. First, it is used for the processes of modification that make organisms better suited to survive and reproduce in a particular environment. Secondly, it is used for the state of being suited to a particular environment. The derogatory adaptationist label is stuck on those who use it in the first sense. However, it is the second sense of current utility that relates to the ethological usage of 'function' and has concerned me here. A functional approach focusses on the job that has to be done here in the present.

In this article I have suggested that the approaches which have worked so well in behavioural ecology will start to furnish us with a set of unifying principles about the mechanisms of behavioural development. On the face of it, knowledge of function ought not to tell us anything about how things work. What is the connection? The steps from function to mechanism may be as follows. First, the correlations observed between behaviour seen in development and the circumstances in which it occurs lead to speculations about current use. These, in turn, can lead in two directions. On the one hand, they may suggest what are likely to be important controlling variables and thence lead to experiment. On the other hand, they may suggest a design for the way in which the mechanism ought to work. Here again,

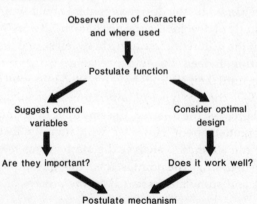

Fig. 2. Proposed steps from studies of function to studies of mechanism.

the proposal can be tested against reality. Either way, we are led to an enhanced understanding of mechanism. The steps are summarised in Fig. 2.

Admittedly, the implication of an inexorable progress towards greater understanding of mechanism is unrealistic. In practice, a great deal more tracking backwards and forwards between steps takes place than is suggested by the scheme shown in Fig. 2. Nonetheless, it serves to make the point that functional considerations can have an important part to play in understanding mechanism. If this argument is correct, we ought to encourage studies that relate development to ecological context as well as good analytical work in the laboratory. Moreover, we should engage in many more comparative studies of behavioural developmental than has been usual, since such studies provide such important insights into function. The rewards of all this work will be reaped in a much broader understanding of ontogenetic processes. If the general style is right, as I believe it is, we shall finally be in a position to solve one of the major problems in biology, namely how animals develop in their own life-times.

Acknowledgements

I am very grateful to Tim Caro, Phyllis Lee and Paul Martin for their comments on earlier drafts of this article.

References

Bateson, P. P. G. (1976) Rules and reciprocity in behavioural development. In *Growing Points in Ethology*, ed. P. P. G. Bateson & R. A. Hinde, pp. 401–21. Cambridge: Cambridge University Press

Bateson, P. P. G. (1979) How do sensitive periods arise and what are they for? *Anim. Behav.*, **27**, 470–86

Bateson, P. P. G. (1983) *Mate Choice*. Cambridge: Cambridge University Press

Bateson, P. P. G. & Reese, E. P. (1969) The reinforcing properties of conspicuous stimuli in the imprinting situation. *Anim. Behav.*, **17**, 692–9

Burtt, E. H. (1977) Some factors in the timing of parent–chick recognition in swallows. *Anim. Behav.*, **25**, 231–9

Caro, T. M. & Bateson, P. [P. G.] (in press) Organisation and ontogeny of alternative tactics. *Anim. Behav.*

Davies, N. B. (1982) Behaviour and competition for resources. In *Current Problems in Sociobiology*, ed. King's College Sociobiology Group, pp. 363–80. Cambridge: Cambridge University Press

Dominey, W. G. (1984) Alternative mating tactics and evolutionary stable strategies. *Am. Zool.*, **24**, 385–96

Dunbar, R. I. M. (1984) *Reproductive Decisions: An Economic Analysis of Gelada Baboons*. Princeton, NJ: Princeton University Press

Evans, R. M. (1970) Imprinting and mobility in young ring-billed gulls, *Larus delawarensis. Anim. Behav. Monogr.*, **3**, 193–248

Fagen, R. (1981) *Animal Play Behaviour*. New York: Oxford University Press

Gould, S. J. & Lewontin, R. C. (1979) The spandrels of San Marco: a critique of the adaptationist program. *Proc. R. Soc. London, B*, **205**, 581–98

Hall, W. C. & Williams, C. L. (1983) Suckling isn't feeding, or is it? A search for developmental continuities. *Adv. Stud. Behav.*, **13**, 219–54

Hamilton, W. J. (1964) The genetical evolution of social behaviour. I & II. *J. Theoret. Biol.*, **7**, 1–52

Hinde, R. A. (1982) *Ethology*. Oxford: Oxford University Press

Kamil, A. C. & Yoerg, S. I. (1982) Learning and foraging behavior. In *Perspectives in Ethology*, Vol. 5, *Ontogeny*, ed. P. P. G. Bateson & P. H. Klopfer, pp. 325–64. New York: Plenum

Krebs, J. R. & McCleery, R. H. (1984) Optimization in behavioural ecology. In *Behavioural Ecology*, 2nd edn, ed. J. R. Krebs & N. B. Davies, pp. 91–121. Oxford: Blackwell

Marler, P. R. & Terrace, H. (1984) *Biology of Learning*. Dahlem Konferenzen. Berlin: Springer

Martin, P. & Caro, T. M. (1985) On the functions of play and its role in behavioral development. *Adv. Stud. Behav.*, **15**, 59–103

Tinbergen, N. (1963) On aims and methods of ethology. *Z. Tierpsychol.*, **20**, 410–33

Tinbergen, N., Broekhuysen, G. J., Feekes, F., Houghton, J. C. W., Kruuk, H. & Szulc, E. (1962) Eggshell removal by the black-headed gull, *Larus ridibundus* L; a behaviour component of camouflage. *Behaviour*, **19**, 74–117

White, L. E. (1977) The nature of social play and its role in the development of the Rhesus monkey. Doctoral dissertation, University of Cambridge

V.2

Parent–offspring conflict: care elicitation behaviour and the 'cry-wolf' syndrome

M. D. HAUSER

Introduction

When interactions between individuals are essentially determined by the nature of the resource at stake, as well as the amount of resource to be allocated, conflicts of interest typically arise. The conflict to be discussed here is the parent–offspring conflict, in which the interests of the offspring with regards to the allocation of care differ markedly from those of the parent. I shall first establish a framework for discussing the parent–offspring relationship and, given this framework, shall describe a model which focuses primarily on the infant's ability and effectiveness at eliciting care. Finally, I shall test some of the major predictions of the model using empirical data on mother–infant interactions in vervet monkeys from Amboseli National Park, Kenya.

The basis of the model presented here starts with the structure built by games theorists (Maynard Smith, 1982). Most recently, Parker (1983) has applied this approach to the parent–offspring relationship and suggested that the interactive strategies of parent and offspring are effectively viewed as an arms race in which the parents are capable of resisting or rejecting the enhanced solicitations of their offspring. With the conceptual formulations of games theory, I shall focus specifically on the exchange of information or signals between parent and offspring and, given this exchange, shall describe how the strategies of parents and offspring come into conflict with regard to interests concerning survivorship and reproductive success.

It should be pointed out that throughout the discussion, reference will be made to *mothers* and offspring rather than *parents* and offspring. The use of a single parent is for simplicity and the model may easily be modified for species where both mother and father partake in the care of young. Thus the model's predictions are not restricted to a particular species or social organization. Finally, although most of the empirical results will focus on the period of infant development from birth to weaning, the model is not restricted to this developmental period.

Mothers, infants and the exchange of information

When infants solicit for care, they are transferring to their mother information which reflects the quality and quantity of investment required. This information can be either 'truthful' or 'deceitful'. Truthful is defined as indicative of actual needs or requests, and deceitful as deviant from actual needs. Given such information, a mother may respond in several ways. Fig. 1 illustrates the matrix which results when mother and infant are set against one another with regards to the exchange of information. The terms selected to represent mother and infant strategies were chosen for ease of description and should not be taken to imply any cognitive reasoning capabilities. In addition, the payoff matrix has been restricted to pluses and minuses rather than numerical values. Thus pluses and minuses are values *relative* to the other interactant and *relative* to other cases within the matrix.

The infant is seen as having two types of signal to solicit care – it may solicit either truthfully or deceptively. In turn, the mother may respond to the infant's solicitations in two ways – the mother can choose either to believe or to doubt the infant's solicitations. The former response is

Fig. 1. Matrix of the parent–offspring conflict. M = mother; I = infant.

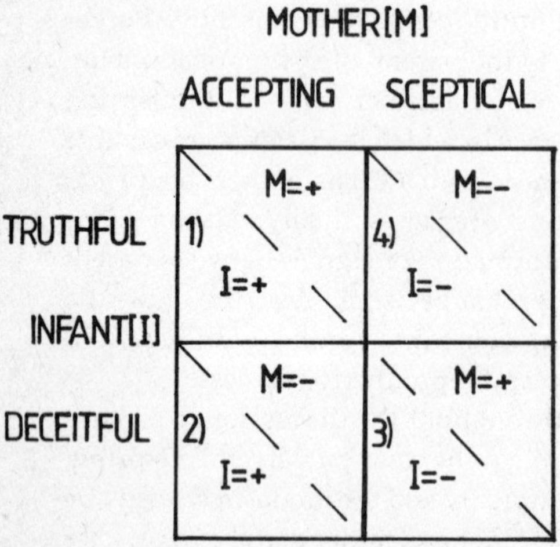

	MOTHER [M]	
INFANT [I]	ACCEPTING	SCEPTICAL
TRUTHFUL	1) M=+, I=+	4) M=−, I=−
DECEITFUL	2) M=−, I=+	3) M=+, I=−

termed 'accepting' and the latter 'sceptical'. An accepting response would indicate that the mother perceives its offspring's signal as indicative of needs, whereas a sceptical response would indicate that the mother perceives the signal as an exaggeration of needs. In describing the matrix, each box will be considered in a counter-clockwise order, beginning in Box 1. There is no biological reason for discussing each box in this order.

In Box 1, if the infant's solicitations are truthful and the mother is accepting, then both mother and infant benefit. For example, if an infant screams to obtain aid from its mother, and in fact the infant is in need of aid, then the mother will benefit by being accepting, and the infant will have benefited by signalling truthfully. Although Box 1 shows benefits for both mother and infant, the infant could gain more by signalling deceptively, as long as its mother remained accepting of the solicitation. This brings us to Box 2 where the infant signals deceptively and the mother continues to accept the information conveyed by the infant. Thus the mother matches the infant's requests and this results in a gain for the infant and a loss for the mother. Due to the imbalance in payoffs, it will now pay the mother to switch response strategies and become sceptical of the infant's solicitations for care. This switch in strategies brings us to Box 3, where the infant continues to solicit deceptively, but its mother is now sceptical of the information conveyed. In this case there are two possible sets of payoffs. In case one, if the mother's allocation of care is less than what is solicited but not less than what is actually needed, then the mother gains but the infant loses relative to the payoffs in Box 2. This is the case illustrated here. In case two, if the allocation of care is below what is actually needed, then both mother and infant may lose. Finally, we come to Box 4 where it no longer pays the infant to solicit deceptively, so it solicits truthfully. However, the mother continues to respond sceptically. As in Box 3, the payoffs depend upon how much care is allocated or, how sceptical the mother is. I have illustrated the general case where if an infant's solicitations represent actual needs and the mother allocates less, then both mother and infant lose.

The payoff matrix shows that the exchange of information between mothers and infants will always involve some level of imperfection or uncertainty. An infant will never know how much care its mother is capable of allocating and conversely, a mother will never know how much care her infant actually needs. The result of such uncertainty is an endless cycling of strategies. It will never pay to remain in one box or with one set strategy indefinitely, because the relative payoffs of changing are continuously fluctuating to the advantage of either the mother or the infant.

Mechanics of the model

The model to be discussed takes the approach described above and addresses one general question: Through the use of various signals which each convey particular bits of information, how can an infant maximize its effectiveness at eliciting care? To address this question, it is necessary to establish a measure of effectiveness which describes an infant's tactics for eliciting care. The equation shown below describes how such a measure of effectiveness can be obtained:

$$E_i = R_i/S_i$$

This equation says that the effectiveness, E_i, of a particular care-eliciting behavior, i, will be a function of the ratio R_i to S_i, where R_i indicates the number of positive responses or consequences given the number of times care-eliciting behavior, S_i, is solicited. For example, if the care-eliciting behavior is 'present to groom' and the respective positive consequence is 'grooming', then if the infant solicits for grooming 10 times and the mother grooms 10 times, the effectiveness of present to groom is 1.00.

The measure of effectiveness described here is based upon *numbers* of care-eliciting behaviors and corresponding *numbers* of consequences. Clearly it is important to consider such factors as the duration of the consequence as well as the interval of time elapsed between successive attempts to elicit care. In discussing some of the infant strategies of care-elicitation, some of these additional factors will be addressed.

Fig. 2 illustrates the hypothetical relation between effectiveness of a given care-eliciting behavior and its change over time. As previously mentioned, the developmental time scale has been left open since no particular period of the parent–offspring relationship is to be emphasized. The curves in Fig. 2 suggest several predictions. First, the effectiveness of care-eliciting behaviors will decline over time, but the rates at which they do so should vary. Secondly, there should be specific periods of time during development when it is most beneficial for infants to switch care-elicitation behaviors, although such a switch does not necessitate the abandonment of former ones. Generally speaking, infants should switch behaviors as a function of diminishing returns. The problem here is comparable to the notion of 'giving-up time' from optimal foraging theory (Krebs, 1978), in that an infant should move on to a new care-eliciting behavior given a decline in effectiveness of a former one. Once a switch has occurred, there will be certain care-eliciting behaviors which initially result in low levels of effectiveness and this will sometimes be due to the mother's inability to understand their function. As illustrated in curve i_3, it is possible that with an increase in the frequency of such care-eliciting behaviors, the mother will come to understand their function and thus, an increase in effectiveness will ensue. Finally, if infants have large care-eliciting behavioral repertoires, then as mothers wean their infants to independence and conflict over the allocation of care peaks, infants will potentially exhaust all means of soliciting care and this will result in a series of curves, packed closely together in time and with, most likely, minimal effectiveness.

By summing the area under these curves, one obtains a statement of how much care an infant receives based upon the mode and tempo of care-eliciting behaviors selected throughout development. Given the amount of care received, it is necessary to examine what types of strategies infants might select and, subsequently, what rules of care allocation mothers might respond with.

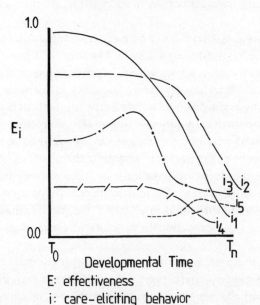

Fig. 2. Hypothetical changes in effectiveness of different care-eliciting behaviors over developmental time.

Infant strategies and mother's allocation rules

Two broad categories of infant solicitation strategies will be examined. These two strategies I have termed 'persistent' and 'try-later'. It should be pointed out that these are by no means the only types of strategies infants may use, nor is a given infant restricted to one type of strategy during development. It is quite likely that some infants will flip from strategy to strategy depending upon various properties of the mother–infant relationship.

The persistent strategy can be described by the case where infants continue attempts to obtain care until successful or until the costs of rejection by the mother become too great. This strategy might involve (1) using different types or sequences of care-eliciting behavior with perhaps varying intensities, or (2) repeatedly using the same care-eliciting behavior. For example, an infant using this strategy whose attempt to gain access to the nipple failed, might *immediately* try to be groomed and if successful, then *immediately* try to gain access to the nipple. Continued attempts to gain access to the nipple might only be stopped by a mother who, for instance, bites her infant. In contrast to the persistent approach, there is the try-later strategy where an infant whose attempt to obtain care failed, would wait for a particular interval of time and then try again. As in the persistent strategy, the infant's subsequent attempts to elicit care may involve either the same or different types of care-eliciting behaviors. What will be particularly important here is the time elapsed between successive attempts to elicit care.

To offset the infant strategies, two rules of care allocation will briefly be discussed. The first rule, suggested by Parker & Macnair (1979), was termed *'pro rata'*. *Pro rata* allocation occurs when the parent gives more than its optimum, but less than the offspring requests. In essence, parental investment will be a compromise between the interests of the parent and those of the offspring. In contrast to *pro rata* allocation, a 'matching rule' is suggested whereby females are assumed to have a finite amount of investment potential or investment threshold. The rule states that: The demands or solicitations by infants will be matched until the threshold value of their mother's investment resources is surpassed. As the threshold is attained, the mother may either cut back to a lower level of care allocation, or become immune to the infant's solicitations.

Before the fit between theoretical predictions and empirical results is assessed, some of the model's main points must be emphasized. First, by concentrating on the repertoire and effectiveness of care-eliciting behaviors and the corresponding rules of care allocation, the parent–offspring relationship emerges as a system where each individual must constantly tune, modify and adjust to the information being exchanged. Secondly, it is important to look at *specific* care-eliciting behaviors and their respective consequences, in addition to the sum total of care received over time. Examining specific care-eliciting behaviors and how they change allows infants to demonstrate their perception of allocation rules as well as their sensitivity to the mother's reproductive and social status.

Vervets as modellers

To assess the various implications of the model, results from observation on free-ranging vervet monkeys in Amboseli National Park, Kenya, will be examined. As a brief introduction, the results to be described are based on observations of five adjacent groups. What is interesting to note about these groups and this sub-population of vervets at large, is that infant mortality is extremely high, of the order of 60%, and part of the cause of this mortality seems to be the unique ecological constraints which confront each group (Cheney *et al.*, 1981, in press; Hauser, unpublished). For example, some groups lack surface water throughout most of the year, whereas other groups have limited access to high-quality food resources.

To illustrate the predictions of the model, two mother–infant pairs will be contrasted. The cases selected are most useful in describing one particularly intriguing outcome of the model's application. This outcome, which I have called the 'cry-wolf syndrome', stems from the maternal allocation rule previously described as the matching rule and states that: if the rate of care-eliciting behaviors escalates beyond a given level, mothers become immune to their infant's solicitations for care and thereby fail to respond. Prior to the study's inception, a series of care-eliciting behaviors and possible consequences were generated. Fig. 3 represents a selective list of care-eliciting behaviors, indicated under the column heading i, and possible consequences, indicated under column headings R_i and *negative* R_i. Although each care-eliciting behaviour has been represented as having a one-to-one correspondence with a particular consequence, this was only done for purposes of illustration. There will be several consequences, both positive and negative, for a given care-eliciting behavior. For example, an infant who presents to groom could be groomed, ignored, bitten or pushed away.

The two mother–infant pairs will be contrasted by focusing on three types of vocalization used by vervet infants to elicit care. The vocalizations have been pooled for purposes of illustration and thus the values of effectiveness represented in Fig. 4 are combined scores for all three vocalizations. Effectiveness has been plotted against the same developmental period for each mother–infant pair. Developmental time is represented in one-week intervals after birth. Thus the first point on each graph indicates the effectiveness of infant vocalizations up to the age of 7 days. Effectiveness was examined up to week 7 for mother–infant pair BA/EG and up to week 5 for AC/BB.

Fig. 4(a) shows that effectiveness for mother–infant pair BA/EG was quite high until week 2 and then dropped to zero in week 3. The decline in effectiveness in week 3 was associated with a substantial escalation or *increase* in the frequency of solicitations. Following this decline in week 3, there was an increase in effectiveness in week 4 and this corresponded to a *decrease* in the frequency of solicitations; this level of effectiveness was reasonably well maintained until week 7. In Fig. 4b, the change in effectiveness for AC/BB is shown. As with BA/EG, effectiveness was quite high early on and then declined to zero in week 4. This decline was also associated with an escalation in the frequency of solicitations. However, in contrast to infant EG, BB continued to escalate the frequency of solicitations following the decrease in effectiveness and, consequently, effectiveness remained

Fig. 3. Examples of different care-eliciting behaviors (i) and potential consequences, both positive (R_i) and negative (negative R_i).

i	R_i
"Scream" vocalization	Retrieved
Accesses nipple	Suckles
Presents to groom	Is groomed
Elicits agonistic aid	Is given aid

Negative R_i
Is bitten
Is pushed away
Is ignored

low. In fact, in week 5, BB died. BB's case can be expanded upon further from ad-lib and focal observations. Three days prior to his death, BB began to escalate the frequency of care-eliciting vocalizations. During a focal sample, BB began to scream while he was struggling to climb up into a tree. AC, his mother, ignored these vocalizations and subsequently BB fell out of the tree and was paralyzed from the waist down. Two days later BB died. Although this represents the most extreme case of the 'cry-wolf syndrome', out of eight infants who died in 1983/84, five deaths were preceded by similar changes or patterns in effectiveness. In addition, the 'cry-wolf syndrome' seems to be a fairly general phenomenon for this sub-population of vervets in Amboseli, since infants in different groups, independently of ecological differences, have suffered from this decline in effectiveness.

In returning to the matrix described in Fig. 1, it should be clear how the results from Amboseli fall into place. In Box 1 we have a stable situation between mothers and infants with the result that effective-

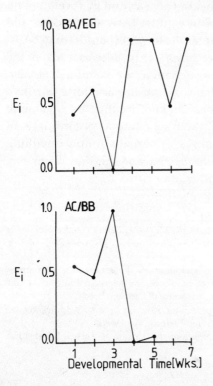

Fig. 4. Effectiveness (E_i) plotted against developmental time (in weeks) for two mother–infant pairs. (top): Mother–infant pair BA/EG; (bottom): Mother–infant pair AC/BB.

ness is high. The infant then escalates the frequency of solicitations and maintains a high level of effectiveness – Box 2. The mother then becomes sceptical of this escalation and this is associated with a decrease in effectiveness – Box 3. For mother–infant pair AC/BB, we arrive in Box 4 and both AC and BB lose; this is the 'cry-wolf syndrome' or, as previously mentioned, one possible outcome of the matching rule. In contrast, for BA/EG, as we arrive in Box 4 the cycle starts up again with Box 1, where effectiveness is high.

Conclusion

In conclusion, the interaction between theoretical predictions and empirical results brings out three main points. First, and as pointed out previously, the analysis of various care-eliciting behaviors and their subsequent effectiveness allows us to understand what type of information infants convey to their mothers about needs and, consequently, how mothers perceive the nature and content of the information proffered. Secondly, by focusing on the changes in effectiveness of different care-eliciting behaviors, it is clear, at least for vervet mothers and infants in Amboseli, that conflict over the amount and quality of care allocation may begin very early on in development. In other words, although the *peak* in conflict between mothers and infants may occur at weaning, more subtle forms of conflict may begin at birth and this type of conflict might only be observed if the effectiveness of various care-eliciting behaviors are examined. Finally, the results from Amboseli show that some infants are quite sensitive to their mother's rules of care allocation and are thereby capable of adjusting the mode and tempo of their solicitations accordingly. In contrast, other infants are less sensitive, or perhaps simply unwilling to compromise in terms of the care requested, and this may ultimately lead to a maladjusted mother–infant relationship.

References

Cheney, D. L., Lee, P. C. & Seyfarth, R. M. (1981) Behavioural correlates of non-random mortality among free-ranging female vervet monkeys. *Behav. Ecol. Sociobiol.*, 9, 153–61

Cheney, D. L., Seyfarth, R. M., Andelman, S. J. & Lee, P. C. (in press) Reproductive success in vervet monkeys. In *Reproductive Success*, ed. T. H. Clutton-Brock. Chicago: University of Chicago Press

Krebs, J. (1978) Optimal foraging: decision rules for predators. In *Behavioural Ecology*, ed. J. Krebs & N. B. Davies. Oxford: Blackwell

Maynard Smith, J. (1982) *Evolution and the Theory of Games*. Cambridge: Cambridge University Press

Parker, G. A. (1983) Arms races in evolution – An ESS to the opponent-independent costs game. *J. Theor. Biol.*, **101**, 619–48

Parker, G. A. & Macnair, M. R. (1979) Models of parent–offspring conflict. IV. Suppression. Evolutionary retaliation by the parent. *Anim. Behav.*, **27**, 1210–35

V.3

Relationships between adult male and infant baboons

D. A. COLLINS

Introduction

The nature of relationships between infants and adult males may vary markedly between species. In some, such as vervet monkeys, males and infants seldom interact (Struhsaker, 1971), while in others, such as baboons and macaques, their interactions may be frequent and complex. This variability can be contrasted with other relationships which tend to be more constant across species (e.g. those with siblings or peers), and raises questions about male–infant relationships concerning their origins, immediate behavioural functions, and contributions (if any) to the social development of the infant and social and reproductive activities of the male. Male–infant relationships are prominent among most savannah baboons, and this paper relates observations on male–infant interactions among yellow baboons (*Papio c. cynocephalus*) from one troop to findings from other studies of baboons in the hope of resolving some of these questions.

Previous studies have highlighted two main features of male–infant relationships. First, it is widely agreed that males often behave protectively to young infants (DeVore, 1963; Ransom & Ransom, 1971; Rhine *et al.*, 1980). It has recently been proposed that males often *carry* infants specifically to protect them from infanticide by immigrant males (Busse & Hamilton, 1981). However, the idea that the males' behaviour is entirely protective is at odds with the second main feature of these relationships, that a male who carries an infant can inhibit his opponent from attacking him (Rowell, 1967; Stoltz & Saayman, 1970; Ransom & Ransom, 1971; Altmann, 1980; Stein & Stacey, 1981; Strum, 1983), and he may even become temporarily dominant to his opponent (Packer, 1980). Such interactions, falling within the definition of agonistic buffering (Deag & Crook, 1971), may result in the risk of injury to the infant and can be considered exploitive rather than protective. Here, I ask how infant-carrying is related to the male–infant relationship, and assess the protective and exploitive contexts in which it occurs in order to determine the immediate and ultimate benefits that might be conferred on participants.

Study troop and methods

The observations presented here were made on a habituated troop of 70 yellow baboons in Ruaha National Park, southern Tanzania. The subjects of the study were 12 males and 19 females (8 of whom had infants below 6 months of age). Both males and females could be ranked in a linear dominance hierarchy on the basis of dyadic supplants and avoidance (excluding all interactions involving escalated aggression, mating partnerships, alliances and infant-carrying) (see data in Collins, 1981, pp. 162–3). The most important distinction among the males was that two were newcomers, young adults who joined the troop 5 months before the study began, and they held the top two ranks throughout the study. In contrast, there were six adult males who had been with the troop at least 1 year; three of these residents were of prime age (ranked 3, 4, and 5) and three were older (ranked 6, 7, and 8). There were four subadults who had been born in the troop, whereas the adults almost certainly were immigrants. The subadults ranked below the adults in accordance with their relative sizes, except for the largest who was ranked between the fifth and sixth adults. Males are referred to by age-class (A for adult, S for subadult) with a suffix indicating their rank within their class. Thus the adults are A1 to A8 and subadults S1 to S4.

There were 14 infants under 1 year old. Those less than 6 months ($n = 6$) are referred to as black infants because of their natal coat colour. Two additional black infants were born, making eight in total. Those between 6 and 12 months were referred to as brown infants ($n = 6$).

Observations were made on foot within the troop, and male–infant interactions were recorded on an *ad libitum* basis. Such observations may be biased towards the animals most often in view or most conspicuous. This potential bias was corrected for by dividing each subject's rate of behaviour by the proportion of time for which it was in view (see Collins, 1981, p. 27), which relatively increased the rates for the animals observed the least. In analyses, results were retained only if they were confirmed both with and without this correction.

A total of 381 interactions between males and infants was recorded during 3 months in 1975–76. The dyadic interactions between males and infants are described using all 381, and a subset of those involving more than one male are examined for triadic male–infant–male relationships.

Dyadic male–infant interactions
Interactions and contexts

In 55% of interactions, a male held the infant while stationary or carried it when moving. Interactions without contact were also relatively frequent (27%) and consisted of interest in an infant (peering, grunting). Handling of an infant while it was in contact with its mother (including trying to pull it off the mother) was the least frequent interaction (17%). It was possible to describe the contexts of interactions from events occurring at the time. Most of the dyadic interactions were relaxed (59.6%), with the male usually sitting with mother–infant clusters and no unusual circumstances evident. Only a few interactions (0.5%) were obviously a protective response to distress on the part of the infant. This may be an underestimate, since the males may have perceived more infant distress than did the observer, who was primarily watching adults. Some interactions appeared to be possessive (1.6%) and others took place when a consort pair was nearby (9.1%). However, the most commonly recognised non-relaxed context (29.4%) was when the male was near to, or involved in, an agonistic exchange or obviously tense situation with other males. Although few of the interactions were overtly protective many could have served this function, as when a male gathered up the infant at the outbreak of nearby aggression. Infants rarely showed distress, struggled or vocalised when handled by males.

Differences between males

The 12 males were compared for five measures of their involvement with infants (Fig. 1). Males who interacted most (measure b) were resident males: A3, A4, A5 and A7. This was in marked contrast to the newcomers A1 and A2 who never interacted with infants. The four subadults had a lower median rate of interaction (1.9 per male per 100 h) than did the adults (8.4). The rate of interaction was not correlated with dominance rank across the 12 males ($r_s = -0.119$; corrected -0.053, NS), but those males who interacted more with infants were more likely to be near mothers of black infants (i.e. as nearest female neighbour within 15 m; measure a, Fig. 1) and therefore

Fig. 1. Twelve males compared for their involvement with infants. Males arranged in order of dominance rank, descending from left to right. Subadults cross-hatched, subjects identified beneath (A = adult, S = subadult), numbered by dominance rank within class. (a) Percentage of neighbour records with a mother as nearest female. (b) Males' rates of interaction with infants per 100 h. (c) Percentage of subject's interactions with other males in which he carried an infant. (d) Percentage of subject's interactions with other males in which opponent carried an infant. (e) Percentage of the 381 male–infant interactions (excluding his own) for which each male was within 25 m of the male–infant pair.

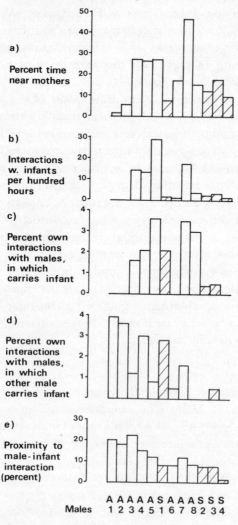

to have infants available to them ($r_s = 0.860$; corrected 0.895; $P < 0.01$). Again, newcomers were the least likely to be near mothers.

This contrast between newcomer and resident males follows the pattern of presumed paternity of the infants concerned. Twelve of the 14 infants were conceived before the newcomers joined the troop. (The two born later were involved in fewer interactions with males since they were observed for a shorter period.) Because interactions with infants took place only with resident males and subadults, the partners were restricted to possible fathers (present during infant's conception) and the natal males, who were potential kin or familiar to the mothers.

Studies elsewhere have found interactions between infants and males to be more frequent between possible father–offspring pairs (e.g. Altmann, 1980; Busse & Hamilton, 1981; Smuts, 1982) and here, evidence of preferences or selectivity between particular male–infant dyads was examined. Males consistently interacted with younger, black infants more than with the older, brown infants in comparisons based on relative availability (Wilcoxon $T = 6$, $N = 10$, $P < 0.05$). Table 1 therefore presents only the number of interactions between the eight black infants and the four resident males who interacted most frequently with infants (86% of all interactions).

Although the infants of F12 and F18 were present as newborns in only 24% and 35% of the study, so low numbers of interactions would be expected, males did not interact equally with all infants even correcting for time available. Each male appeared to focus his interactions on one or more particular infants. To examine these preferences, the eight mothers were ranked on two measures relating to each

Table 1. *The numbers of interactions seen between each of four adult males and eight black infants. For each male are compared two rankings of the eight mothers, in order of (1) infants' interaction rate, and (2) mothers' association, with the male. Probabilities:* * < 0.05, ** < 0.01

	Infants named by maternal rank								Correlation of rankings (Spearman r_s)
	1	2	3	8	12	14	16	18	
Males									
A3	7	8	5	2	2	2	3	10	0.139
A4	13	–	15	1	–	2	–	–	0.944**
A5	2	40	1	6	1	20	7	–	0.826*
A7	3	–	20	3	7	10	4	5	0.922**

male. These were the percentage of time spent within 15 m (from point-sample, nearest-neighbour data), and the rate of interactions between their infants and each male. For three of the four males, the two rankings were significantly correlated (Table 1: rates for infants of F12 and F18 were corrected for availability). Access to infants reflected associations with mothers for these males. These partnerships between males and mothers often included high rates of grooming. The exceptions to this mainly involved males A3 and A7 and the three youngest infants, over whom some competition was observed. The general pattern to male–infant interactions was that males had access to the infants of females with whom they associated and groomed, but they also tried to interact with other, very young, infants.

This may explain why the two newcomers did not interact with infants. They had no association or grooming relationships with mothers. Due to the limited time of this study, I was unable to determine whether the observed male–mother–infant triads reflected previous mating partnerships (and thus probable fathers). Such persistent partnerships have been observed elsewhere (Altmann, 1980; Smuts, 1983). However, Smuts' data led to the conclusion that the male's selectivity for infants was determined more by his relationships with mothers than by the probability of paternity. If this is the case, male–infant relationships may coincide with paternity only to the extent that long-term male–female partnerships coincide with matings. While many do, exceptions have been noted (e.g. Seyfarth, 1978).

Benefits of male–infant relationships

A number of studies have suggested that if a female has persistent affiliative relationships with an adult male, her infant may gain protection from others as a result of his proximity (e.g. Stein & Stacey, 1981), or as a result of his active defence (e.g. DeVore, 1963; Busse & Hamilton, 1981). Infants also may have an increased chance of survival if separated or orphaned (e.g. DeVore, 1963; Smuts, 1982), and feeding interference may be reduced when associating with adult males who are dominant to most of the infant's potential competitors (e.g. Altmann, 1980). Such potential enhancement of infant survival will also contribute to the inclusive fitness of the mother, and possibly to that of the male if he is the father. These benefits are discussed further by Smuts (1983).

Triadic male–infant interactions

Although most interactions between males and infants occur in the context of the male's relationships with particular mother–infant pairs, some interactions are related to the proximity or behaviour of other males and may even put the infant at risk. The following section examines this subset of interactions.

Contexts

In Table 2, the 381 male–infant interactions are assigned (*post hoc*) contexts according to those behaviours exchanged between the infant's male partner and other males in the preceding, same, or following minute as that of the male–infant interaction. In the majority of cases, there was no interaction between males. However, in 13.9%, the interactions with the infant occurred shortly after the male had interacted with another male. In most cases, the male who interacted with the infant had received a close pass, supplant or aggression from another male. In a further 10.8%, the male first contacted the infant then, while holding it, interacted with another male, again mainly receiving passes, supplants or aggression. These amounted to 41 male–infant interactions. In some of these, more than one male opponent was involved so there were 57 male–infant–male triadic interactions in total.

Table 2. *Classification of male–infant interactions according to male–male interaction context.* $N = 381$

Interactions between carrying male and other males, in the same, preceding, or following minute as the male–infant interaction	
None	75.3%
Male gives interaction to another (2.1% aggressive, 0.9% not), then interacts with infant in 1 min	3.0%
Male receives interaction from another (3.4% aggressive, 7.5% not), then interacts with infant in 1 min	10.9%
Male interacts with infant, then while holding or carrying it receives interaction from another male (4.2% aggressive, 3.2% not)	7.4%
Male interacts with infant, then while carrying it gives interaction to another male (1.4% aggressive, 2.0% not)	3.4%

The 57 cases where a male held or carried an infant during interactions with another male are classified in Table 3 into different types of male–male interactions. Infants were involved most frequently in interactions consisting of close passes between males, supplants and dyadic aggression. An important distinction is made between passive carrying, in which the infant-carrier received an interaction from another male, and active carrying, where the male held the infant and then moved towards or initiated aggression to another male. Such active carrying may increase the risks to an infant, and seldom appears to be protective.

Protection or exploitation: comparisons between males

The two previously mentioned hypotheses of protection or exploitation are compared to interpret these triadic interactions (those involving two males and an infant). Cases of infant-carrying involving a second male were analysed in terms of who carried infants against whom, separating males on the basis of class, possible paternity and dominance rank.

Comparisons of measures b and c in Fig. 1 show that the males who carried infants against other males in a high proportion of their male–male interactions tended to be those males who interacted more with infants at other times ($r_s = 0.897$, corrected 0.886, $P < 0.01$). This emphasises the observation that males could not be separated into protectors and exploitative carriers. The males A3, A4, A5 and A7 were prominent on both measures, while the high proportions for males S1 and A8 were based on low absolute numbers of interactions. Again, newcomers who never had access to infants did not carry them in this way.

Table 3. *Types of male–male interactions showing the total number of such interactions observed through the study, the percentage with infant carrying, and the number of active and passive carryings observed*

Interaction type	Overall total	Percentage in which infant was carried	Ratio of active:passive carrying
Close pass	107	10.3	6:5
Contact	232	0.4	0:1
Avoidance (>1 m)	146	0.7	1:0
Supplant (<1 m)	539	2.0	2:9
Aggression (solo)	458	7.0	11:21
Aggression (with allies)	245	0.4	1:0

Relationships: adult males and infant baboons

The class of the opponents during each male's infant carrying is compared with the relative availability of males in that class in Fig. 2. The contrast between resident males as possible fathers, and newcomers as non-fathers who might gain by infanticide, allows a test of the two most extreme hypotheses of protection or exploitation. If infants are carried to protect them from potentially infanticidal immigrants (Busse & Hamilton, 1981), then resident males should carry infants against newcomers. This was the case in 42% of the 57 interactions. If, however, non-fathers carry infants as hostages to inhibit the aggression from possible fathers, as has been suggested by Popp & DeVore (unpublished, cited with permission), then newcomers should carry infants during agonistic interactions with resident males. This was never seen.

A majority of the triadic carrying took place during interactions between residents and/or subadults and was not directed at potentially infanticidal immigrants. This cannot rule out infant carrying that may have been protective against lesser risks than infanticide. Nor does it eliminate the possibility of hostage carrying or otherwise exploitative carrying among residents and subadults (see also Packer & Pusey, 1985).

Neither of the two most extreme hypotheses explains more than a minor proportion of the interactions observed. An alternative is to examine the pattern in terms of dominance rank: a higher-ranking

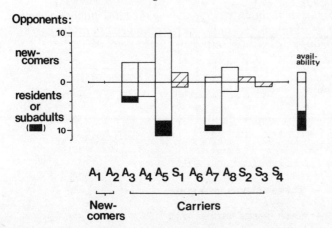

Fig. 2. The number of occasions on which each male, when carrying an infant, interacted with newcomer, resident, or subadult male opponents; on the right is shown the availability of each class. Interactions of subadult carriers cross-hatched. Males named beneath as in Fig. 1.

opponent may pose a greater risk to both the carrying male and to the infant, since high-ranking males are often young males with unworn canines (Packer, 1980). If newcomers were opponents more than expected from their high rank, there would be evidence that carrying is protective against immigrants. The distribution of infant carrying by males of each dominance rank is presented in Fig. 3. Interactions received by males holding infants (passive carrying) tended to be initiated by males of higher rank (83% downrank) and differed little from the majority of male–male dyadic interactions (89% downrank). However, when males holding infants approached or initiated aggression to others, they tended to do so against males of higher rank than themselves (62% of active carrying, Fig. 3). Pooling passive and active carrying, opponents tended to be of higher rank than the carriers in 79% of interactions. Males of higher rank thus tended to have infants carried against them in a higher proportion of their interactions than did males of lower rank (Fig. 1(d); $r_s = 0.836, N = 12, P < 0.01$).

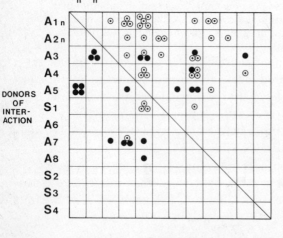

Fig. 3. The distribution of the 57 triadic interactions, according to identity of donor (on left) and recipient (above), with identity of infant carrier assigned as donor (closed circles) or recipient (open circles). Subjects arranged in order of dominance rank decreasing from top and from left, the newcomers being A1 and A2, indicated as 'n'.

In addition to being involved in more interactions where infants were carried against them, high-ranking males also tended to be found within 25 m of dyadic male–infant interactions more frequently than were lower rankers (Fig. 1(e); $r_s = 0.937$, $N = 12$, $P < 0.01$), despite the fact that the high-ranking males did not spend more time near other males in general (Collins, 1981). It occasionally appeared that a nearby male was, by his behaviour, responsible for another male interacting with an infant, for example by behaving aggressively. High-ranking males were responsible for a higher proportion of those male–infant interactions to which they were near than were low-ranking males ($r_s = 0.911$, $N = 12$, $P < 0.01$). The newcomers, in particular, were responsible for a far greater proportion of such cases (31%) than were any of the residents (highest 19%).

Opponents of high rank appeared, by these measures, to pose the greatest risk to the infants, their carriers or both. Because the risks to infants cannot be separated from those to carriers, the protective and exploitive aspects of these interactions cannot be determined. The above effects of rank were not due solely to the newcomers' actions. The correlations were all significant when newcomers were excluded. However, newcomers occupied the highest ranks, were 'responsible' for more male–infant interactions when nearby, and were more likely than were residents to have infants carried against them by subordinates (see Fig. 3). The newcomers, perhaps by virtue of both their high rank and their unfamiliarity, posed the greatest perceived risk to resident males.

Discussion

Male–infant interactions among these baboons were relatively common and most were neutral or benign in effect on the infant. The presumed benefits appeared to take the form of protection for infants and possible paternal investment for the male. However, 11% of the male–infant interactions involved another male. Infants may have been protected from the other male in some cases, but in others the infant appeared to be put at risk. Although Hrdy (1976) suggested that males may protect some infants but exploit others, the data presented here support previous observations that males may protect and exploit the same infant at different times (Ransom & Ransom, 1971; Altmann, 1980; Packer, 1980; Smuts, 1982; Strum, 1983). This may result in part because only familiar infants cling sufficiently to the male to allow their exploitation (Altmann, 1980; Nicolson, 1982; Strum, 1983). Access to infants appears to benefit the male in his relationship with other males (see also Collins, 1981). In the short term, the carrying and handling of infants enabled males to withdraw from conflicts without defeat, and it also appeared to allow males to direct agonistic interactions to males of higher rank, and thus may have enhanced their effectiveness in male–male agonistic interactions.

This leaves open the question of why infant carrying should in any way inhibit the carrier's opponent. That it does so seems widely established (see references in introduction; also Gilmore, unpublished; Shopland, 1982; Smuts, 1982). Strum (1983) suggests that the presence of the infant is instrumental in ending the conflict before the carrier concedes defeat, either because the infant alters the focus of attention between the males, or because contact with it alters the carrier's motivational state. Others have suggested that the conspicuous natal coat of the black infant with its pink skin acts as a signal which inhibits other baboons from attack, although there are cases of wounding of such infants which suggest that the inhibition is not absolute (Collins, Busse & Goodall, 1984). Packer & Pusey (1985) view the inhibition as a conflict asymmetry between males, the asymmetry being the greater likelihood that the carrier will escalate the contest in defence of the infant. This is plausible when the carrier may be the father, but there is also evidence that signs of distress by the infant are likely to bring concerted defence by other troop-members, not only adult males but also females and immatures (Gilmore, unpublished; Rhine et al., 1980; Smuts, 1982; Strum, 1983), which might constitute greater asymmetry than the carrier's defence of the infant. This is especially likely to be effective against newcomers because troop residents, notably lactating females and immatures, discriminate against them in particular (Packer, 1979; Busse & Hamilton, 1981; Smuts, 1982; Strum, 1983). The two newcomers were the most frequent targets of mobbings in this study (Collins, 1981).

Although it may be possible to explain particular behaviours, such as protectiveness to particular infants, as innate behavioural tendencies favoured through natural selection, yet males' interactions with infants are so varied that learning must also be involved. For example, the males here differed greatly in their frequency of carrying infants, and in their proportions of active and passive carrying. Learning may explain the more complex interactions, and the effectiveness of agonistic buffering, in the following way. Although protectiveness may not be general to all male–infant relationships, yet it is sufficiently intense that others would inevitably learn to be cautious of any male protecting an infant, or merely even close to one. From this, the protector would learn the increased effectiveness of his agonistic actions when paired with a black infant, and could use it *passively*, to avoid or withdraw from conflicts, or *actively*, by carrying an infant when initiating interaction with another male. This would explain why protectiveness and exploitation appear so closely linked, and suggests that studies of the ontogeny of infant carrying, from male juvenile life through to adulthood, would be of great value. For the infant, such experiences might also have longer-term consequences; if, for example, it influences its tendency to form alliances (Ransom & Ransom, 1971), it could provide advantage in mating competition, and might influence reproductive success in adulthood.

References

Altmann, J. (1980) *Baboon Mothers and Infants*. Cambridge, Mass.: Harvard University Press

Busse, C. & Hamilton, W. J. III (1981) Infant carrying by male chacma baboons. *Science*, **212**, 1281–3

Collins, D. A. (1981) Social behaviour and patterns of mating among adult yellow baboons (*Papio c. cynocephalus* L 1766). Ph.D. Thesis, University of Edinburgh. Ann Arbor: University Microfilms International, 8270026

Collins, D. A., Busse, C. D. & Goodall, J. (1984) Infanticide in two populations of savanna baboons. In *Infanticide: Comparative and Evolutionary Perspectives*, ed. G. Hausfater & S. B. Hrdy, pp. 193–215. New York: Aldine Publishing Co.

Deag, J. M. & Crook, J. H. (1971) Social behaviour and 'agonistic buffering' in the wild Barbary macaque *Macaca sylvana* L. *Folia primatol.*, **15**, 183–200

DeVore, I. (1963) Mother–infant relations in free-ranging baboons. In *Maternal Behavior in Mammals*, ed. H. L. Reingold, pp. 305–35. New York: Wiley

Hrdy, S. B. (1976) Care and exploitation of nonhuman primate infants by conspecifics other than the mother. *Adv. Stud. Behav.*, **6**, 101–58

Nicolson, N. (1982) Weaning and the development of independence in olive baboons. Ph.D. thesis, Harvard University

Packer, C. (1979) Inter-troop transfer and inbreeding avoidance in *Papio anubis*. *Anim. Behav.*, **27**, 1–36

Packer, C. (1980) Male care and exploitation of infants in *Papio anubis*. *Anim. Behav.*, **28**, 512–20

Packer, C. & Pusey, A. (1985) Asymmetric contests in social mammals: respect, manipulation and age-specific aspects. In *Evolution: Essays in Honour of John Maynard Smith*, ed. P. J. Greenwood, P. H. Harvey & M. Slatkin, pp. 173–86. Cambridge: Cambridge University Press

Ransom, T. W. & Ransom, B. S. (1971) Adult male–infant relations among baboons (*Papio anubis*). *Folia Primatol.*, **16**, 179–95

Rhine, R. J., Norton, G. W., Roertgen, W. J. & Klein, H. D. (1980) The brief survival of free-ranging baboon infants (*Papio cynocephalus*) after separation from their mothers. *Int. J. Primatol.*, **1**, 401–9

Rowell, T. E. (1967) A quantitative comparison of the behaviour of a wild and a caged baboon group. *Anim. Behav.*, **15**, 499–509

Seyfarth, R. M. (1978) Social relationships among adult male and female baboons. II. Behaviour throughout the female reproductive cycle. *Behaviour*, **64**, 227–47

Shopland, J. M. (1982) An intergroup encounter with fatal consequences in yellow baboons (*Papio cynocephalus*). *Am. J. Primatol.*, **3**, 263–6

Smuts, B. B. (1982) Special relationships between adult male and female olive baboons (*Papio anubis*). Ph.D. Thesis, Stanford University

Smuts, B. B. (1983) Special relationships between adult male and female olive baboons: selective advantages. In *Primate Social Relationships: an Integrated Approach*, ed. R. A. Hinde, pp. 262–71. Oxford: Blackwell Scientific

Stein, D. M. & Stacey, P. B. (1981) A comparison of infant–adult male relations in a one-male group with those in a multi-male group for yellow baboons (*Papio cynocephalus*). *Folia Primatol.*, **36**, 264–76

Stoltz, L. P. & Saayman, G. S. (1970) Ecology and behaviour of baboons in the Northern Transvaal. *Ann. Transvaal Mus.*, **26**, 99–143

Struhsaker, T. (1971) Social behaviour of mother and infant vervet monkeys (*Cercopithecus aethiops*). *Anim. Behav.*, **19**, 233–50

Strum, S. C. (1983) Use of females by male olive baboons (*Papio anubis*). *Am. J. Primatol.*, **5**, 93–110

V.4

The role of alliances in the acquisition of rank

S. B. DATTA

Introduction

It is well established that in several species of macaques and baboons (Sade, 1967; Bernstein, 1969; Gouzoules, 1975; Hausfater, 1975; de Waal, 1977; Lee & Oliver, 1979; Dunbar, 1980; Walters, 1980), offspring inherit maternal rank relative to others in the group hierarchy. In such a hierarchy one might expect to find factors, associated with maternal rank, which set limits on how high an immature can rise.

A potentially crucial factor (though considered secondary by some researchers (e.g. Walters, 1980) and relatively infrequent and therefore less influential by others (e.g. Lee & Oliver, 1979)) is the interventions by third parties in agonistic disputes between immatures and other group members.

In this paper, I shall show that the pattern and nature of interventions in disputes involving male and female immature rhesus monkeys conforms to what would be expected if alliances were important in the acquisition of rank.

Interventions expected to change the direction of dominance–subordinance

What kind of interventions might change the direction of dominance–subordinance? Consider two individuals, A and B, with A clearly dominant to B. In disputes between them A is the aggressor while B submits. A third individual, C, intervenes in their disputes to threaten A, thus supporting B. (C may be referred to as the interferer, with A the 'target' of the interference and B the 'beneficiary' of the interference.) In one case (top line, Fig. 1) C is dominant to A, so A submits to C; in the other (bottom line, Fig. 1) C is subordinate to A and intervention results in A attacking C and C submitting. Both these kinds of intervention have been observed among primates (e.g. Kaplan, 1977). In the first case B receives what will be referred to as 'effective' support. Intuitively, we might expect that B, given such support, would in time challenge A's dominance, and with continued effective support, eventually reverse dominance with A. No such progression would be expected in the second case.

Study site and animals

Data come from a 16-month study of rhesus macaques in the dominant matriline of Group J on Cayo Santiago. Subjects were 11 related adult females and their pre-adult offspring. Dyadic agonistic interactions were used to determine the direction of dominance, using Sade's (1967) criteria. Focal and *ad lib.* data on these and on interventions were combined for this analysis.

Dominance–subordinance relations during the study

Rhesus females on Cayo Santiago maintain highly stable linear dominance hierarchies and offspring inherit maternal rank relative to other females and their offspring (Missakian, 1972; Sade, 1972). Throughout this paper the offspring of a female will be referred to as high-born (HB) relative to individuals their mother outranks. (The latter are of course low-born – LB – relative to the former.)

The direction of dominance was known in 532 inter-family dyads at the end of the study. In 170 (32%) of these the offspring of higher-ranked females did *not* dominate individuals low born to themselves.

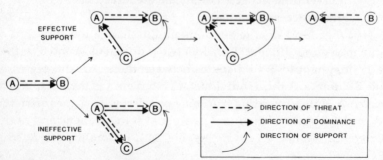

Fig. 1. The effect of support on dyadic dominance–subordinance. A and B are the original contestants, C the interferer in their dispute. B is the beneficiary and A the target of the interference.

Fig. 2 suggests why: relative age/size of opponents is a determinant of dyadic dominance relations. Thus, while HBs typically dominate all LB opponents their own age or younger, the proportion of older LBs dominated declines as the discrepancy in age/size increases.

From Fig. 2 the importance of the other determinant of dyadic dominance relations – maternal dominance status – is also apparent: LBs often do not dominate even those HB opponents the same age as or younger than themselves. The pattern of dominance–subordinance is strongly skewed in favour of HBs.

This is further illustrated by the pattern of challenges and dominance reversals seen during the study. In all 76 dyads where the subordinate began to challenge the dominant during the study the former was younger and higher born than the latter. This was also the case in 30 out of 32 reversals seen (30 reversals were preceded by a period of challenge).

Interventions in disputes involving immatures

As described above, HB subordinates were more likely than LB subordinates to challenge dominants and to reverse dominance

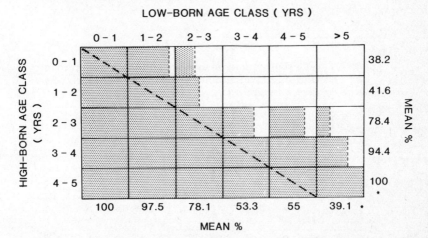

Fig. 2. Median percentage of low-born opponents dominated by high-born individuals. Each cell represents 100%. The proportion of the cell that is shaded gives the median percentage of their low-born opponents dominated by individuals in a given age-class of high-borns. The median number of dyads per cell was 14, with a range of 2–47. The median number of high-borns per age-class was 7 (range 2–9) and of low-borns was 11 (range 6–14).

* $p < 0.05$, 2 tailed Spearman

with them. To discover whether HB subordinates received a different kind of support from LB subordinates, those interventions in disputes during the period *before* challenge (if challenge occurred at all) were examined.

Both HB and LB subordinates received support from third parties. Fig. 3 shows the frequency with which support to HB and LB subordinates (a) caused submission in the target of the interference (and hence was 'effective'), and (b) halted the original dispute. These were the two most common consequences of interference. Consequences were determined by asking questions (e.g. was the dispute halted or not?) of each triad (a unique combination of interferer, target, and beneficiary) and taking the most frequent sequelae as the consequence of interference in that triad. Thus if a dispute was halted two out of three times in a given triad the consequence was 'dispute halted'. Of course an interference could have more than one consequence – it could halt the dispute *and* cause submission in the target.

It is clear that support to both HB and LB subordinates usually terminated the original dispute (Fig. 3a) and may thus have prevented further or serious injury. However, while the supporters of HB sub-

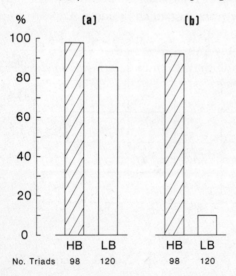

Fig. 3. Consequences of support to high-born (HB) and low-born (LB) subordinates. (a) Percentage of triads with interference in which dispute is halted. (b) Percentage of triads with interference in which support causes submission in the target of the interference. The difference between HB and LB subordinates is significant for both (a) and (b): $p \leq 0.05$, 2-tailed, comparing B-distributions (Phillips, 1973).

ordinates usually elicited submission from the (LB) dominants who had threatened the HB subordinates, the supporters of LB subordinates were rarely able to elicit submission from the (HB) dominants who had threatened LB subordinates (Fig. 3b). This was largely because the supporters of HB subordinates were dominant to their targets while those of LB subordinates were subordinate to their targets (Datta, 1983).

Discussion and conclusions

Those subordinates (HB) who do typically go on to challenge dominants and to reverse dominance with them receive a different kind of support – support that is effective in the sense of causing submission in the target of the interference – than those subordinates (LB) who do not. Among rhesus monkeys on Cayo Santiago, 'effective' support is strongly associated with relative maternal status, hence maternal dominance is a reliable predictor of offspring rank. In other groups of these or other primates, effective support may not be so associated. One would expect nevertheless that the availability of effective support – no matter who it is from, as long as it is persistent and consistent – would encourage subordinates to rebel. In other words, the 'knowledge' that effective support *is* available against a particular opponent may be crucial in the development of challenging behaviour against that opponent.

Walters (1980), however, suggests that whether or not an immature 'decides' to 'target' (that is, challenge) a dominant or not depends on the 'birth-rank' of the immature relative to the dominant. Thus, if (at the immature's birth) its mother outranked the dominant the immature would, at adolescence, be likely to challenge the dominant, even if the latter now outranked the immature's mother.

There are two difficulties with this rather deterministic view. First, since Walters does not present data on alliances *before* challenge it is difficult to know how important effective support against particular opponents was in encouraging immatures to rebel: immatures might have chosen targets according to who they could count on getting powerful support against, rather than according to 'birth-rank' as such. Secondly, if the concept of birth-rank is to be consistent, we must expect immatures whose mothers have risen in rank since the immatures' birth *not* to challenge those individuals previously dominant to their mother that she now outranks. This seems unlikely and is indeed not supported by the available evidence (e.g. Marsden, 1968; Koyama, 1970; Chance, Emory & Payne, 1977).

The alternative view (assuming that individuals would prefer to have high rank) is that immatures will challenge whoever, in their assessment, they are capable of defeating without undue cost to themselves. In other words they behave opportunistically. As discussed elsewhere (Datta, 1983) and outlined in Fig. 2, assessment may be based not only on the nature of the alliances of self and opponent, but on factors related to the relative age/size of opponents. On this view it would be predicted not only that immatures would switch rank when they could (not just at some absolute age or point in the life-cycle such as adolescence), but that, depending on circumstances, among them the acquisition or loss of powerful allies, individuals may throughout their lives be capable of, or susceptible to, changes in dominance status.

References

Bernstein, I. S. (1969) Stability of the status hierarchy in a pigtail monkey group (*Macaca nemestrina*). *Anim. Behav.*, **17**, 452–8

Chance, M. R. A., Emory, G. R. & Payne, R. G. (1977) Status referents in long-tailed macaques (*Macaca fascicularis*): precursors and effects of a female rebellion. *Primates*, **18**, 611–32

Datta, S. B. (1983) Relative power and the acquisition of rank. In *Primate Social Relationships – An Integrated Approach*, ed. R. A. Hinde, pp. 93–102. Oxford: Blackwell

Dunbar, R. I. M. (1980) Determinants and evolutionary consequences of dominance among female gelada baboons. *Behav. Ecol. Sociobiol.*, **7**, 253–65

Gouzoules, H. (1975) Maternal rank and early interactions of infant stumptail macaques (*Macaca arctoides*). *Primates*, **16**, 405–18

Hausfater, G. (1975) *Dominance and Reproduction in Baboons* (Papio cynocephalus). Contrib. Primatol., 7. Basel: Karger

Kaplan, J. R. (1977) Patterns of fight interference in free-ranging rhesus monkeys. *Am. J. Phys. Anthropol.*, **47**, 279–88

Koyama, N. (1970) Changes in dominance rank and division of a wild Japanese monkey troop in Arashiyama. *Primates*, **11**, 335–90

Lee, P. C. & Oliver, J. I. (1979) Competition, dominance and the acquisition of rank in juvenile yellow baboons (*Papio cynocephalus*). *Anim. Behav.*, **27**(2), 576–86

Marsden, H. M. (1968) Agonistic behaviour of young rhesus monkeys after changes induced in the social rank of their mothers. *Anim. Behav.*, **16**, 138–44

Missakian, E. A. (1972) Genealogical and cross-genealogical dominance relations in a group of free-ranging rhesus monkeys on Cayo Santiago. *Primates*, **13**(2), 169–80

Phillips, L. D. (1973) *Bayesian Statistics for Social Scientists*. London: Nelson

Sade, D. S. (1967) Determinants of dominance in a group of free-ranging monkeys. In *Social Communication among Primates*, ed. S. Altmann, pp. 99–115. Chicago: Aldine

Sade, D. S. (1972) A longitudinal study of the social behaviour of rhesus monkeys. In *Science and Psychoanalysis*, vol. 12, ed. J. Masserman, pp. 18–38. New York: Grune & Stratton

Waal, F. B. M. de (1977) Organisation of agonistic relations within two captive groups of Java monkeys (*Macaca fascicularis*). *Z. Tierpsychol.*, **44**, 225–82

Walters, J. (1980) Interventions and the development of dominance relationships in female baboons. *Folia Primatol.*, **34**, 61–89

V.5

Environmental influences on development: play, weaning and social structure

P. C. LEE

Introduction

Development in the social milieu has been one main focus of studies on immature primates. As a result, research has tended to concentrate on individual differences (or personality, e.g. Stevenson-Hinde, 1983), on isolates or deviant behaviour (e.g. Harlow, 1969; Anderson & Mason, 1974; Suomi, 1974), and on details of interactions and relationships between immatures and other group members (see Hinde, 1983). The physical environment of immatures has only infrequently been considered as a variable affecting immature social development. Studies concentrating on the effect of the habitat on development have tended to examine specific juvenile specialisations such as weaning foods (e.g. Rhine & Westlund, 1978). Only recently has the habitat been examined with respect to its effects on inter-individual relationships and development (see Altmann, 1980).

Furthermore, much previous research on development has tended to emphasise those aspects of development where elements of adult behaviour are present, but the repertoire is incomplete. In this paper, I will attempt to demonstrate that the behavioural flexibility exhibited by immatures can be a direct response to ecological pressures and in some cases may have important consequences for the life history of individuals. I will focus on three aspects. First, I will examine how some interactions of immatures, such as play, are directly affected by ecological conditions. Secondly, I will look at how the energetics of the mother are affected by ecological variables and how this in turn affects the mother–infant relationship. Thirdly, I will briefly discuss how population parameters, in themselves a function of ecological conditions, affect group composition and thus alter the availability of partners for interactions. The physical environment can thus be a major influence on the types, frequencies and qualities of relationships formed during development.

Methods

The data presented here are based on a 22-month study of vervet monkeys (*Cercopithecus aethiops*) in Amboseli National Park, Kenya. A total of 53 immature animals, ranging in age from birth to 5 years old, were observed on a focal animal basis (Altmann, 1974) from three adjacent groups. Details of observational techniques and statistical analyses are presented elsewhere (Lee, 1983*a*,*b*; 1984*a*,*b*).

Habitats and diet

The general habitats of vervet monkeys in Amboseli consist of: (1) *Acacia xanthophloea* woodlands located near permanent swamps; (2) *A. tortilis* woodlands on well-drained soils away from swamps; (3) open alkaline grasslands and bushlands; and (4) the swamps and swamp-edge herb and *Cynodon dactylon* grass understory associated with dense immature *A. xanthophloea*. Rainfall is highly seasonal and the majority falls between November and May. The seasonal distribution of the rain greatly affects the productivity of shrubs, herbs and grasses, while the tree species also have a yearly cycle of production. As a result, different food species and food products are available in different seasonal periods. During the years of this study, rainfall was above average (>400 mm).

Measures of diet were based on point samples at 60-s intervals during 1200 h of focal animal samples. Food species and food type or plant part were recorded when an animal placed food in its mouth, or was in the process of acquiring food from the ground or trees. Seasons were defined from rainfall records and were considered as dry (May–October) or wet (November–April). These seasons were further divided into an early and a late half of 3 months each.

Results

The yearly diet of the immature vervets in all three groups showed a marked reliance on the products of the two tree species (*Acacia xanthophloea* and *A. tortilis*). These two species alone contributed over 50% of the diet in each of the groups. However, each group lived in adjacent but different micro-habitats, and there were differences between the groups in the food species and food types eaten throughout the year (Fig. 1). Two of the groups had access to swamps both for water and for food, while the third group (A) had no standing water available during the dry season. This group lived primarily in the *A. tortilis* woodlands and bushlands and had little access to the dense stands of *A. xanthophloea* near the swamps.

Inter-group differences in the food species eaten were largely a function of the overall availability of these species within each of the territories of the three groups. Availability was determined from total counts of food species (see Lee, 1981). There was a strong relation between the canopy availability ($N \times$ mean canopy volume per species per hectare) of the different species and their proportional representation in the diet of each group (Fig. 2).

The immatures in each group were thus exposed to a different set of ecological pressures and opportunities and responded in a different way from group to group. These inter-group differences were further affected by seasonal variation in the availability of food and water. The activity budgets of the immatures were thus seasonally variable and differed between the groups (Lee, 1983b; 1984b). During the dry season, when foods were clumped and of relatively low quality, the time spent feeding was high and the time spent resting and interacting

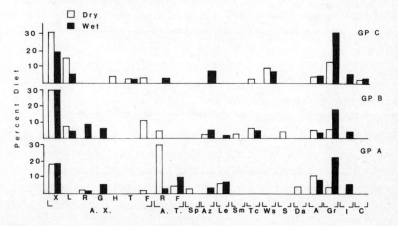

Fig. 1. The percentage of each food species and food type (plant part) for the 10 most frequently eaten foods, which accounted for over 80% of the yearly diet. The dry and wet season diets of each of the three groups are presented. A.X. = *Acacia xanthoploea**, A.T. = *Acacia tortilis** (X = exudate, L = leaf, R = ripe fruit, G = green fruit, H = thorn, T = twig leaf tip, F = flower); Sp = *Salvadora persica**; Az = *Azima tetracantha**; Le = *Lycium europaeum**; Sm = *Suaeda monoica**; Tc = *Trianthema ceratocepala**; Ws = *Withania sominifera**; S = *Solanum* sp.; Da = *Dicliptera albicaulis*; A = *Abutilon* sp.; Gr = grass species; I = invertebrates; C = creepers (2 species). Asterisks are species censused in Fig. 2.

was low. During the wet season, foods were more dispersed and evenly distributed, and the time spent resting and interacting was high while the time spent feeding dropped. The seasonal differences were reflected in inter-group differences such that the group with the poorest-quality habitat spent the most time feeding and the least time interacting of the three groups in both the dry and wet seasons (Lee, 1984*b*).

Social interactions among the immatures

Play has been considered to be one of the most ubiquitous behaviours of immature animals, and its form and function have been continually debated (see Fagen, 1981). The suggestion that play frequencies vary as a result of energetic constraints or environmental harshness has frequently been put forward (e.g. Baldwin & Baldwin, 1976). Among the immature vervets in these groups, play was a relatively rare event, especially during the dry season (Lee, 1983*a*). Play rates changed from a mean of less than one bout per hour during the dry season to over 20 bouts per hour during the wet season. Although there was a tendency for rates of play to decline with increasing age, the seasonal changes in play rates were found for animals of all ages and sexes, including adults (Lee, 1983*a*). The wet/dry season cycles in play have been observed over several years

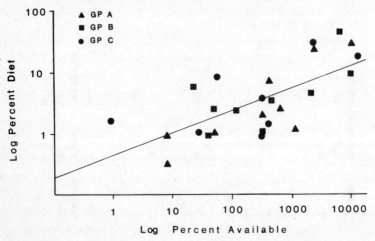

Fig. 2. The double log plot of canopy availability (m^3/ha) of each censused species (starred in key to Fig. 1) against the proportion of that food in the yearly diet of each of the different groups. Line drawn from least squares regression.

and were strongly related to changes in the availability and quality of foods rather than to developmental trends or group composition (Lee, 1984b).

Play thus appears to be a sensitive indicator of dietary quality and of the energetics of immatures. The differences in caloric intake required for an immature who plays and one who does not are estimated to be 5–10% over resting metabolic requirements (Lee, 1981; Martin, 1982). In most experiments or observations where play has been eliminated from the immature repertoire, food has been reduced by as much as 50% (Loy, 1970; Baldwin & Baldwin, 1976). Animals and humans have been reported to be listless and apathetic under these conditions (Hall, 1963; Loy, 1970; Rutishauer & Whitehead, 1972). Zimmerman et al. (1975) specifically reduced the protein in the diet of rhesus monkeys and found low play rates, low rates of investigation, reduced attention to tasks, and fearful responses to novel stimuli. The protein-deprived monkeys showed persisting effects in that they were generally less sociable as adults and puberty was delayed.

The quality and quantity of the diet can thus have long-lasting effects on social and physical development. Lack of play or at least seasonal constraints on play rates have not yet been shown to have a major effect on development, in isolation from that of the diet. The immature vervets appeared to be responding to ecological pressures directly in an 'adaptive' way, by reducing rates of play during times of stress. They may have compensated to some degree by playing at high rates when food availability and quality were higher, but during the dry season they were, of energetic necessity, forced to find other ways to gain whatever social knowledge or skills can be derived from play. However, they were also energetically constrained from engaging in some other types of interactions.

Rates of competitive approach–retreat and aggressive interactions showed the same seasonal changes as those of play (see Lee, 1983b, 1984b). Although it could be expected that the frequency of active competition for food would increase during times of shortage (e.g. Dittus, 1977), this was not the case among these vervets. Rates of competition, measured by supplants over food, rose significantly during the wet season over those during the dry season, as did rates of aggression (Lee, 1983b). During the dry season, immatures dispersed and foraged in isolated clumps, and thus their opportunities for engaging in competition or aggression were limited. As for grooming, while the percentage of time spent grooming did not change significantly between the seasons, and the majority of bouts took place between mothers and offspring, the bouts between immature peers did appear to be somewhat more frequent and longer in the dry season than in the wet. During times of little peer–peer interaction in the form of play, some contact may have been replaced through grooming.

Weaning and mother–infant interactions

Seasonal and yearly variation in food availability and distribution has been shown to affect infant survival and the probability of conception among these vervets (Cheney et al., 1981; Cheney, Seyfarth, Andelman & Lee, in press). During harsh or extended dry seasons, or when major staple foods fail or are reduced in availability, mothers are less likely to conceive or if they do conceive, to carry the foetus to term. This was seen in the overall pattern of fecundity among the vervets. In Cheney et al.'s (in press) analysis, the interbirth interval of these vervets was longer than one year on average, and Group A with the poorest-quality habitat had a mean interbirth interval of close to 2 years. While the other two groups did not differ significantly from each other in interbirth intervals, less than half of the females in either group were able to conceive in subsequent years. The typical pattern of reproduction among these vervets appears not to be that of subsequent years, but rather a variable pattern of giving birth every 2 years under average (poor) conditions, and giving birth in subsequent years only during exceptionally good conditions.

This tendency to produce infants every 2 years had important implications for mother–infant relationships. Suckling success was lower earlier for those infants whose mothers gave birth the following year, declining to zero by the time of the sibling's birth (Fig. 3A). On the other hand, infants whose mother did not give birth were able to suckle successfully into their second year (Fig. 3B). The earlier termination of suckling for infants whose mothers gave birth appeared to affect infant survivorship. Infants in Group A with the poorest-quality habitat had lower suckling success and higher mortality at younger ages than did infants from the other groups (see Lee, 1984b). Since infant survival during the first year was one of the major factors affecting the interbirth interval (see Cheney et al., in press), mothers appeared to be using different tactics in their relationship with an infant, which depended at least in part on their levels of nutrition and activity budgets.

Thus, habitat variables appeared to play an important role in the timing and structure of mother–infant conflict over weaning. Mothers may have been able to assess their physiological condition around the time of conception, and if in poor condition, they diverted additional resources to their still dependent infant, prolonging their investment in that infant into a second year and reducing infant mortality. In addition, mothers who prolong suckling investment may also perpetuate a grooming and supportive relationship with their offspring that could ultimately affect dominance relations and alliance structures in the group (e.g. Seyfarth, 1980).

Demography, partner availability and development

Since ecological factors influence the probability of mothers bearing infants at yearly intervals, there may be several consequences for social development. At unpredictable intervals, depending on rainfall, food availability and tree production cycles, there may be

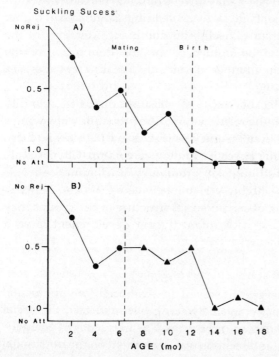

Fig. 3. Suckling success measured as the mean ratio of the number of rejections from the nipple to the number of attempts per hour for infants between 0 and 18 months old. (A) Infants whose mothers gave birth in month 12 ($N = 5$). All infants have been combined up to the point of mating (6 months; $N = 16$). (B) Infants whose mothers did not give birth until at least month 24 ($N = 10$). ● = All with conception possible (<6 months) and actual (>6 months). ▲ = Those whose mothers did *not* conceive (>6 months only).

either a very small or a large number of infants born in a seasonal cohort. Thus the number of peers available for interactions such as play and grooming or with whom dominance relations can be established can vary enormously from year to year and from group to group. For example, in 1977 there were seven infants born in Group B, in 1978 there was one infant, and in 1979 there were six infants. The lone 1978 infant had no other age-mates and was forced to interact exclusively with older, larger immatures. Large or small peer cohorts may affect the frequency with which some types of interactions occur, such as those particular types of play where partners need to be relatively well matched in size and strength. At the same time, a single playful and attractive individual may dominate the partner preferences within a group, minimising the appearance of age- or sex-mate preferences during interactions such as play.

Other interactions, such as the allomothering of newborns by juvenile females, may not be seen as frequently in years when fewer infants are born, or, alternatively, some juvenile females may be excluded from access to the infants (see also Lee, 1983c). It was interesting that the main allomothering partner of the lone infant born in Group B was a nulliparous female of highest dominance rank. This infant, who was the son of the lowest-ranking female, was subsequently able to supplant and to defeat yearlings larger and of higher maternal dominance than himself. The dominance relations of the immatures appeared to be influenced by the dominance of the allomother as well as the number of immatures of different ages, sex and maternal dominance.

Another potential effect of ecologically based variation in fecundity was on the nature of sibling relationships. The probability of having a sibling of the same sex in subsequent years was low (<0.40) and these probabilities were further affected by low survivorship (Cheney et al., 1981). Thus patterns of sibling play, grooming, and dominance acquisition were likely to be highly variable between years and between groups. As a result, the observed social structure of the groups may have been different, and each infant during development faced a unique social environment.

Discussion

Among the vervets observed in Amboseli, environmental variation, seasonal, yearly and between groups, affected the social development of immatures in three ways. The frequency of energetically demanding social interactions was dependent on the nutritional content of the diet, and during seasonal periods of low food availability and quality, some interactions were eliminated from the immatures' repertoire. A relation between play and food intake or general environmental characteristics has been observed in many species of primates and other mammals (see Fagen, 1981). However, studies of development have only rarely considered the question of how such environmentally induced variation can affect behaviour over the long term. If we are to approach the study of development through examining those behavioural responses to local and immediate conditions (e.g. juvenile specialisations to food shortage; see Bateson; Chapter V.1) and those responses that are part of a consistent developmental sequence leading to adult behaviour (e.g. scaffolding; Bateson; Chapter V.1), we must begin to incorporate aspects of the physical environment.

The time courses of weaning and mother–infant relationships were also affected by the seasonal and inter-group variations in food and water. Altmann (1980) and Berger (1979) have both demonstrated the importance of ecological constraints on levels of maternal investment. Here again, we can emphasise different questions. First, what are the implications for mothers of constraints on their ability to invest in either the present infant or in subsequent infants. Secondly, how is the mother–infant relationship affected as a result (see also Hauser; Chapter IV.2). As Hauser notes, we can find mothers who are unwilling or unable to respond to their infants' needs and infants who are poor at assessing their mothers' abilities to provide under different ecological conditions. Additionally, we can ask how the infant's own development is affected by its relationship with a mother under different ecological constraints. What are the implications for an infant in terms of growth, survivorship, and dominance, or in its ability to form persisting beneficial relationships (e.g. Seyfarth, 1980)?

Finally, as a result of environmental influences on mortality and fecundity, the demographic structure of each group was unique at any point in time. We cannot as yet assess how aspects of a group such as the availability of siblings, peers and other social partners of different ages and sex affect the social development of immatures, but as Collins (Chapter V.3) and Datta (Chapter V.4) point out, relationships between immatures and other group members have important consequences for both the others' and the immatures' abilities to manipulate their social environment.

We are thus left with the suggestion that our understanding and interpretation of both the specifics and generalities of social development need to be placed within a framework that incorporates the physical environment of the immature. In speculating about the adaptive function and the current utility of immature behaviour, the environmental variation influencing the expression of behaviour must also be observed.

Acknowledgements

I would like to thank the Office of the President, Republic of Kenya, and the Ministry of Tourism and Wildlife for permission to carry out research. I would like to thank the Wardens of Amboseli for great assistance and the Institute of Primate Research, Nairobi, for sponsorship. This research was funded by the L. S. B. Leakey Trust Studentship at New Hall, Cambridge. I am grateful for their financial support. Finally, I would like to thank D. L. Cheney and R. M. Seyfarth for sharing the vervets and their data with me, and Profs R. A. Hinde and P. Bateson for guidance and encouragement during writing.

References

Altmann, J. (1974) Observational study of behaviour: sampling methods. *Behaviour*, **49**, 227–65

Altmann, J. (1980) *Baboon Mothers and Infants*. Cambridge, Mass.: Harvard University Press

Anderson, C. O. & Mason, W. A. (1974) Early experience and complexity of social organisation in groups of young rhesus monkeys. *J. Comp. Physiol. Psychol.*, **87**, 681–90

Baldwin, J. D. & Baldwin, J. I. (1976) Effects of ecology on social play: a laboratory simulation. *Z. Tierpsychol.*, **40**, 1–14

Berger, J. (1979) Weaning conflict in desert and mountain bighorn sheep (*Ovis canadensis*): an ecological interpretation. *Z. Tierpsychol.*, **50**, 188–200

Cheney, D. L., Lee, P. C. & Seyfarth, R. M. (1981) Behavioural correlates of non-random mortality among free-ranging female vervet monkeys. *Behav. Ecol. Sociobiol.*, **9**, 153–61

Cheney, D. L., Seyfarth, R. M., Andelman, S. J. & Lee, P. C. (in press) Reproductive success in vervet monkeys. In *Reproductive Success*, ed. T. H. Clutton-Brock. Chicago: University of Chicago Press

Dittus, W. P. J. (1977) The social regulation of population density and age-sex distribution in the toque macaque. *Behaviour*, **53**, 281–322

Fagen, R. M. (1981) *Animal Play Behavior*. Oxford: Oxford University Press

Hall, K. R. L. (1963) Variations in the ecology of the chacma baboon *Papio ursinus*. *Symp. Zool. Soc. London*, **10**, 1–28

Harlow, H. F. (1969) Age-mate or peer affectional systems. *Adv. Study Behav.*, **2**, 333–83

Hinde, R. A. (1983) *Primate Social Relationships: An Integrated Approach*. Oxford: Blackwell

Lee, P. C. (1981) Ecological and social influences on development of vervet monkeys. Ph.D. thesis, University of Cambridge

Lee, P. C. (1983a) Play as a means for developing relationships. In *Primate Social Relationships*, ed. R. A. Hinde, pp. 81–9. Oxford: Blackwell

Lee, P. C. (1983b) Ecological influences on relationships and social structure. In *Primate Social Relationships*, ed. R. A. Hinde, pp. 225–30. Oxford: Blackwell

Lee, P. C. (1983c) Caretaking of infants and mother–infant relationships. In *Primate Social Relationships*, ed. R. A. Hinde, pp. 146–51. Oxford: Blackwell

Lee, P. C. (1984a) Early infant development and maternal care in free-ranging vervet monkeys. *Primates*, **25**, 36–47

Lee, P. C. (1984b) Ecological constraints on the social development of vervet monkeys. *Behaviour*, **91**, 245–62

Loy, J. (1970) Behavioral responses of free-ranging rhesus monkeys to food shortages. *Am. J. Phys. Anthropol.*, **33**, 263–72

Martin, P. (1982) The energy costs of play: definition and estimation. *Anim. Behav.*, **30**, 292

Rhine, R. J. & Westlund, B. J. (1978) The nature of a primary feeding habit in different age-sex classes of yellow baboon (*Papio cynocephalus*). *Folia Primatol.*, **30**, 64–79

Rutishauer, I. H. E. & Whitehead, R. G. (1972) Energy intake and expenditure in 1–3 year old Ugandan children in a rural environment. *Br. J. Nutr.*, **28**, 145–52

Seyfarth, R. M. (1980) The distribution of grooming and related behaviours among female vervet monkeys. *Anim. Behav.*, **28**, 1070–94

Stevenson-Hinde, J. (1983) Individual characteristics: a statement of the problem. In *Primate Social Relationships*, ed. R. A. Hinde, pp. 28–30. Oxford: Blackwell

Suomi, S. J. (1974) Social interactions of monkeys reared in a nuclear family environment vs monkeys raised with mothers and peers. *Primates*, **15**, 311–20

Zimmerman, R. R., Strobe, D. A., Steere, P. & Geist, C. R. (1975) Behaviour and malnutrition in the rhesus monkey. In *Primate Behavior*, Vol. 4, ed. L. A. Rosenblum, pp. 241–65. New York: Academic Press

Part VI

Social interactions: development and maintenance

Editors' introduction

In Part V, the development of behaviour was considered from a functional perspective. This section begins with specific issues in social development and ends by considering these same patterns among adults. The main questions raised by the papers in this section are those of mother–offspring relationships and the nature of dominance, its acquisition and maintenance.

The first and one of the most important relationships among primates is that between a mother and her infant. Due to the persisting influence of this relationship on the socialisation of individuals, primate research continues to focus on its nature, variability and later effects on behaviour. We begin with a problem common to many studies of early development. Tartabini and Simpson (Chapter VI.1) explore techniques for reliably assessing the consistency of an individual's behaviour in different contexts. They use as an example interactions between mother and infant rhesus monkeys and relate these to the infant's interactions with other members of the social group. The procedure used to demonstrate reliability in measures of maternal behaviour and infant responses should be useful for those working with small sample sizes, as is often the case in studies of primate social development.

Mendoza and Mason (Chapter VI.2) have examined the allocation of care between the different sexes of parents in a monogamous primate. While this may be the 'unusual' case among primates, this study demonstrates the different principles involved in care-taking, such as the division of labour between parents and the nature of parent–infant bonding. The roles played by the mother, father and infant in initiating and controlling the allocation of care are distinguished through several experimental manipulations.

Hendy (Chapter VI.3) examined the interactions of infant baboons with group members other than the mother. She concludes that the age of the infant was the most important variable affecting the amount of interactions with others, while sex affected the distribution of interactions throughout the range of potential partners. She found little effect of any maternal variables such as dominance rank or parity on an infant's interactions with others. Infant independence and the nature of socialisation with different physical and social experiences are explored.

Quiatt (Chapter VI.4) compares the contribution of different ages and sexes of juveniles to group activities in free-ranging rhesus monkeys. He attempts to determine the applicability of a concept of 'household' and principles of energy exchange to the analysis of the behaviour of non-human primates. His paper develops definitions and methodology for assessing these ideas that may help clarify some of our ideas about the nature of social roles among primates.

The next series of papers deal with dominance relationships and focus specifically on the acquisition of rank, the relations between rank and other friendly and aggressive interactions, and on a cross-species comparison of aggressive interactions. These issues have been of interest and controversy over the past 20 years. The papers that follow illustrate the differences between the mother–infant relationship, with underlying regularities to its development, and other types of relationships where rare events and interactions may have consequences that alter the nature of the relationships themselves. We can ask in what cases are the rules regulating relationships similar, and when are they different? Furthermore, these papers help to refine our understanding of dominance, of its definition and description, and of its variety of functions in primate societies.

Netto and van Hooff (Chapter VI.5) analyse interventions by third parties during aggressive interactions in order to explain how rank is acquired by immature macaques. They find that interventions are often made when the risks to the supporter are low, and that these serve to consolidate existing family rank relationships. They also suggest that adult males can be highly influential in the maintenance of rank relations among families.

Messeri and Giacoma (Chapter VI.6) investigated the role played by dominance as an intervening variable in friendly and aggressive interactions among female pigtail macaques. Dominance ranks were shown to be consistent across different types of interactions, but the frequency of interactions depended more on the hierarchical distance between females than on absolute rank. They conclude that dominance acted to regulate the behaviour considered.

In Chapter VI.7, Thierry compares the frequency and nature of inter-individual aggression between different species of macaques in captive groups. He finds that each species appears to have different rates of aggression, different intensities of interactions and different responses. He concludes that the rates, intensities and responses to aggression are characteristics of the species studied. He notes that little is known of the behaviour of at least one of these macaques in the wild, and conclusions about the adaptiveness of these inter-specific differences are premature.

VI.1

The use of confidence intervals in arguments from individual cases: maternal rejection and infant social behaviour in small groups of rhesus monkeys

A. TARTABINI AND M. J. A. SIMPSON

Introduction

When an individual's behaviour varies from one observation session to the next, the overall score produced by that individual through one or several observation sessions must always be an uncertain quantity. In other words, if any of the subjects in a particular field study or experiment had been observed at different times through the observation period, such as in 10 sessions beginning at half past the hour rather than on the hour, their overall scores would have been different. Given a large group of comparable individuals to work from, intra-individual variability from session to session may be a relatively small problem. In many primate studies, however, we are forced to work with just a few individuals, and a number of consequences can follow.

One individual's score can contribute substantially to a group result. Few of us have not felt grateful to the outlier on a scatter plot that 'strengthens' an otherwise 'weak' correlation coefficient, or impatient about the single wildly 'exceptional' individual that prevents its group from differing from a comparison group. We would therefore like some explicit method, that is not too arbitrary or *post hoc*, that lets us assess the consistency with which each individual is likely to score. Would the outlier or exception have repeated its performance in sessions starting at half past the hour rather than on the hour?

An individual's overall score in, say, the rate at which its mother rejected it, will not necessarily reflect a continuous God's-eye view of the rate. Thus, to study the effects of an 'independent' variable, such as maternal rejection rate, by selecting two groups of high- and low-scoring individuals is a less reliable procedure than controlling the variable experimentally. But in practice we are often unable to experiment, or we may fear unwanted side effects, which are specially likely when we attempt to control just one aspect of one of the partners in a mother–infant dyad.

If a number of scores is used for selecting individuals for making up a sub-group, for example of infants often rejected and often restricted by their mothers and spending long out of contact with them (e.g. Simpson, 1985), we multiply the risk of wrongly classifying individuals. Given a 4/5 (0.8) chance that an individual scoring 'high' on any one of the three behaviours really is high, we have about an even chance of correctly classifying an individual on the basis of three behaviours simultaneously (0.8^3).

Unless we know how consistent and inconsistent individuals have contributed, pooling results from two or more studies can be an uncertain procedure. For example, two studies contributing all their subjects could fail to show some effect when used separately or in combination, while the few consistent individuals available from both might together have shown the effect.

In this paper we show how judgements about individual consistency can be made in studies using small sample sizes. With Thiemann & Kraemer (1984) we believe that a research design can be strengthened if subjects can be observed repeatedly, so long as irrelevant factors affecting scores do not persist at the same level from one sampling session to the next. For example, sampling an individual chimpanzee's grooming partners at 1-minute intervals could result in scores reflecting their reluctance to leave comfortable sitting places within minutes of settling down. Accordingly one would sample sessions separated by enough time or appropriate interventions to ensure the available partners were 'reshuffled'. Application of common sense to the problem at hand should result in reasonable independence. Absolute statistical independence will never be attained, although degrees of independence can sometimes be assessed (e.g. Haccou, Dienske & Meelis, 1983).

Our approach involves estimating the range within which an individual's score is likely to lie, given its variability from session to session. We call the range within which 95% of the scores based on comparable series of sessions should lie the confidence interval (CI) (Moroney, 1956; Phillips, 1973), and we calculate a separate CI for each individual. In this paper we use β-distributions of ratio scores (Phillips, 1973; Bernard, Blancheteau & Rouanet, 1985). In general, any series of observations provides one basis for estimating CIs, and with normally distributed (or normalized) data we would use standard deviations. β-distributions are appropriate when we deal with 'success or failure' or 'yes or no' kinds of events. For example, a coin whose bias was in question could have yielded 7 heads in a series of 10 tosses ('observation sessions'). While intuition would not lead us to expect exactly 7 more heads in a second series of 10 tosses, we might feel that series with 4 or fewer heads would be rare. A β-distribution of the distribution of numbers of heads in series of 10 tosses, given a 7:3 ratio, shows that 95% of future series should yield at least 4 heads. A 70% bias in a larger series, say of 50 tosses, would give us more statistical 'power', and 95% of repeat series would be expected to yield at least 28 (and fewer than 41) heads, and we would then have concluded that the coin was biased.

A formal statistical method is only one way of judging an individual's CI on some measure: any acceptable explicit method could be used. For example, for most purposes single weighings of rhesus monkeys are enough, and informal 'confidence intervals' can be guessed from our views of the likely ranges of weights of contents of food-pouches, bladders and guts. Similarly, in assessing rhesus mother–daughter dominance relationships, one or two observations of a mother supplanting her daughter may be enough to confirm us in our view of this expected bias, given our experience embodied in the Kawamura Hypothesis. Of course, our decision to use some particular statistical technique for estimating CIs is no less open to judgements of confidence than our decision to invoke prior knowledge about dominance relationships or whatever.

This chapter aims to show how explicit procedures can be used to calculate CIs, rather than to present some perfect recipe. As an illustration we refer to data addressing the question of whether maternal rejection of young rhesus monkey infants' attempts to make body and nipple contact promotes or inhibits social activity, especially as they interact with potentially alarming older sisters who sometimes 'kidnap' infants (Spencer-Booth, 1968; Hooley & Simpson, 1983). Maternal rejecting behaviour considered here is 'passively preventing' the infant's attempt to make body or nipple contact, usually by raising an arm between the chest and the approaching infant, or by turning away from the infant. Acts of rejecting by hitting or avoiding the infant as it approaches, included in scores of rejection used by Hinde (1974),

Berman (1980) and Simpson & Howe (1986) are not considered here.

Methods

Four male and three female rhesus monkey infants and their mothers were observed through 4 weeks, starting in week 11. At this age all infants are approaching their maximum rates of social interaction with others and rejection by their mothers (Hinde, 1974), and their mothers have almost ceased to restrain them from leaving. Between thirty-eight and fifty 5-minute records were made for each infant (Fig. 1). Up to two records per infant were made daily, usually separated by at least 45 minutes. Observation was begun as soon as the infant had moved more than 60 cm from its mother to ensure that all observations were of 'active' infants (Simpson, 1979a; Dienske et al., 1980).

A computer-compatible keyboard-operated event-recording system ('MICRO') comparable to the 'WRATS' system (White, 1971; Simpson, 1979b) was used for recording.

We recorded *'mother passively prevents'* her infant's access to her nipple by putting her arm up in front of the infant, or turning her body away as it approaches, and *'social interaction involving companions other than mother'*. Such social contacts could be simple physical contacts, or they could be affiliative or mildly aggressive.

The infants and their mothers lived in six small social groups. All except R's group had a resident adult male. The mothers of H, R and B were dominant to the other females in their groups. All infants had young companions, but R, Z and H did not have companions born in the same year. B and U lived in the same group and had each other as young companions. K was in the largest group, with six young and three adults. The six groups lived in separate indoor–outdoor runs, in a building like that described by Anderson & Simpson (1979) but larger and with more places of refuge.

All results are presented as percentage scores, indicated by ($>$ $<$) on the Figs, with the vertical bars showing 95% CIs based on β-distributions from Phillips (1973) (and also available from the Numerical Algorithm Group, NAG Central Office, Mayfield House, 256 Banbury Road, Oxford, OX2 7DE, UK).

For example, a β-distribution for the ratio 15 records with $\geqslant 1$ passive prevention to 34 without, for U in Fig. 1(a), shows that we can be confident that 95% of comparable sets of 49 records from U would produce percentage scores in the range 18% to 44%. Such intervals facilitate visual judgements about which sets of records are likely to be similar, and which might contrast reliably. Another CI with a maximum of 18% would be seen as not overlapping and 'significantly different' (as would another CI with a minimum of 44%). The chance that the scores from comparable sets of records from U and the other monkey would be equal or in the reverse order is $\leq(2.5/100)^2$ or ≤ 0.001. This relatively stringent criterion is necessary when we argue from single contrasts among many (see Results and Discussion) or when we classify individuals with reference to several measures (see Introduction).

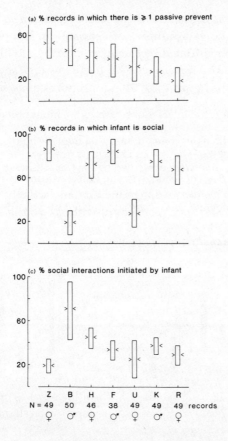

Fig. 1. Percentage of records in which: (a) mother passively prevents her infant's access to her nipple at least once; and (b) the infant has at least one social interaction with a companion other than the mother. (c) Percentage of all social interactions observed with companions other than the mother that were initiated by the infant. In all, the vertical bars indicate 95% confidence intervals. Numbers of records used for each infant are shown at the foot.

Results

The CIs in Fig. 1(a) suggest that Z's mother passively prevented her infant in at least half the 5-minute observation periods, while the mothers of K and R did so in fewer than half the records. The figure also shows that B's mother passively prevented in significantly more records than R's, according to our criterion that CIs should not overlap. Whether the rates received by the remaining infants were different from each other or from those of Z, K and R must remain in question. Fig. 1(b) shows that B and U stand out as the only infants interacting with companions other than mother in fewer than half their records. The contrasts between four infants (Z, B, K and R) show that rejection and social interaction are not related in any simple way. Thus, contrasts in passive preventions can occur without contrasts in rates of social interaction (Z was passively prevented more often than R, but was not more socially active), and contrasts in rates of social interaction can occur where there are none in rejection (Z *versus* B, and U *versus* K).

Every social interaction was initiated either by the infant or a social companion, and Fig. 1(c) shows a number of contrasts in the proportions of social interactions initiated by the infants, with both high and low proportions initiated by infants with high and low total social interaction scores (Fig. 1b). To produce these proportions, every

Fig. 2. Percentage of all social interactions with companions other than the mother initiated by the infant, with infants born in the same year (same age) and with older companions other than the mother (black bars). For each of the infants F and K, results from records without passive prevention are represented by the two left-hand bars, and results from records in which the mother passively prevented at least once by the two right-hand bars. For infants without same-age or older partners, bars are replaced by 'ND' (no data).

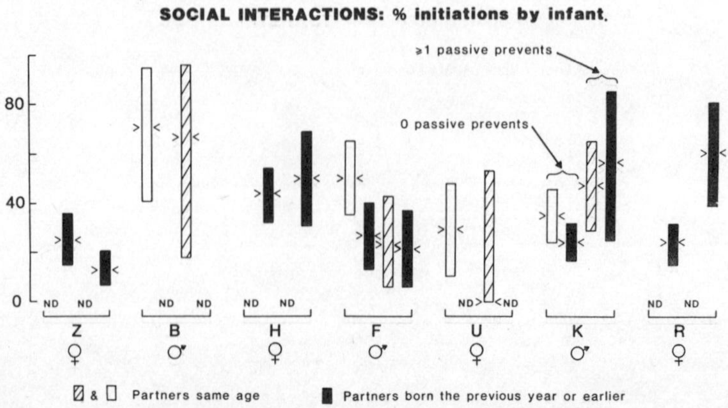

occurrence of a social act between an infant and a social companion other than the mother was counted, successive occurrences being regarded as independent. Moreover, contrasts in proportions of infants' social initiatives can accompany contrasts in passive preventing scores in both directions (e.g. Z *versus* R, B *versus* R in Fig. 1c), providing strong evidence against any simple association between the measures.

Associations between rates of rejection and proportions of social initiatives can also be tested when infants are used as their own controls. For each infant, records without passive prevention and records having at least one were grouped (Fig. 2), and compared for proportions of initiatives with same-age and older partners. R initiated a greater proportion of her interactions in observation sessions in which she was rejected, while F just failed to show an association in the opposite direction. At this stage, the view that rejected infants initiate higher proportions of their social encounters must be based on the single case provided by R. Here we are not following the traditional approach of appealing to the null hypothesis that 'chance' could have produced R's result in Fig. 2. Instead we wait for significant contrasts in the opposite direction to appear in future studies.

Finally, we focus on the proportions of total social interactions that infants have with their oldest nulliparous sisters (Fig. 3), which are

Fig. 3. Percentage of all social interactions with companions other than the mother initiated by three infants with their oldest nulliparous sisters in all records in which their mothers passively prevented at least once (black bars), and in records without passive preventions.

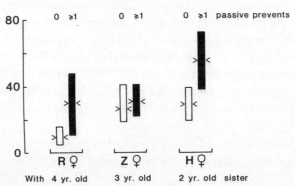

potentially alarming because of their tendencies to 'kidnap' infants. Contrasts provided by H and R approach our criterion of 'significance', and suggest that any strong inhibiting effect of maternal rejection on social interaction is unlikely. It could even be the case that some rejected infants seek comfort from their oldest nulliparous sisters.

Discussion

Because we found contrasts in two measures of the infants' social activity with companions other than mother in the absence of accompanying contrasts in maternal rejection rate, we could refute the view that maternal rejection has any simple effect on social activity. In taking each individual as its own control and comparing records with and without maternal passive preventions, we found one case with a higher proportion of social initiatives in the records with passive preventions (R in Fig. 2). More complex relations could be tested if future studies yielded more cases providing useable contrasts. For example, Fig. 3 suggests that when they are likely to be rejected, R and H but not Z interact more with their oldest nulliparous sisters. Z's mother had a somewhat tense relationship with an unpredictably aggressive adult male, making the position of Z and her mother less secure in their social group. To test this view we need cases providing significant contrasts in some measure of the rates at which mothers and infants are disturbed by aggression in their groups, as well as contrasts in infant social initiative and maternal rejection.

Once we know the ranges within which individuals are likely to score, we may feel less committed to the traditional strategy of working against the 'null' hypothesis that 'chance' explains our results. This strategy must be chosen, however, when 'chance' rather than some specific biologically meaningful hypothesis really does seem the best alternative explanation, as when within-subject variability from observation session to observation session, and/or inter-subject variability due to uncontrollable, irrelevant or unidentifiable factors, is high. In following the null hypothesis testing strategy, it will often be supported when sample sizes (Ns) are small (e.g. 7 subjects), because of the poor statistical power. With large enough Ns, the null hypothesis almost always fails, even against relatively weak effects, and it can therefore become too easy to support any other hypothesis (Meehl, 1967). More formal critical discussion of these problems is available in Phillips (1973).

Another approach to hypothesis testing becomes possible when we can exclude the less-consistent subjects, using information about intra-individual variability. In this 'classification approach' (Simpson, 1985), just one contrast where CIs do not overlap supports our initial hypothesis, while one such contrast in a direction opposite to that predicted by the hypothesis is sufficient grounds for refutation. On the strength of that one exception, however, we may set up a modified hypothesis. Our new view would have to stand or fall on our view of the sense it made, its testability, and ultimately the results of further testing as more cases became available. In following out such a strategy, exceptions can 'improve' the rule.

Of course we will want to accumulate as many examples as we can, if only to improve our confidence in the generality of our rules. As our sample size builds up, applying the null hypothesis testing approach to our significant contrasts can be instructive. For example, with eight cases following the rule that the male twin is heavier in different-sex twin births, and a ninth exception, the sex bias stands against the null hypothesis ($P = 0.040$, two-tailed binomial test (Siegel, 1956)), and the exception could be seen as a result of 'chance'. If, however, we were then told that the confidence intervals of all our twins' weights overlapped with each other, we might suspect that we had been 'lucky' to refute the 'chance' hypothesis on this occasion, and we should also expect to replicate such a 'significant' result only once in 25 times. We might also ask how often we depend on this kind of luck every time we test group results in ignorance of the individuals' confidence intervals. If, however, none of the CIs of the male twins had overlapped with those of their female twins, we might instead view the single exception as meaningful in the context of some more complex hypothesis.

Conclusions

We have shown how to assess the limits within which an individual's score can be seen as an uncertain quantity, with reference to a study of the effect of maternal rejection on the social behaviour of seven rhesus monkey infants. Such limits or confidence intervals let us recognize those individuals most likely to make group results and comparisons uncertain. Identifying the more 'fuzzy' individuals may help us to judge the contributions of extreme individuals to group results, to select subjects whose sampled behaviour is most likely to reflect their behaviour and experience at non-sampling times, to classify individuals with reference to several variables, and to improve our ability to pool subjects from more than one study.

By reducing the uncertainty associated with sampling the behaviour of individuals, we may gain in the confidence with which we work with hypotheses vulnerable to single exceptions and provisionally resting on single cases. Out of this strategy, biologically plausible rules of moderate complexity will sometimes emerge. At other times, of course, application of this strategy will reveal that the 'rules' are becoming progressively complicated by *post hoc* modifications depending partly on *ad hoc* reference to currently unmeasurable variables. Then a 'null' hypothesis, that many variables work together in a complex and unspecifiable way to produce 'chance' effects, is supported after all.

Acknowledgements

We are grateful to the Italian C.N.R. for supporting A. Tartabini's visit to work on the rhesus colony, and to the Medical Research Council for supporting M. J. A. Simpson and the colony. We also thank Les Barden for preparing the figures, David Rayment for his care of the monkeys and Brian Styles and Graham Titmus for constructing the MICRO recorder and the computer programs for it.

References

Anderson, D. M. & Simpson, M. J. A. (1979) Breeding performance of a captive colony of rhesus macaques (*Macaca mulatta*). *Lab. Anim.*, **13**, 275–81

Berman, C. M. (1980) Mother–infant relationships among free-ranging rhesus monkeys on Cayo-Santiago: a comparison with captive pairs. *Anim. Behav.*, **28**, 860–73

Bernard, J.-M., Blancheteau, M. & Rouanet, H. (1985) Le comportement prédateur chez un forficule, *Euborellia moesta* (Géné). II: Analyse séquentielle au moyen de méthodes d'inférence bayésienne. *Biol. Behav.*, **10**, 1–22

Dienske, H., Metz, H. A. J., van Luxemburg, E. & de Jonge, G. (1980) Mother–infant body contact in macaques. II: Further steps towards a representation as a continuous time Markov chain. *Biol. Behav.*, **5**, 61–94

Haccou, P., Dienske, H. & Meelis, E. (1983) Analysis of time-inhomogeneity in Markov chains applied to mother–infant interactions of rhesus monkeys. *Anim. Behav.*, **31**, 927–45

Hinde, R. A. (1974) *Biological Bases of Human Social Behaviour*. New York: McGraw Hill

Hooley, J. M. & Simpson, M. J. A. (1983) Influence of siblings on the infant's relationship with the mother and others. In *Primate Social Relationships: An Integrated Approach*, ed. R. A. Hinde. Oxford: Blackwell Scientific

Meehl, P. E. (1967) Theory-testing in psychology and physics: a methodological paradox. *Philos. Science*, **34**, 103–15

Moroney, M. J. (1956) *Facts from Figures*. Harmondsworth: Penguin

Phillips, L. D. (1973) *Bayesian Statistics for Social Scientists*. London: Nelson

Siegel, S. (1956) *Nonparametric Statistics for the Behavioral Sciences*. New York: McGraw Hill

Simpson, M. J. A. (1979a) Daytime rest and activity in socially living rhesus monkey infants. *Anim. Behav.*, **27**, 602–12

Simpson, M. J. A. (1979b) Problems of recording behavioral data by keyboard. In *Social Interaction Analysis: Methodological Issues*, ed. M. E. Lamb, S. J. Suomi & G. R. Stephenson. Wisconsin: University of Wisconsin Press

Simpson, M. J. A. (1985) Effects of early experience on the behaviour of yearling rhesus monkeys (*Macaca mulatta*) in the presence of a strange object: classification and correlation approaches. *Primates*, **26**, 57–72

Simpson, M. J. A. & Howe, S. (1986) Group and family differences in the behaviour of rhesus monkey infants. *Anim. Behav.*, **34**, 444–59

Spencer-Booth, Y. (1968) The behaviour of group companions towards rhesus monkey infants. *Anim. Behav.*, **16**, 541–57

Thiemann, S. & Kraemer, H. C. (1984) Sources of behavioral variance: implications for sample size decisions. *Am. J. Primatol.*, **7**, 367–75

White, R. E. C. (1971) Wrats: a computer-compatible system for automatically recording and transcribing behavioural data. *Behaviour*, **40**, 135–61

VI.2

Parenting within a monogamous society

S. P. MENDOZA AND W. A. MASON

Introduction

The study of parent–infant attachment in nonhuman primates has focused on such species as the squirrel monkey or rhesus macaques, in which primary care giving is a female activity. In these species interactions between mother and infant constitute a complex and reciprocal dyadic relationship that extends well beyond the nutritional, thermoregulatory or transportational requirements of the infant. This relationship is characterized by the formation of an attachment or emotional bond between parent and infant which functions, in part, to maintain proximity between mother and infant, thereby establishing one of the essential conditions for providing basic care and creating opportunities for the infant to acquire basic social and nonsocial skills. One of the attributes of such a bond is that the mother and infant respond adversely to disruption of the relationship; another is that each animal buffers the other from experiencing the full physiological effects of environmental perturbation (Mason, 1971; Mendoza *et al.*, 1980; Swartz & Rosenblum, 1981).

For a few primate species, parenting is a shared activity in which both mother and father participate in providing routine care of the young. One such species, *Callicebus moloch*, shows an interesting division of care-taking behavior in which the mother usually has the infant during periods surrounding active nursing and the father carries it during the remaining time (Fragaszy, Schwartz & Shimosaka, 1982). Furthermore, once the infant has achieved locomotor independence, it will generally respond to disturbance by returning to the father in preference to the mother if both are equally available.

The differentiation of attachment and nurturing functions within a triadic parent–offspring relationship presents an unusual research opportunity. The nature of each individual's contribution to the development and maintenance of this relationship may, in principle, assume a variety of different forms. For example, each parent could be specifically motivated to perform the role that it characteristically plays, or the chief contributor to the parental division of labor could be the infant, who, for whatever reason, is disposed to treat one parent as a source of transport and the other as a source of food. Similar possibilities exist with respect to emotional attachments within the triad. It is clear from previous research that *Callicebus*, which are monogamous in nature, form a specific attachment to their mates, as reflected in behavior and physiological responses to separation (Mason, 1966, 1974, 1975; Cubicciotti & Mason, 1975, 1978). One might ask whether a similar attachment forms to the infant. Are the parents equivalent in this respect? If so, what is the basis for the apparent differential attachment of the infant to its father? The present research was designed to investigate such questions. Specifically, we examined the behavioral and physiological responses of each parent to separation from the pairmate and the infant, assessed the relative preference of each animal for the other members of the triad, and traced the development of parental retrieval response during the first 2 months of life.

Experiment 1: Response of parents to separation

Involuntary separation from an attachment figure has been shown to elicit clear behavioral and physiological responses indicative of distress (e.g. Mendoza *et al.*, 1978; Mineka & Suomi, 1978; Coe & Levine, 1981). Conversely, distressful events typically induce approach and contact with an attachment figure and such contact tends to reduce the behavioral and physiological responses that would otherwise be shown (Mason, 1971; Mendoza *et al.*, 1980). This pattern of a strong response to separation, proximity-seeking in distressful circumstances, and the mitigation of reactions to distress by presence of the attachment figure has been observed in a variety of species; indeed, it is commonly considered diagnostic of an attachment or emotional bond (Ainsworth, 1972). Accordingly, in this experiment we used the response to separation as a means of assessing the relationship of the parents to each other and to their infant. Each parent was separated from the infant, from the mate, and from both infant and mate. Furthermore, to assess the relative effectiveness of infant and mate in modifying responses to an unfamiliar and therefore potentially distressful setting, these procedures were carried out in the familiar living cage and in a novel environment. The primary measure of responsiveness was alteration in concentrations of plasma cortisol. This endocrine measure has been shown to be sensitive to a variety of environmental disturbances, including novelty and separation from an attachment figure (Hennessy & Levine, 1977; Mendoza et al., 1980). Because the cumulative amount of blood required to complete this assay for all conditions was considered potentially hazardous for the infant, only adults were sampled.

Subjects

Subjects for this and subsequent experiments reported here were opportunistically recruited into each study as viable young were born. Adult subjects were a combination of wild-born and laboratory-born monkeys. Laboratory-born animals had all been raised in stable family units. A total of seven triads was tested. All animals were housed in standard indoor/outdoor cages and maintained according to established laboratory protocol.

Methods

Testing started when the infants were 2.5 months of age. Each parent was tested in the indoor portion of the home cage under each of the following conditions. (1) *Base*. Behavioral observations and blood samples were collected prior to any disturbance of the subjects. (2) *Infant separation*. All three animals were caught, placed in separate transport cages, both adults were returned to the home cage, and the infant was taken to a remote location. One hour later the parents were recaptured and their blood was sampled. (3) *Mate separation*. The same procedures were followed as with infant separation, except that one parent and the infant were returned to the home cage. (4) *Mate and infant separation*. Same as infant separation, except that only one parent was returned to the home cage. (5) *Disturbance control*. Capture procedures were followed as in the separation conditions, but all three animals were simultaneously returned to the home cage. Conditions (3)–(5) were repeated in an unfamiliar environment using a cage which was identical to the indoor portion of the home cage. Order of presentation of conditions was balanced across pairs and at least 1 week intervened between successive presentations. All blood samples (1.0 ml) were collected via femoral puncture from unanesthetized animals within 3 min of entry into the test cage. Behavioral data were collected to determine the quality and extent of parent–infant interactions during the test hour for those conditions in which the infant was present. Two types of behavioral data recording were employed. The first method sampled duration of time spent carrying the infant by each parent in alternate 5-min intervals, yielding 30 min of observation per condition. The second method recorded the frequency of behaviors (using a predefined list) directed by each parent toward the infant throughout the hour. Behavioral categories included: exploratory behavior – close visual manual or olfactory investigation of the infant; supportive behavior – retrieval or attempted retrieval of the infant, or providing it manual or postural support; and punitive behavior – biting, hitting, or grabbing the infant, rejecting or attempting to reject the infant being carried.

Results

Analysis of plasma cortisol levels obtained in the home cage tests indicated that both sexes differentiated among conditions ($F(4,48) = 10.184, p < 0.001$). *Post hoc* evaluations of these data indicate that separation from the mate elicited significant elevations in plasma cortisol in both males and females, in comparison to disturbance or basal values ($p < 0.05$, Newman–Keuls test). Neither parent showed any evidence of such a response to separation from the infant only (see Fig. 1).

Fig. 1. Mean plasma cortisol levels (±S.E.) of father and mother *Callicebus* in the home cage tests.

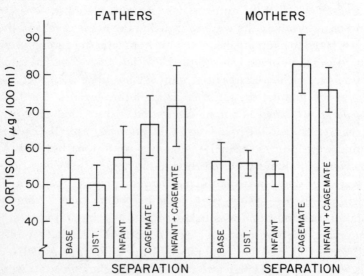

In the novel environment both sexes showed a substantial plasma cortisol elevation in response to the situation, regardless of separation conditions ($F(3,36) = 16.248$, $p < 0.001$). For females, but not males, some differentiation among the test conditions was apparent, and they showed the highest cortisol response to the condition in which they were in the novel environment with their infant, but without their mate (see Fig. 2).

Analysis of time spent carrying the infant showed that the father transported the infant significantly more than did the mother ($F(1,12) = 9.061$, $p < 0.05$). When both parents were present in the home cage tests, the father carried the infant an average of 44.5% of the test period, as compared to 10.9% for the mother. When only one parent was present with the infant in the home cage the difference between parents was attenuated, owing to the increase in time spent on the mother (carry time: fathers $\bar{X} = 52.7\%$, mothers $\bar{X} = 35.9\%$). In the novel environment the father spent substantially more time carrying the infant, as compared to the same conditions in the home cage, whether he was alone with the infant (78.8%), or the mother was also present (76.6%); mothers, however, showed no such increase ($\bar{X} = 9.6\%$ with the father present; $\bar{X} = 44.5\%$ with father absent).

Fig. 2. Mean plasma cortisol levels (±S.E.) of father and mother *Callicebus* in the base condition and following exposure to a novel environment.

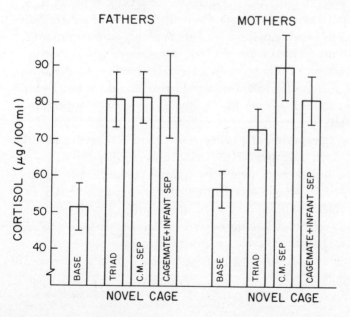

In all conditions the incidence of infant-directed behaviors shown by the father was relatively low. He did not groom the infant, cradle it, inspect it, or reject it. The same was true for the mother, except for those conditions in which she was alone with the infant. Under these circumstances, whether in the home cage or novel environment, she frequently inspected the infant and frequently rejected or attempted to reject it, as evidenced by aggressive grabbing, pushing away, biting, or strenuous efforts to dislodge the clinging infant. Thus, even though the mothers spent more time carrying the infant in the father's absence, this effect seemed largely to be the result of the infant's persistent efforts to cling, and it was accompanied by strong indications that the mother sometimes resisted these efforts, thereby limiting the time the infant spent in contact with her.

Experiment 2: Preference testing

Results of the previous experiment showed that neither parent exhibited a reliable physiological response to separation from its infant, as measured by elevations in corticosteroid concentrations, whereas they showed an unequivocal response to separation from each other. By this measure of attachment it appears that *Callicebus* parents are bonded to each other, but show little evidence of attachment to their infant. The parents did differ, however, in that fathers appeared to be tolerant of contact with their infants, whereas mothers were relatively intolerant, a conclusion supported by both the behavioral and physiological data. These findings suggest that relative tolerance of male and female for extended contact with the infant is a more important factor in the parental division of labor shown by *Callicebus* than any differences in parental attraction to the infant. Presumably, differential tolerance would also lead the infant eventually to prefer the father over the mother as an attachment figure or security object.

The following experiment was designed to assess these possibilities by offering each member of the triad an opportunity to choose between the other two members (paired-comparison); because paired-comparison tests do not distinguish between lack of preference for the nonpreferred stimulus *versus* active avoidance, each member of the triad was also tested with the other members presented singly. The experiment was conducted with infants that were 6 months old, an age when locomotor abilities and differentiation between parents are firmly established.

Subjects

Five parent–infant triads were tested in this situation. Living arrangements were as described. Testing started when the infants (four males, one female) were 6 months of age.

Methods

Preferences were assessed using a Y-maze. A cage located at the end of each arm presented the stimuli and a similar cage at the distal end of the stem of the Y constituted the start-cage from which both stimulus-cages were in clear view. The distance between start-cage and stimulus-cage was 1.5 m. Each member of a triad was tested with each of the other members of the triad presented singly, and with both presented simultaneously. All stimulus conditions were presented twice on each test day, resulting in 20 tests per condition per subject.

Results

Given a choice between mate and infant, both parents tended to prefer the mate. This pattern was shown by all males and by 4/5 females. Although the effect was significant by t tests for the combined sexes, based on total scores ($p < 0.05$), it should be noted that, with the exception of one male, parents of both sexes approached the infant on at least some trials (35% and 40% for males and females, respectively). In the same situation, all infants preferred the father, selecting him over the mother on 68% of trials ($p < 0.05$, t test). The single-stimulus tests provided no clear evidence of active avoidance among the members of the triad. It is of interest, however, that each infant more consistently chose the male over the empty stimulus-cage, as compared to the female in the same condition (78% *versus* 60% of trials, $p < 0.05$, t test).

Experiment 3: Infant retrieval

Results of the the previous experiments, based on infants between 2.5 and 6.5 months of age, indicate that *Callicebus* parents of both sexes prefer the mate over their infant, and show reliable increases in cortisol response to separation from each other, but not to separation from the infant. Furthermore, the data suggest that an influential element in the parental division of labor characteristic of this species is the extent to which male and female are tolerant of prolonged contact with the infant. This difference may help to account for the eventual emergence of the father as the infant's primary attachment figure or security object.

During the age-range when these experiments were performed, however, the infants had considerable prior experience with their parents and had developed the motor skills to be fully capable of responding to them differentially. In earlier stages of development when the pre-ambulatory infant is being carried continuously, either by its parents or an older sibling, it seems likely that the parents play a more direct and active part in creating the distinctive structure of the parent–infant triadic relationship. The present experiment focused on this early period. Parents were tested at weekly intervals from the first to the eighth week following the birth of an infant, a phase when the infant is seldom if ever observed out of contact with an older animal, and when it lacks the locomotor capabilities to re-establish contact on its own initiative. To assess differences between mother and father in tendency to approach and make contact with the infant, a simple retrieval test was used in which the infant was separated from its parents and placed on the floor of the home cage.

Subjects

Fourteen parent–infant triads were the subjects for this study. Some parents were included more than once, with infants born in subsequent years. The data represent contributions from nine different fathers and eight different mothers and from eight male infants and six female infants. Testing for each triad was initiated when the infants were a few days old and continued at weekly intervals throughout the next 8 weeks. Older siblings which were present in the home cage of five pairs were removed prior to testing.

Methods

The general procedure involved the capture of the parent carrying the infant, removal of the infant from the parent, return of the parent to the home cage and presentation of the infant to both parents by placing it on a terry-cloth towel on the floor of the home cage. Identity of parent carrying the infant prior to initiation of the test was noted. Latencies to first contact and to successful retrieval (i.e. infant carried by parent) were measured, and the identity of the contacting and retrieving parents was noted. If the infant was not retrieved within 10 min, the test was terminated and the infant was placed on the parent that had been carrying it at the start of the test.

Results

As can be seen in Table 1, there was a definite tendency for fathers to carry the infant more often than mothers (73.2% vs 26.8%, $\chi^2 = 12.07$, $p < 0.001$). This difference was suggested by the second week of life and was clear and consistent from the fourth week until the end of the study. Although some variability occurred between triads (the observational data indicated that in three triads the infant was carried equally or more often by the mother than by the father), it is clear that the infant generally was more likely to be found on the father. Because our sample included individuals with varying amounts of prior experience with infants we examined the influence of this factor on the extent to which parents differed in the tendency to carry the infant. The results indicated that fathers with no prior parental experience carried the infant substantially less often ($N = 5$ pairs, male carried on 52.5% of observations), than did fathers who had had two or more prior infants ($N = 9$ pairs, male carried on 84.7% of observations). Thus, two variables are associated with an increase in the amount of time fathers carry infants, the age of the infant and previous parental experience by the male.

Both parents were generally responsive when confronted with their separated infant. In this test the infant invariably emitted high-pitched distress vocalizations which continued until it was retrieved.

Table 1. *Results of retrieval test across weeks*

	Infant age (weeks)								Combined weeks
	1	2	3	4	5	6	7	8	
Carry									
Father	7	11	6	9	13	11	13	12	82
Mother	7	3	8	5	1	3	1	2	30
Contact									
Father	6	6	6	6	8	7	6	8	53
Mother	8	8	7	8	6	7	8	6	58
$\bar{X}$ Latency (s)	12.9	19.0	10.6[a]	25.2	15.5	16.9	27.4	18.7	18.6[a]
Retrieval									
Father	5	5	6	5	7	6	6	8	48
Mother	7	8	7	9	7	8	8	6	60
$\bar{X}$ Latency (s)	35.5[a]	46.9[a]	20.6[a]	34.6	24.5	24.2	28.6	19.6	30.6[a]

[a] Balks excluded

The parents typically oriented toward the infant, showed signs of agitation, and eventually one or both moved to the cage floor, approached and made contact with the infant. On only one occasion (from a total of 112 tests) was the infant not contacted within the 10-min test period. Mean latencies to contact (shown in Table 1) were about 20 s, although the range was broad (1–178 s). Males and females were equally likely to contact the infant. Retrieval was most often the result of the infant grasping the contiguous parent and seldom included any specific action by the parent, such as scooping up the infant. For this reason several successive contacts were sometimes made before retrieval was accomplished. On six tests the parent initiating first contact did not retrieve the infant and on three of these occasions retrieval did not occur. Although mothers tended to retrieve the infant somewhat more often than did fathers, this difference was not statistically significant.

Discussion

By the time their infant is about 1 month old, and usually before this age, *Callicebus* parents show a clear-cut division of parental labor in which the female has the infant for brief periods of nursing and the male is mainly involved in transporting the young. Informal observations suggest that the male also becomes the primary attachment figure, in the sense that he is sought out by the infant when it is distressed.

These experiments examined some of the variables influencing the structure of the parent–infant triad, with particular reference to the role each individual plays in contributing to the division of labor in care-taking activities. The first experiment found no evidence that attachment to the infant was stronger in the father than in the mother, as measured by cortisol response to separation. In fact, neither sex showed a significant elevation in cortisol concentrations when separated from the infant, whereas both did so when separated from each other. Results from the preference test (Experiment 2) were consistent with these findings, in that both parents generally preferred each other over the infant in direct choice tests. In this same test, however, the infants clearly preferred the father over the mother.

Some of the results of Experiment 1 may indicate how the infant's preference for the father comes about. When the male was separated from the mother and infant, the time the infant spent in contact with the mother increased. This seemed to result largely from the infant's efforts to establish social contact; the mother's response to this was an increase in cortisol to a level slightly above that in any other condition, and various behaviors indicative of rejection. This lack of tolerance by the mother is relative, of course, and the results of the retrieval test (Experiment 3) indicate that both parents are responsive to the presence of their physically separated infant, and are about equally likely to approach and make contact with it.

Based on our observations during the retrieval test, and on other occasions, we believe that the distress vocalizations of the infant and the parental reactions to these sounds may be a critical element in the genesis of the *Callicebus* triadic relationship. Most mothers appear to be tolerant of contact with the infant for only relatively brief periods, as during a nursing bout. As the mother grows increasingly restive she attempts to dislodge the infant by pushing at it, biting its hands, feet or tail. The infant invariably vocalizes in response to this treatment, which causes the father to approach the mother and permits the infant to transfer. The father is relatively more tolerant of the infant (in fact we have rarely seen any attempts at rejection before the infant is ambulatory), and continues to carry it until it is hungry, at which time it begins to vocalize, drawing the mother toward the father.

If this construction is essentially correct, as we believe it to be, the infant plays a major role in coordinating the division of care-taking activities within the triad. Its ability to accomplish this smoothly and efficiently improves as its locomotor skills develop and as it becomes better able to distinguish between its parents. The general tendency of the parents to remain close to each other, and to approach the source of infant vocalizations are important enabling conditions which permit the infant to negotiate between its source of food and its principal source of transportation and emotional security.

Acknowledgements

The authors would like to express their appreciation to Ann Butler, Julia Mead, and Judy Mikols for their assistance in the execution of this research. This work was supported by the National Research Service Award 1-F32-HD06127 (to S. P. Mendoza) and National Institute of Health, Division of Research Resources # RR00169.

References

Ainsworth, M. D. S. (1972) Attachment and dependency. In *Attachment and Dependency*, ed. J. L. Gewirtz, pp. 97–137. Washington, D.C.: V. H. Winston & Sons

Coe, C. L. & Levine, S. (1981) Normal responses to mother–infant separation in nonhuman primates. In *Anxiety: New Research and Changing Concepts*, ed. D. F. Klein & J. Rabkin, pp. 155–77. New York: Raven Press

Cubicciotti, D. D. III & Mason, W. A. (1975) Comparative studies of social behavior in *Callicebus* and *Saimiri*: male–female emotional attachments. *Behav. Biol.*, **16**, 185–97

Cubicciotti, D. D. III & Mason, W. A. (1978) Comparative studies of social behavior in *Callicebus* and *Saimiri*: heterosexual jealousy behavior. *Behav. Ecol. Sociobiol.*, **3**, 311–22

Fragaszy, D. M., Schwarz, S. & Shimosaka, D. (1982) Longitudinal observations of care and development of infant titi monkeys (*Callicebus moloch*). *Am. J. Primatol.*, **2**, 191–200

Hennessy, M. B. & Levine, S. (1977) Effects of various habituation procedures on pituitary-adrenal responsiveness in the mouse. *Physiol. Behav.*, **18**, 799–802

Mason, W. A. (1966) Social organization of the South American monkey, *Callicebus moloch*: a preliminary report. *Tulane Stud. Zool.*, **13**, 23–8

Mason, W. A. (1971) Motivational factors in psychosocial development. In *Nebraska Symposium on Motivation*, ed. W. J. Arnold & M. M. Page, pp. 35–67. Lincoln, Nebr.: University of Nebraska Press

Mason, W. A. (1974) Comparative studies of social behavior in *Callicebus* and *Saimiri*: behavior of male–female pairs. *Folia Primatol.*, **22**, 1–8

Mason, W. A. (1975) Comparative studies of social behavior in *Callicebus* and *Saimiri*: strength and specificity of attraction between male–female cagemates. *Folia Primatol.*, **23**, 113–23

Mendoza, S. P., Coe, C. L., Smotherman, W. P., Kaplan, J. & Levine, S. (1980) Functional consequences of attachment: a comparison of two species. In *Maternal Influences and Early Behavior*, ed. R. W. Bell & W. P. Smotherman, pp. 235–52. New York: Spectrum Press

Mendoza, S. P., Smotherman, W. P., Miner, M. T., Kaplan, J. & Levine, S. (1978) Pituitary-adrenal response to separation in mother and infant squirrel monkeys. *Develop. Psychobiol.*, **11**, 169–75

Mineka, S. & Suomi, S. J. (1978) Social separation in monkeys. *Psychol. Bull.*, **85**, 1376–1400

Swartz, K. B. & Rosenblum, L. A. (1981) The social context of parental behavior: a perspective on primate socialization. In *Parental Care in Mammals*, ed. D. J. Gubernick & P. H. Klopfer, pp. 417–54. New York: Plenum Press

VI.3

Social interactions of free-ranging baboon infants

H. HENDY

Introduction

A number of baboon investigators have provided general descriptions of infant social interactions with other troop members (Bolwig, 1959; Hall & DeVore, 1965; Kummer, 1968a; Altmann & Altmann, 1970; Ransom & Rowell, 1972; Nash, 1974). The purpose of the present study was to examine quantitatively the social interactions between free-ranging baboon infants less than a year old and all age/sex categories of troop members. The few other quantitative reports that are available for free-ranging baboons focus primarily on mother/infant interaction (Ransom & Ransom, 1971; Nash, 1978; Altmann, 1980) or they are concerned with a later time in the young baboon's life (Owens, 1975, 1976; Cheney, 1977, 1978a,b; Lee & Oliver, 1979; Chalmers, 1980).

Past research on both baboons and the closely related macaques has suggested a number of variables that may be related to the amount and type of interaction infants have with troop members (Seay, 1966; Hinde & Spencer-Booth, 1967; Rosenblum & Kaufman, 1967; Mitchell & Brandt, 1970; Harlow & Lauersdorf, 1974; Gouzoules, 1975; Cheney, 1978a,b; Altmann, 1980). The present study thus examined the effects of age, sex, parity, dominance, and troop differences on the social interactions that infant baboons have with their mothers and other troop members during the first few months of life. The present study was part of a cooperative effort among students from the University of California, Riverside (P. Ender, D. Rasmussen, K. Rasmussen, and the present author), the first group of researchers at Mikumi National Park, Tanzania.

Methods

Eleven infant subjects (five males, six females) were from a 106-member troop of *Papio cynocephalus* living in the grassy floodplain of Mikumi National Park, Tanzania. Five infant subjects (four males, one female) were from a 31-member troop of *Papio anubis* living in the mountainous, woodland lakeside of Gombe National Park, Tanzania. Both troops were selected because they were relatively undisturbed by humans.

The Gombe baboons had been observed for over 6 years and had a flight distance of approximately 2 m (i.e. they ran away when approached to within 2 m). Their only known predators were chimpanzees, although little predatory activity on the baboons had been seen in the 3 or 4 years before this study (unpublished Gombe Research files). The Mikumi troop had been observed for a period of 8 months and had a flight distance of approximately 5 m at the time of the present study. The Mikumi baboons were threatened by leopards in their sleeping trees and lions in the grassy flood-plain where they foraged during the day (P. Sayellel, D. Rasmussen & K. Rasmussen, personal communication).

The Gombe subjects were studied during the wet season from November, 1974, through May, 1975. The woodland habitat of the Gombe baboons was dense and lushly green with many types of fruit, grasses, leaves, and insects available for food. Water was obtained two or three times a day from Lake Tanganyika or from the streams that cut through the troop's home range and ran into the lake. The Mikumi baboons, on the other hand, were studied during the dry season from June, 1975, through November, 1975. Much of their grassy range at this time was dry, blackened stubble from brush fires, but this made for excellent observation conditions. The baboons ate almost exclusively tamarinds, the bulbous part of sedges, and a few grasshoppers. Once a day they visited one of the three waterholes available in their home range.

Five types of social interaction were recorded for the 16 infant subjects:

1. *holding* the infant either ventrally or dorsally
2. *momentary touching* of the infant with either the hand or mouth
3. *grooming* through the hair of the infant using either the hand or mouth
4. *contact playing* with the infant during which grappling or wrestling occurred

5. *aversive touching* of the infant in which the infant squealed or jerked its body away after being touched.

These five types of infant interaction were recorded for four categories of troop member:

1. *The infant's mother.* For all infant subjects the identity of the mother was certain because she had been individually recognized and followed for some months prior to the infant's birth.
2. *Adult and subadult males.* Males were considered as adults if they had a full-sized body, mane, and fully erupted canines, and subadults if they were were at least the size of an average adult female.
3. *Adult and subadult females other than the mother.* Females in this category were of full body size and had regular estrous cycles and were either nulliparous, primiparous, or multiparous.
4. *Immatures.* This category included all troop immatures from birth to approximately 4 years of age (Altmann, Altmann & Hausfater, 1981). It was further subdivided into three immature age groups:

   $1 =$ *a black infant* having at least one-half its head still with this natal coloring. (Immatures from birth to about 6 months of age were in this category.)

   $2 =$ *a small juvenile* still under 30 cm in height at the shoulder. (Immatures from about 6 months to 2 years old were in this category.)

   $3 =$ *a large juvenile* between 30 cm in height at the shoulder and the size of an average adult female. (Immatures from about 2 years to 4 years were in this category.)

The infant's social interactions were recorded only for troop members who were individually recognized throughout all observations on all infant subjects. At Gombe, recognized troop members included 3 adult males, 12 adult males, and 12 immatures. At Mikumi, recognized troop members included 22 adult males, 26 adult females, and 19 immatures.

Scores of agonistic dominance were calculated each month for the adult males and adult females who were under weekly observation for D. Rasmussen's study being conducted at Gombe and Mikumi. Only these regularly observed adults were considered to avoid obtaining inaccurately low dominance scores for those individuals who were encountered less often by observers. Separate dominance scores were calculated among adult males and adult females and they were based on spontaneously occurring interactions in which an adult approached and supplanted a same-sexed adult from a stationary position. Each individual's dominance score was then defined as the difference between the proportion of individuals he/she supplanted and the proportion of individuals that supplanted him/her (with such scores then averaged over the months of infant observation).

Each of the 16 infant subjects was sampled for a 2-month span during its first 8 months of life (Table 1). Observations were made while the infant was within 10 m of its mother. Approximately 152 h of observation were made, with an average of 9.5 h per infant. Sampling periods for each infant were from 1 to 3 h long, during which data were collected for successive 5-min intervals. During each 5-min interval, every dyad of infant-subject/other-troop-member received a score for each of the five behaviors of either a 'one', if that behavior had been observed at any time during the interval, or a score of 'zero', if the behavior had not occurred during the interval. These dyad scores were summarized over all observations for the infant subject and prorated as the number of 5-min intervals in which each behavior occurred per 100 observed 5-min intervals. Finally, means for these prorated scores were calculated across the particular infant or interactant category being considered in each analysis. Despite controversy over the use of one/zero samples (J. Altmann, 1974; Simpson & Simpson, 1977), they

Table 1. *Infant subject information.* N = 1830 5-min intervals (152.5 h)

Subject	Troop	Average age (days)	Sex	Mother's dominance	Number of 5-min intervals observed
060	Mikumi	145	F	−0.04	144
064	Mikumi	206	M	−0.29	142
068	Mikumi	181	F	missing	35
078	Mikumi	186	M	+0.38	129
088	Mikumi	116	F	−0.18	131
089	Mikumi	58	M	+0.08	154
090	Mikumi	68	F	−0.15	136
091	Mikumi	30	F	−0.40	121
092	Mikumi	33	F	+0.05	154
097	Mikumi	21	M	−0.15	96
098	Mikumi	19	M	−0.38	60
171	Gombe	256	M	−0.29	23
172	Gombe	178	M	−0.43	40
173	Gombe	52	M	+0.03	188
174	Gombe	22	M	−0.14	192
175	Gombe	9	F	−0.29	85

do provide a measure of the *relative* amount of behavior, and as long as the same one/zero technique is used consistently for all comparison groups, it is an effective behavioral measurement for analyses that examine the behavior's relationship to other variables such as age, sex and dominance (Rhine & Flanigon, 1978; Rhine & Linville, 1980; Rhine & Ender, 1983).

The data for the present study were collected by Dennis Rasmussen and the present author, who each had over 2 years' experience collecting systematic samples of captive macaque behavior. To check interobserver reliability, the two observers simultaneously sampled 11 h of infant interactions. The proportion of observed 5-min intervals in which the observers' scores were in agreement was over 97% for all five behaviors.

Results

In the analyses, Pearson correlation coefficients were used to examine the relation of the infant's age to the amount of social interaction (defined as the number of observed intervals in which the interaction occurred per 100 observed 5-min intervals). The three immature age groups were compared in the amount of interaction they had with infants by using analysis of variance. Spearman correlation coefficients were used to consider the relation of dominance scores to the amount of social interaction with infants. Sex differences and parity differences were examined with t-tests.

The overall pattern of age/sex classes with which the infants had particular types of interaction was very similar to the generalized reports of previous investigators. Infants showed holding and grooming interactions primarily with their mothers, aversive touching interactions primarily with adult males and adult females, momentary touching interactions primarily with adult females and immatures, and play interactions primarily with immatures (Fig. 1). However, the patterns of how variables such as age, sex, dominance, and parity were related to infant social interaction were in some cases quite different from previous findings.

Infant variables

Unlike previous reports for baboons and macaques (Missakian, 1972; Gouzoules, 1975; Cheney, 1977, 1978b; Lee & Oliver, 1979; Silk, Rodman & Samuels, 1979; Altmann, 1980; Berman, 1980; French, 1981), no relation was found between the mother's dominance and the social interactions her infant had with either the mother herself or with other troop members (with social interaction again defined as the number of observed intervals in which the interaction occurred per 100 observed 5-min intervals). This is different from the finding of Altmann (1980) that high-ranking females restrained their infants less and that their infants received more grooming from other troop members than did the infants of low-ranking females. The present study examined infant interactions that took place within 10 m of the mother. Perhaps it is only beyond this distance that we see a difference in the ability of low- and high-ranking mothers to affect their infants' social interaction. In addition, the effects of the mother's dominance might have been more apparent if dominance had been calculated from the more violent (though less frequent) behaviors of chase or attack, rather than from supplant interactions.

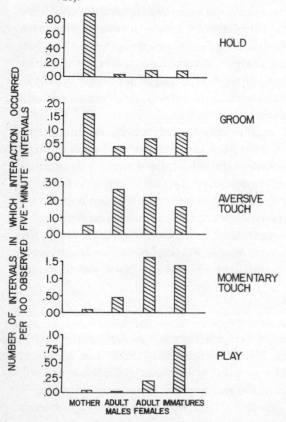

Fig. 1. The overall pattern of five types of social interaction which baboon infants showed with four categories of troop member ($N = 16$).

Also, unlike previous reports in macaque studies and in some baboon studies (Jensen, Bobbitt & Gordon, 1968; Kummer, 1968a; Ransom & Rowell, 1972; Harlow & Lauersdorf, 1974; Norikoshi, 1974; Rosenblum, 1974; Owens, 1975, 1976; Cheney, 1978a), no sex differences were found in infant social interactions for the five types of behavior received from the four categories of troop member, except that adult females other than the mother held female infants more than they held male infants ($t = 2.46$, $d.f. = 14$, $p < 0.05$). Wheeler (1983) has suggested that adult females tend to form bonds with the female infants because they will probably stay in the troop their entire lives, whereas the male infants will probably leave the troop at adolescence.

Some baboon studies suggest that early sex differences in behavior may be less consistent for baboons than for macaques (Rowell, Din & Omar, 1968; Young & Bramblett, 1977; Nash, 1978; Altmann, 1980). The baboon studies that do report sex differences in infant social interaction are either studies in which infant behavior was *not systematically* sampled (Kummer, 1968a; Ransom & Rowell, 1972), they are studies on infants *older* than 6 months (Owens, 1975, 1976; Cheney, 1978b), or they are studies on *captive* animals (Coelho & Bramblett, 1981; Young, Coelho & Bramblett, 1982). All these factors can affect the findings of infant sex differences. For example, qualitative descriptions of infant behavior are at risk for selective perception and memory of *expected* results, especially if the observers are familiar with the consistent findings of early sex differences in macaque behavior. Also, if sex differences in behavior do not appear in baboons until after 6 months of age, the young infants of the present study would not show such sex differences in social interaction. Finally, captive infants may show earlier sex role differentiation than free-ranging infants who face more immediate problems of learning to find sufficient food, water, and shelter before they can invest as much energy in social interactions (Kummer & Kurt, 1965; Baldwin & Baldwin, 1976).

Of the three infant variables (age, sex, mother's dominance) considered in the present study, age seemed to be the variable most related to the patterns of infant social interaction. Mothers held younger infants more than older infants ($r = -0.80$, $n = 16$, $p < 0.01$). Both adult males and adult females showed more momentary touching and aversive touching with younger infants (for adult males, $r = 0.65$, $n = 16$, $p < 0.01$; $r = -0.44$, $n = 16$, $p < 0.05$: for adult females, $r = -0.48$, $n = 16$, $p < 0.05$; $r = -0.59$, $n = 16$, $p < 0.01$, respectively). Immatures also had more momentary touching interactions with the younger infants ($r = -0.47$, $n = 16$, $p < 0.05$) but they directed more aversive touches to the older infants ($r = 0.46, n = 16, p < 0.05$). In general, then, younger infants received the most contact from other troop members. These findings are similar to those of Altmann (1980), who noted that the first 3 months was the time when infants were most in contact with their mothers and when they had the most troop members nearby.

Other troop member variables

Besides the infant variables of age, sex, and mother's dominance, variables concerning the troop members themselves might also be related to the amount and type of infant social interaction. The effects of age, sex, parity, and dominance of the other troop members were therefore examined. Although the sex of the infant subject was not a significant factor for infant interactions with most troop members, the sex of the immature troop member was significant; female immatures gave more grooming and momentary touching interactions to infants than did male immatures ($t = 3.88, d.f. = 29, p < 0.01; t = 2.61, d.f. = 29, p < 0.05$; respectively). This finding is similar to that of many other macaque and baboon investigators, who often suggest that the female immatures are practicing the skills they will later use with their own infants (Lancaster, 1971, 1976; Poirier & Smith, 1974; Cheney, 1978b; Quiatt, 1978; Silk et al., 1979).

The three immature age groups (birth to 6 months, 6 months to 2 years, and 2 years to 4 years) were found to differ in the amount of play they had with the infant subjects, with the oldest age group showing the most play ($F = 8.03, d.f. = 2, 16, p < 0.01$). This finding is similar to that of other studies of baboons and macaques showing an increase in play as the juvenile matures (Owens, 1975; Caine & Mitchell, 1979).

The dominance of adult males and adult females was significantly related to the amount of infant social interaction they showed. The higher the male's dominance, the more momentary touching interactions he showed with infants ($r^s = 0.46, n = 15, p < 0.05$), and the higher the female's dominance the more aversive touching interactions she showed with infants ($r^s = 0.65, n = 20, p < 0.01$). A number of authors (Smith & Peffer-Smith, 1979; Packer, 1980; Berenstain, Rodman & Smith, 1981; Stein & Stacey, 1981) have suggested that dominant males, or males most likely to be the fathers of the infants, show the most gentle interaction with infants. Other authors (Silk et al., 1979; Altmann, 1980) have also found that dominant females are most likely to direct aggressive behaviors to the infants of other females, which may enhance their own reproductive fitness by putting at risk the reproductive investments of their rival females (Fedigan, 1983).

No parity differences were found when experienced, multiparous females were compared to inexperienced, nulliparous or primiparous females in the amount of interaction they showed with infants not their own. Although the adult female's parity was unrelated to her interaction with the infants of other females, it was found that females who currently had an infant under a year of age in the troop showed more holding and momentary touching interactions with other troop infants than did females who did not currently have an infant under a year old ($t = 3.23$, $d.f. = 36$, $p < 0.01$; $t = 2.53$, $d.f. = 36$, $p < 0.05$; respectively). This finding is similar to that of Altmann (1980), who indicated that females with young infants tended to cluster together, perhaps speeding the social maturation of the infants or increasing protection from predators.

Troop differences

In addition to an examination of the above infant variables and other-troop-member variables that might be related to the amount and type of social interaction received by infants, troop differences were considered. Gombe infants had more aversive touching interactions from their mothers and they had more play interactions with both adult females and immatures in their troop (Fig. 2). The infant subjects of the Gombe and Mikumi troops did not differ significantly in age or mother's dominance, so these factors cannot account for the troop differences. The Gombe troop did have a higher proportion of male infants than did the Mikumi troop, but the same pattern of results appears even when only the male infants of each troop are considered in the analyses, so sex differences in the troops cannot account for these troop differences either.

The pattern of troop differences suggests a greater independence of infants at Gombe. Gombe infants had more play interactions with other troop members and they had more of the aversive interactions with the mother which many authors have suggested push the infant into independence (Altmann, 1980; Negayama, 1981). Perhaps this greater independence of infants at Gombe is due to its reduced predator pressure and increased availability of food and water that allowed greater social time investments than in the harsh dry season conditions of Mikumi. Certainly a number of primate investigators (Crook, 1970; Kummer, 1971; Eisenberg, Muckenhirn & Rudran, 1972;

S. Altmann, 1974) have suggested such relations between primate social behavior and ecology. Also, closely related species have been found before to differ in particular patterns of social interaction (Rosenblum & Kaufman, 1967; Kummer, 1968b; Small, 1982), so it is hardly surprising that the *Papio anubis* and *P. cynocephalus* subjects in the present study were found to differ in their behavior.

A final explanation for the troop differences in infant social interactions could be a difference in the degree of habituation of the two troops. The Gombe troop had been closely observed for over 6 years and had an average flight distance of 2 m, whereas the Mikumi troop had been observed for only 8 months and had an average flight distance of 5 m at the beginning of the present study. Perhaps at Mikumi the presence of humans inhibited the aversive interactions with which mothers encourage their infant's independence, and the

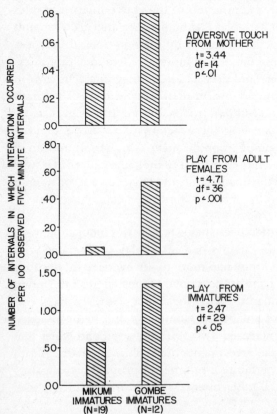

Fig. 2. Troop differences in interactions which infants had with other troop members.

play interactions of infants and other troop members that are more likely under relaxed, nonthreatening conditions. Only further observation of these and other well-habituated baboon troops can answer these questions of troop differences in infant social development.

Conclusions

To summarize, the age of the baboon infant was the variable most related to the amount of social interaction it had with other troop members, with younger infants generally receiving the most interaction. Unlike previous studies, the sex of the infant and its mother's dominance were unrelated to its interactions with most troop members. Female immatures in the troop had more interactions with the infant subjects than did male immatures, and immatures older than about 2 years showed more interaction with the infants than did the younger age groups. Adult males and females of high dominance had more interactions with infants than did males and females of low dominance. Although an adult female's parity was unrelated to her interaction with the infants of other females, females who currently had a young infant of their own showed more interaction than did females without their own infant. Finally, troop differences were found in infant social interaction, suggesting greater independence of the Gombe infants than the Mikumi infants.

Future study of a number of well-habituated troops could answer questions about other variables that might be related to patterns of infant social development (such as kinship, immigration, and habitat change). In the present study the author was fortunate to study in two baboon field research centers. Only further cooperative use of the few available field sites can help us gain understanding of all aspects of primate behavior.

Acknowledgements

The present study was conducted while the author held a National Science Foundation Graduate Fellowship at the University of California, Riverside. The author is grateful to Jane Goodall, Derek Bryceson, and Philip and Ara Sayellel for their generous help in arranging the baboon research in Tanzania's Mikumi National Park. Thanks are also extended to Philip Ender, Dennis Rasmussen and Kathlyn Rasmussen for the exhaustive months spent habituating the baboon troop, and to Dennis Rasmussen for help in data collection. Salum Kemo, Charles Kidungho, and Felix Muroto gave assistance in finding and identifying the baboons; Debra Forthman-Quick and Horace F. Quick, IV, offered suggestions and friendship during the final months of the field study; Lorraine Kimble did the graphics; and Diane Evans typed the manuscript. Most of all, appreciation is felt for Ramon J. Rhine who gave advice and encouragement throughout the many phases of the project.

References

Altmann, J. (1974) Observational study of behavior: sampling methods. *Behav.*, **49**, 227–67

Altmann, J. (1980) *Baboon Mothers and Infants.* Cambridge, Mass.: Harvard University Press

Altmann, J., Altmann, S. & Hausfater, G. (1981) Physical maturation and age estimates of yellow baboons, *Papio cynocephalus*, in Amboseli National Park, Kenya. *Am. J. Primatol.*, **1**, 389–99

Altmann, S. (1974) Baboons, space, time, and energy. *Am. Zool.*, **14**, 221–48

Altmann, S. and Altmann, J. (1970) Baboon ecology. *Biblio. Primatol.*, No. 12

Baldwin, J. D. & Baldwin, J. I. (1976) Effects of food ecology on social play: a laboratory simulation. *Z. Tierpsychol.*, **40**, 1–14

Berenstain, L., Rodman, P. S. & Smith, D. G. (1981) Social relations between fathers and offspring in a captive group of rhesus monkeys. *Anim. Behav.*, **29**, 1057–63

Berman, C. M. (1980) Early agonistic experience and rank acquisition among free-ranging infant rhesus monkeys. *Int. J. Primatol.*, **1**, 153–70

Bolwig, N. (1959) A study of the behaviour of the chacma baboon, *Papio ursinus*. *Behaviour*, **14**, 136–63

Caine, N. & Mitchell, G. (1979) A review of play in the genus *Macaca*: social correlates. *Primates*, **20**, 535–46

Chalmers, N. P. (1980) The ontogeny of play in feral olive baboons (*Papio anubis*). *Anim. Behav.*, **28**, 570–85

Cheney, D. L. (1977) The acquisition of rank and the development of reciprocal alliances among free-ranging immature baboons. *Behav. Ecol. Sociobiol.*, **2**, 303–18

Cheney, D. L. (1978a) Interactions of immature male and female baboons with adult females. *Anim. Behav.*, **26**, 389–408

Cheney, D. L. (1978b) The play partners of immature baboons. *Anim. Behav.*, **26**, 1038–50

Coelho, A. M. & Bramblett, C. A. (1981) Effects of rearing on aggression and subordination in *Papio* monkeys. *Am. J. Primatol.*, **1**, 401–12

Crook, J. H. (1970) The socio-ecology of primates. In *Social Behavior of Birds and Mammals*, ed. J. H. Crook, pp. 103–66. London: Academic Press

Eisenberg, J. F., Muckenhirn, N. A. & Rudran, R. (1972) The relation between ecology and social structure in primates. *Science*, **176**, 863–74

Fedigan, L. M. (1983) Dominance and reproductive success in primates. *Yearbook Phys. Anthropol.*, **26**, 91–129

French, J. A. (1981) Individual differences in play in *Macaca fuscata*: the role of maternal status and proximity. *Int. J. Primatol.*, **2**, 227–46

Gouzoules, H. (1975) Maternal rank and early social interactions of infant stumptail macaques. *Primates*, **16**, 405–18

Hall, K. R. L. & DeVore, I. (1965) Baboon social behavior. In *Primate Behavior*, ed. I. DeVore, pp. 53–110. New York: Rinehart & Winston

Harlow, H. F. & Lauersdorf, H. E. (1974) Sex differences in passion and play. *Perspect. Biol. Med.*, **17**, 348–60

Hinde, R. A. & Spencer-Booth, Y. (1967) The behavior of socially-living rhesus monkeys in their first two and a half years. *Anim. Behav.*, **15**, 169–96

Jensen, G. D., Bobbitt, R. A. & Gordon, B. N. (1968) Sex differences in the development of independence of infant monkeys. *Behaviour*, **30**, 1–14

Kummer, H. (1968a) *Social Organization of Hamadryas Baboons.* Chicago: University of Chicago Press

Kummer, H. (1968b) Two variations in the social organization of baboons. In *Primates: Studies in Adaptation and Variability*, ed. P. C. Jay, pp. 293–312. New York: Rinehart & Winston

Kummer, H. (1971) *Primate Societies: Group Techniques of Ecological Adaptation*. Chicago: Aldine

Kummer, H. & Kurt, F. (1965) A comparison of social behaviors in captive and wild hamadryas baboons. In *The Baboon in Medical Research*, vol. 1, ed. H. Vagtborg, pp. 65–80. Austin, Texas: University of Texas Press

Lancaster, J. B. (1971) Play-mothering: the relation between juvenile females and young infants among free-ranging vervet monkeys. *Folia Primatol.*, 15, 161–82

Lancaster, J. B. (1976) Sex roles in primate societies. In *Sex Differences: Social and Biological Perspectives*, ed. M. S. Teitelbaum, pp. 22–61. New York: Anchor Press

Lee, P. C. & Oliver, J. I. (1979) Competition, dominance, and the acquisition of rank in juvenile yellow baboons (*Papio cynocephalus*). *Anim. Behav.*, 27, 576–85

Missakian, E. A. (1972) Genealogical and cross-genealogical dominance relations in a group of free-ranging rhesus monkeys on Cayo Santiago. *Primates*, 13, 169–80

Mitchell, G. & Brandt, E. M. (1970) Behavioral differences related to experience of mother and sex of infant in the rhesus monkey. *Devel. Psychol.*, 3, 149

Nash, L. T. (1974) Social behavior and social development in baboons at the Gombe Stream National Park, Tanzania. Doctoral Dissertation, University of California, Berkeley

Nash, L. T. (1978) The development of the mother–infant relationship in wild baboons (*Papio anubis*). *Anim. Behav.*, 26, 746–59

Negayama, K. (1981) Maternal aggression to its offspring in Japanese monkeys. *J. Hum. Evol.*, 10, 523–7

Noriskoshi, K. (1974) The development of peer-male relationships in Japanese macaque infants. *Primates*, 15, 39–46

Owens, N. W. (1975) Social play behavior in free-living baboons. *Anim. Behav.*, 23, 387–408

Owens, N. W. (1976) Development of sociosexual behaviour in free-living baboons. *Behaviour*, 57, 241–59

Packer, C. (1980) Male care and exploitation of infants in *Papio anubis*. *Anim. Behav.*, 28, 512–20

Poirier, F. E. & Smith, E. O. (1974) Socializing functions of primate play. *Am. Zool.*, 14, 275–87

Quiatt, D. (1978) Aunts and mothers: adaptive implications of allomaternal behavior of nonhuman primates. *Am. Anthropol.*, 81, 310–19

Ransom, T. W. & Ransom, B. S. (1971) Adult male–infant relations among baboons. *Folia Primatol.*, 16, 179–95

Ransom, T. W. & Rowell, T. E. (1972) Early social development of feral baboons. In *Primate Socialization*, ed. F. E. Poirier, pp. 105–44. New York: Random House

Rhine, R. J. & Ender, P. B. (1983) Comparability of methods used in the sampling of primate behavior. *Am. J. Primatol.*, 5, 1–15

Rhine, R. J. & Flanigon, M. (1978) An empirical comparison of one/zero, focal-animal and instantaneous methods of sampling primate social behavior. *Primates*, 19, 353–61

Rhine, R. J. & Linville, A. K. (1980) Properties of one/zero scores in observational studies of primate social behavior: the effect of assumptions on empirical analyses. *Primates*, 21, 111–22

Rosenblum, L. A. (1974) Sex differences in mother–infant attachment in monkeys. In *Sex Differences in Behavior*, ed. R. C. Friedman, R. M. Richart & R. L. VandeWiele, pp. 123–41. New York: John Wiley & Sons

Rosenblum, L. A. & Kaufman, I. C. (1967) Laboratory observations of early mother–infant relations in pigtail and bonnet macaques. In *Social Communication among Primates*, ed. S. A. Altmann, pp. 33–42. Chicago: University of Chicago Press

Rowell, T. E., Din, N. A. & Omar, A. (1968) The social development of baboons in their first three months. *J. Zool.*, **155**, 461–83

Seay, B. (1966) Maternal behavior in primiparous and multiparous rhesus monkeys. *Folia Primatol.*, **4**, 146–68

Silk, J. B., Rodman, P. S. & Samuels, A. (1979) Kidnapping and spite in female infant relations of bonnet macaques. *Am. J. Phys. Anthropol.*, **50**, 481–2

Simpson, M. J. A. & Simpson, A. E. (1977) One/zero and scan methods for sampling behaviour. *Anim. Behav.*, **25**, 726–31

Small, M. F. (1982) Comparative social behavior of adult female rhesus macaques and bonnet macaques. *Z. Tierpsychol.*, **59**, 1–6

Smith, E. O. & Peffer-Smith, P. G. (1979) Adult male–immature interactions among stumptail macaques. *Am. J. Phys. Anthropol.*, **50**, 483

Stein, D. M. & Stacey, P. B. (1981) A comparison of infant–adult male relations in a one-male group with those in a multi-male group for yellow baboons. *Folia Primatol.*, **36**, 264–76

Wheeler, R. L. (1983) The influence of infant sex on mother and infant relationships with other adult females in *Macaca nemestrina*. *Am. J. Primatol.*, **4**, 358

Young, G. H. & Bramblett, C. A. (1977) Gender and environment as determinants of behaviour in infant common baboons. *Arch. Sex. Behav.*, **6**, 365–85

Young, G. H., Coelho, A. M. & Bramblett, C. A. (1982) The development of grooming, sociosexual behavior, play and aggression in captive baboons in their first two years. *Primates*, **23**, 511–19

VI.4

Juvenile/adolescent role functions in a rhesus monkey troop: an application of household analysis to non-human primate social organization

D. QUIATT

Introduction

A human household can usually be described in terms of spatial and material features as well as social organization, a fact allowing prehistorians to take an interest in the household as 'the initial and crucial point at which to study variability in the archaeological record' (Willey, 1982). For our own species, at least, the household appears to be the appropriate unit in which to observe the articulation of 'social groups . . . with ecological processes' (Wilk & Rathje, 1982).

Hominization may have involved gain in reproductive fitness via regular reduction of the inter-birth interval in conjunction with systematic provision for allomaternal care (Lovejoy, 1981). If so, the nature of the primary system of sociospatial organization that facilitated those gains must be a matter of concern to paleoanthropologists (Isaac, 1978; Quiatt & Kelso, 1983). In this instance departure from speculation can begin with consideration of energetic contributions to the group by young animals, neither wholly dependent infants nor wholly mature adults, across a range of primate species. Questions as to the nature and importance of such contributions are implicit in the observation that reduction of the birth interval must increase numbers of immature animals both absolutely and relative to numbers of breeding adults in a given generation. An approach comparing different species seems appropriate, keeping the focus on subgroupings in which mechanisms of regular association and interaction can work to promote behaviors such as alloparenting. I have begun with an examination of rhesus monkey subgroupings in which kin are behaviorally recognized, asking whether these serve functions in a way (similar to human households) that smaller or larger groupings do not, and how one might relate individual activities to household system functions. It is not important that such subgroupings be called *households*, but it is helpful to frame analytic concepts in a way that will facilitate comparisons across species.

The initial consideration is of young animals; the basic issue is what they may contribute to the group. Important preliminary questions are: How shall individuals' contributions be related to group functions?; and, At what level of inclusiveness shall the group be defined? In this first report I am most interested in problems of method and will pay particular attention to the question of definition.

Subjects and methods

Data are drawn from focal animal records of the behavior of juvenile rhesus monkeys (*Macaca mulatta*) born on Cayo Santiago, Puerto Rico in the winter and spring of 1981, compared with adolescents born in 1979 and 1980 and with mothers. Focal animals included 6 juveniles (36 h of scheduled observations), 12 adolescents (18 h), and 6 mothers of juvenile subjects (17 h). Behavior categories are presented in Table 1, along with a provisional list of household functions and interactant classes.

Videotape records were also made of individual, subgroup, and group activities. Review of videotapes and qualitative assessment of behavior should afford a more definitive list of group functions and indicate how these are served by individual actions. Here I report a first attempt at separating basic units of spatial/interactional associates that appear to operate like little households. I began in a conventional way with comparisons within and between subject groups of orientation toward family (mother and siblings) and wider matriline (mother and siblings and other matrilineal relatives) indicated by proportions of total social behavior directed by subjects toward those classes of interactant.

Results

Table 2 summarizes behavior initiated by focal animal subjects in dyadic social interactions according to category of interactant. To see how subjects of different age/sex categories might differ in

Table 1. *Categories of observation and analysis*
(a) *Behavior categories: observational units of individual action*

Lone action	Social action	
Alert/Alarm	Approach	Huddle
Alone	Bite	Lipsmack
Feed	Chase	Mount
Inspect	Chatter	Near
Investigate	Contact/Proximity	Play with
Lookout	Displace/Replace	Present
Object play	Enlist Support	Solicit
	Support	Threat
	Follow	Touch
	Grimace	Ventral clasp/Carry
	Groom	Retrieve/Restrain
	Hit	

(b) 'Household' functions served by individual action in one or more of the above categories of behavior[a]	(c) Relation to subject (actor) of recipient of social action
	Interactant categories
Food activity	Family
Predator defense	Matriline
Reproduction	Non-matriline
Thermal activity	Unidentified
Acquisition of information	
Infant care (distinguished from other social system maintenance)	
System maintenance: 1. Within household social support that involves affectional behavior only	
System maintenance: 2. Within household social support with agonistic components	
Network activity. Behavior which unites households in larger systems	
External Competition	

[a] Intuited functions of durable groupings intermediate between the single individual and the largest system of continuous long-term association, i.e. Group L itself.

behavior directed toward family, toward matriline, and toward the larger Group L, I put a number of questions:

1. Do juveniles differ by sex in orientation toward family (mother and siblings)?
2. Do juveniles differ by sex in orientation toward matriline?
3. Do adolescents differ by sex in orientation toward family?
4. Do adolescents differ by sex in orientation toward matriline?
5. Sex disregarded, do juveniles as a class differ from adolescents as a class in orientation toward family?
6. Sex disregarded, do juveniles as a class differ from adolescents as a class in orientation toward matriline?
7. Do juvenile females differ from adolescent females in orientation toward family?
8. Do juvenile females differ from adolescent females in orientation toward matriline?

Table 2. *Social action initiated by focal animal subjects. Behavior in the several social action categories listed in Table 1 is here summarized by class of recipient. Proportionate expressions are of total behavior including lone action*

	Subject as actor		
Juvenile subjects			
	Females	Males	Both sexes
Family	526 (0.667)	307 (0.529)	833 (0.608)
Matriline	87 (0.110)	84 (0.145)	171 (0.125)
Non-matriline	49 (0.062)	120 (0.207)	169 (0.123)
Unidentified	127 (0.161)	69 (0.119)	196 (0.143)
Subtotal	789 (0.756)	580 (0.669)	1369 (0.717)
Adolescent subjects			
	Females	Males	Both sexes
Family	50 (0.714)	94 (0.433)	144 (0.502)
Matriline	11 (0.157)	32 (0.147)	43 (0.150)
Non-matriline	6 (0.086)	62 (0.286)	68 (0.237)
Unidentified	3 (0.043)	29 (0.134)	32 (0.111)
Subtotal	70 (0.560)	217 (0.687)	287 (0.597)
Mothers of juvenile subjects			
	Mothers of females	Mothers of males	males and females
Family	85 (0.582)	80 (0.468)	165 (0.521)
Matriline	24 (0.164)	51 (0.298)	75 (0.237)
Non-matriline	16 (0.110)	23 (0.135)	39 (0.123)
Unidentified	21 (0.144)	17 (0.100)	38 (0.120)
Subtotal	146 (0.685)	171 (0.750)	317 (0.719)

Juvenile/adolescent role functions in rhesus

9. Do juvenile males differ from adolescent males in orientation toward family?
10. Do juvenile males differ from adolescent males in orientation toward matriline?

In each case age/sex classes were compared in terms of actions initiated toward (as appropriate): A = family members, including mother and siblings only; B = non-family, including wider matriline, non-matriline, and unidentified individuals; C = matriline; and D = non-matriline, including unidentified individuals. For each comparison, the chi-squared test was applied to data drawn from Table 2. The answer to every question but number 7 was *Yes*. Relevant comparisons and levels of significance of difference were as follows:

1. Juvenile males and females compared with regard to A and B (χ^2, $p < 0.01$)
2. Juvenile males and females compared with regard to C and D (χ^2, $p < 0.01$)
3. Adolescent males and females compared with regard to A and B (χ^2, $p < 0.01$)
4. Adolescent males and females compared with regard to C and D (χ^2, $p < 0.01$)
5. Juveniles of both sexes and adolescents of both sexes compared with regard to A and B (χ^2, $p < 0.01$)
6. Juveniles of both sexes and adolescents of both sexes compared with regard to C and D (χ^2, $p < 0.01$)
7. Juvenile and adolescent females compared with regard to A and B (χ^2, $p > 0.05$)
8. Juvenile and adolescent females compared with regard to C and D (χ^2, $p < 0.01$)
9. Juvenile and adolescent males compared with regard to A and B (χ^2, $p < 0.01$)
10. Juvenile and adolescent males compared with regard to C and D (χ^2, $p < 0.01$)

It is not surprising that behavior should appear to differ by age and sex. The one exception (Item 7) reflects a more interesting pattern that begins to emerge with a closer focus on within-sex comparisons across age classes. Juvenile males appeared to differ from adolescent males, directing more behavior toward family members than toward wider matriline relatives and non-relatives, while with adolescent males (evidently moving out of that tight family circle) the reverse obtained, although adolescent males still exhibited a preference for related over unrelated associates. Where young females were concerned, the situation differed. Both juvenile and adolescent females interacted with family and non-family associates in much the same ratio (about 2:1); differences were insignificant (Item 7). However, where comparison was of matrilineal *vs* unrelated associates, differences were significant (Item 8), perhaps because juvenile females were more likely to initiate interaction with unrelated infants and yearlings, while adolescent females were more likely to engage in solitary activities when not interacting with relatives.

Discussion

The questions posed above are something like little hypotheses about behavior. One object of doing summary analyses can be to derive such little hypotheses to verify or discard after close further study of line-item differences. That would lead naturally to the formation of subsidiary hypotheses about which group functions might be served by which individual social actions, but I hesitate to extend this line of analysis until I have satisfied myself that conventional groups of orientation and comparison (i.e. family, matriline, the larger Group L) are in fact appropriate referent groups. I have some doubts about that and will therefore turn to the issue of at what level of inclusiveness to define the referent group.

In analysis of a young animal's energetic contribution to a group of associates (for instance, mother plus siblings), as reflected in social behavior, that animal's behavior must be compared with behavior of individuals in other age classes – primarily in this instance the mother). However, difficulties arise from the fact that a reproducing female may herself have a surviving mother and siblings. In actual socio-spatial groupings, there is frequent carryover between generations. Compare the first two sets of associates in Fig. 1.

Subgroup A consists of juvenile female D16, two sisters, and mother. Group B consists of juvenile male C78, two brothers, mother, and mother's younger brother. The one is, from a household perspective, a simple family household, the other an extended family household. Rhesus monkey family groupings of course are commonly extended along the matriline. My point is that this introduces difficulties in making comparisons from the perspective of different individuals, even within a kin unit as simple and as seemingly natural as mother-plus-offspring. Nor would it help much to suggest that since rhesus family ties are extended down the matriline we forget about lesser groupings and make all comparisons within and between matrilines.

The effect of matrilineal relationships is real, but human observers' records of genetic relatives may not reveal just which relatives are going to be treated behaviorally as kin.

The matriline and the mother-plus-offspring unit are groupings which are broadly useful in record-keeping but more limited in their analytic applications. A problem in trying to use them to relate individual activities to group functions is that these static models of genetic kin relationships may not represent functioning households, which quite commonly form and reform themselves on the basis of behavioral kin recognition and at a level of inclusiveness somewhere between the parent–offspring unit and the wider cohort of matrilineal

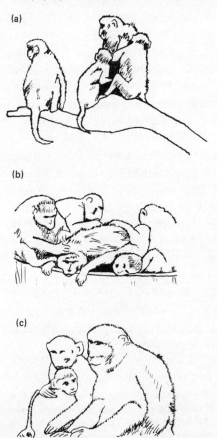

Fig. 1. Non-transient socio-spatial subgroupings. (A) Juvenile female D16 (grooming mother) with mother and siblings. (B) Juvenile male C78 (far right) with mother, siblings, and mother's brother (center rear). (C) Juvenile/adolescent female C75 with male F4 and infant.

relatives. A real such system of spatial-interactional associates may include non-relatives; it may, as in subgroup C, consist entirely of unrelated individuals, here female C75, lately arrived at first mense and consorting with male F4, hence in close company also with an orphaned infant (in ventral cling with C75) that has attached itself to F4.

Conclusions

There are certainly more than enough problems in the way of measuring energetic contributions of young animals to the group or, more generally, relating individual social actions to presumed group functions. The initial problem, as it seems to me, is simply that subgroups of primary analytic concern have much to do with but are not identical with descent groups. The weight placed on relatedness in genetic theories of behavior should not lead us to forget that preferential association is more likely to stem from familiarity than from relatedness *per se* (Holmes & Sherman, 1983; Frederickson & Sackett, 1984). Subgroups of customary associates are not easy to reconstruct with precision from data form records of dyadic interactions; but, given their likely significance as units of day-by-day (learned) adaptation to habitat and of behavioral recognition of kin, it is important that they be studied as a category in their own right – whatever the term applied to them – and not simply equated with genetic kin groupings.

Moving-image data provided by film or (with increasing practical advantage) video is clearly appropriate for such work. Video records can be designed for qualitative assessment of social action and for quantitative analysis of socio-spatial processes in systems which include several individuals, which are characterized by dynamic accommodation of individual action (and, less frequently, synchrony of behavior), hence not readily accessible to conventional dyadic description and interpretation. Videotape records were collected in conjunction with the focal animal studies which provided the primary body of data for this report. The discussion has been based in part on review of those records but has been necessarily provisional, with final conclusions awaiting more complete analysis.

Acknowledgments

Lineages for Group L were prepared by Matt J. Kessler, DVM. Work was supported in part by NIH grant RR01293 to the Caribbean Primate Research Center, University of Puerto Rico School of Medicine, San Juan, P. R., and by a Grant-in-Aid from The University of Colorado Graduate School.

References

Frederickson, W. T. & Sackett, G. P. (1984) Kin preferences in primates (*Macaca nemestrina*): relatedness or familiarity? *J. Comp. Psychol.*, **98**, 29–34

Holmes, W. G. & Sherman, P. W. (1983) Kin recognition in animals. *Am. Scientist*, **71**(1), 46–55

Isaac, G. (1978) Food sharing and human evolution: archaeological evidence from the Plio-Pleistocene of East Africa. *J. Anthropol. Res.*, **34**, 311–25

Lovejoy, C. O. (1981) The origin of man. *Science*, **211**, 341–50

Quiatt, D. & Kelso, J. (1983) Household economics in the Pliocene. *Anthropos (Athens)*, 126–42 (Proceedings of the 3rd European and 1st Panhellenic Anthropological Congresses)

Wilk, R. R. & Rathje, W. L. (1982) Household archaeology. *Am. Behav. Scientist*, **25**, 617–39

Willey, G. (1982) Foreword to special issue of *ABS* devoted to Archaeology of the household. *Am. Behav. Scientist*, **25**, 613–16

VI.5

Conflict interference and the development of dominance relationships in immature *Macaca fascicularis*

W. J. NETTO AND J. A. R. A. M. VAN HOOFF

Introduction

A matrilineal hierarchy in which offspring, particularly females, acquire ranks adjacent and usually below that of their mother, has been found in a number of cercopithecines. Two main types of rank acquisition have been proposed. First, supportive interventions in conflicts by the mother (e.g. Kawai, 1958; Sade, 1967; de Waal, 1977; Cheney, 1977; Watanabe, 1979) or other close relatives (e.g. Kaplan, 1977, 1978; Kurland, 1977; Massey, 1977; Berman, 1980), both of which can offer effective help against opponents of lower family rank. Secondly, other and ontogenetically early influences that are dependent on family rank may determine later social interactions. Identification with and imitation of the mother (e.g. Kawai, 1965; Sade, 1978; Altmann, 1980) or treatment by the group members which is dependent on the rank of the mother (Gouzoules, 1975; Walters, 1980; Datta, 1983; Horrocks & Hunte, 1983) may determine the offspring's (later) social interactions.

The purpose of our study is to investigate the pattern of interventions in conflicts in order to establish to what extent it is compatible with the acquisition and maintenance of dominance–subordinance relationships. By means of triadic analysis of interventions we hope to answer the following questions:

Can the pattern of interventions explain family rank conservatism?

Does it offer opportunities enabling an offspring to escape from the family rank constraints?

Does support by relatives differ from that by non-relatives (better adjusted to the need of the juvenile; greater involvement in strategically risky conflicts, i.e. more 'altruistic'; cf. Watanabe, 1979)?

Study group and methods

Social interactions of 16 immature offspring in a well-established captive group of *Macaca fascicularis* were recorded by means of focal animal (10 h/individual) and *ad libitum* behaviour-dependent sampling (170 h) methods in 1982.

The group consisted of 2 adult males, one of which was a clear α-male, 8 subadult males, 13 adult females, 16 juveniles between 6 and 24 months old and 3 infants less than 4 months old which were not included in the analysis. Altogether 10 mothers, 4 with 1 and 6 with 2 offspring were treated in the analysis. The animals lived in a spacious outer enclosure of 220 m², in which the observations were conducted, and an inner compound of 50 m².

Analysis and non-parametric statistics have been done at individual levels. For the sake of clarity the data in the figures have been combined into totals for all individuals of a given class.

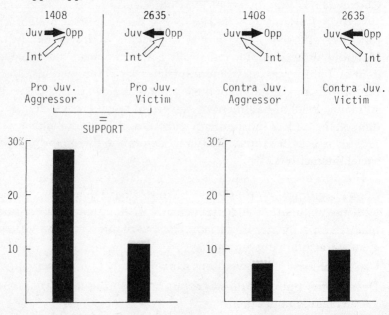

Fig. 1. Types of intervention. Black arrow = direction of aggression; white arrow = direction of aggressive intervention. Juv = juvenile; Opp = opponent; Int = intervener.

Results
Types of intervention

Four types of fight intervention were distinguished (Fig. 1). A juvenile may be involved in a conflict either as an initiative aggressor ($n = 1408$) or as a victim of aggression ($n = 2635$). The intervener (mother, other relatives or non-relatives) may make a support choice PRO the juvenile (in 27.8% of the 1408 conflicts in which the juvenile was the aggressor – 'aggressor support' – and in 10.6% of the 2635 conflicts when it was a victim – 'victim support'). Or the intervener may choose CONTRA the juvenile (in 6.5% of juvenile aggressor cases and in 9.5% of juvenile victim cases). The juveniles were mostly supported rather than intervened against (Wilcoxon test, two-tailed: $p < 0.01$). Moreover, they obtained relatively more support when they acted as aggressors than when they were the victims ($p \ll 0.01$).

Rank dependency of support choices

In the case of choices PRO, the question is how supporters choose in regard to the rank relationship between the opponent and the juvenile (i.e. its 'family rank', not its actual rank) and in regard of the relationships of the supporter with both of these. Fig. 2 presents the results for those interventions in which the juvenile and its opponent were of different family.

Both supporters ranking lower than the family of the juvenile (LF supporters) and supporting relatives (by definition of equal family rank) ally with a juvenile almost exclusively against opponents who are both of lower family rank than the juvenile (LF opponents) and of lower rank than themselves! The juvenile is more likely to receive support, moreover, when it is the aggressive party (Wilcoxon: $p < 0.01$). Thus, not only LF non-family, but also family engage predominantly in support with minimal strategic risk, and they do not differ significantly in this respect. This support pattern clearly reinforces existing family rank relationships.

Remarkably, supporters of higher family rank than the juvenile offer a considerable amount of support against HF opponents (*N.B.* again the opponents are almost always of lower rank than themselves). Only this type of support might offer opportunities to the child to break through the family rank constraints. Since such breakthroughs are rare, one wonders why this large amount of support against HF opponents (c. 30% of all support) is not effective as a rank tilt lever. To answer this we need a closer look at the distribution of the support against these HF opponents.

294 W. J. Netto and J. A. R. A. M. van Hooff

Support against HF opponents by HF supporters

There were 35 interactions in which a juvenile, when it acted as aggressor against HF opponents, received support of one or more HF supporters, ranking higher than the opponent (total $n = 48$). We analyzed, per interaction, the highest-ranking supporters, because their support was considered, in general, to be the most effective. Of these support actions, 13 were executed by the α-male, 15 by adult females ($n = 13$), 3 by some of the subadult males ($n = 9$) and 4 by the other juveniles ($n = 15$). In other words, the α-males excelled in support against HF opponents. This was even more outspoken when we looked at victim support: out of 53 cases, 32 were executed by the α-male, whereas the remaining 37 group members accounted for only 21 cases (females 13, subadult males 3, other juveniles 5). Of those 32

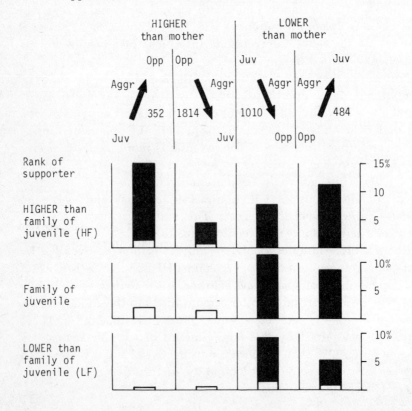

Fig. 2. Interventions in conflicts of juveniles with non-relative opponents. Black bars = intervener of *higher* rank than juvenile's opponent; white bars = intervener of *lower* rank than juvenile's opponent.

cases of victim support by the α-male, 18 were in support of the three lowest ranking of the 16 juveniles! Qualitative evidence suggests that the victim support by the α-male has as its motive the provision of help to the juveniles in distress.

We doubt whether most of the other interferers have this motive as well. If we look at the HF opponents against which the support was given, we see, for instance, that the juveniles received 21 out of the 35 cases of aggressor support against one of the 9 subadult males, whereas the 13 adult females accounted for only 1 case and the other 15 juveniles for 13 cases. The difference between the high percentage of support opportunities utilised against subadult males as compared with the low percentage utilised against adult females (Mann–W: $p < 0.02$) and other juveniles (Mann–W: $p < 0.01$) is remarkable. Although our data are too few to be conclusive on this, it seems that the adult females were particularly involved in the support against subadult males. Because these support actions sometimes formed part of massive alliances against the subadult males, we hypothesized that this support was part of a more wide-ranging conflict with the subadult males. These behaved in a pestering and provocative manner and gradually rose above the adult females. The aggression of a juvenile towards a subadult male may therefore act as an opportunity to start an alliance against him. The motive of these support actions may therefore not be to help the juvenile.

We must conclude then that support against HF opponents cannot be very effective as a means for the juvenile to break the constraints of the matrilineal hierarchy. Against HF females this support is received almost solely when the juvenile is the victim (30 times out of 354 opportunities) rather than the aggressor (1 time out of only 9 opportunities). This only offers relief from distress, therefore, and is hardly a reinforcer of rank improvement strategies. By contrast, supported aggressive initiatives which could act as reinforcers are directed not against HF adult females, but primarily against HF subadult males. But subadult male long-tailed macaques tend to emigrate in the wild (van Noordwijk and van Schaik, 1985), and, in our group, become dominant over the females and peripheralise somewhat; so this is not effective either.

Support of the mother

The offspring received more support from the mother than from other relatives (Wilcoxon: $p < 0.01$). The mother used more occasions to support the offspring than did the average non-relative

($p \ll 0.001$). This applies particularly to support against LF opponents: whereas mothers (Fig. 3) utilized 10.0% of the opportunities (all agonistic interactions in which immature offspring was victim or start aggressor) against adult females, 15.0% against subadult males and 3.5% against other juveniles, non-relatives on the average utilized c. 1.3%, 1.9% and 0.6%, respectively, of these opportunities (all differences: $p < 0.01$). However, with respect to HF opponents the figures are: for the mother, respectively, 0.5%, 2.4% and 1.1%, and for the average non-relative, respectively, 0.3%, 1.5% and 0.1%. These are not significantly different.

In other words: although mothers do invest considerably more energy in supporting offspring than do non-relatives, they *do not* invest comparatively more in strategically risky support!

Fig. 3. Percentage (average per juvenile) of support opportunities utilized by the mother (white bars) and by non-relatives (black bars) against different opponent classes.

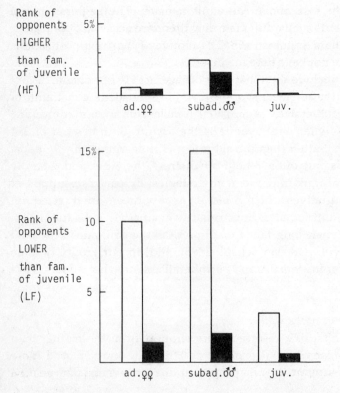

The ratio of interventions PRO and CONTRA

The foregoing concerned support, i.e. the interventions PRO the juvenile. But as a reinforcement for its social involvements, the interventions that it experiences CONTRA are equally important. So far, this aspect has received little attention in the literature. No doubt it is the ratio of choices PRO and CONTRA the juvenile, which encourages or discourages it and its opponents. Fig. 4 shows that a juvenile experienced almost no interventions CONTRA in conflicts with LF opponents, irrespective of whether it was the aggressor or the victim and irrespective of the sex/age class of the opponent.

In conflicts with HF opponents there was comparatively little intervention if the juvenile was the victim. In so far as interventions did occur, the juvenile tended to be supported rather than opposed if HF males victimised it ($p < 0.02$). However, if victimised by HF juveniles it tended to receive interventions CONTRA ($p < 0.02$). If the juvenile aggressed against a HF opponent, then the balance tended to be negative if the opponent was a female (values too low for testing). The balance tended to be positive if the opponent was a male; this reflects the 'battle' against the subadult males again!

Except for the last case, therefore (but see above), the intervention ratios consolidate the existing family rank system.

It is important to note that relatives and non-relatives did differ strongly with respect to this PRO/CONTRA ratio. In conflicts of a

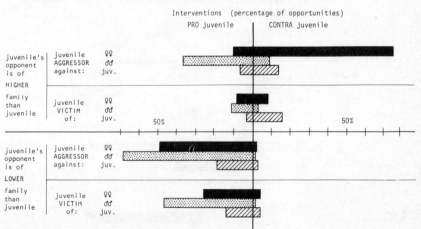

Fig. 4. Interventions in conflicts of juveniles with different opponent classes: hierarchical relationships of the different parties.

juvenile with non-family (!), relatives almost never allied against the juvenile.

The juvenile as an ally

The interventions in conflicts of the juvenile will discourage aggressive initiatives that disregard the matrilineal rank system, especially against HF adult females. The only other possibility that juveniles might have to break through this constraint is when they themselves ally against HF adult females that are the victims in a conflict. Remarkably, however, Fig. 5 shows that choices CONTRA HF victims were comparatively rare. So this possibility is not really utilized, presumably because the risks of redirected counter-attack are prohibitive. For the rest, juveniles intervened opportunistically, taking the side of the involvee which was the aggressor rather than the victim ($p \ll 0.001$). They more often chose PRO the HF involvee and CONTRA the LF involvee if this was a female ($p < 0.05$) or another juvenile ($p < 0.001$), but not if this was a sub-adult male!

Conclusion

Interventions are mainly in the strategically least risky directions, even in the case of the mother, and clearly consolidate and perpetuate the existing family rank relationships, especially with respect to the stable female nucleus of the group (cf. Datta, 1983). Although interventions against HF opponents are not uncommon, it is

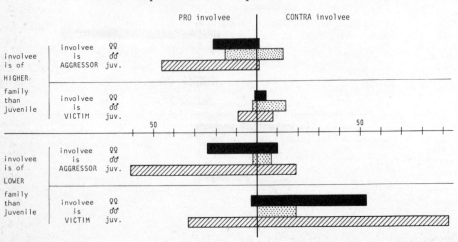

Fig. 5. Interventions by juveniles themselves in conflicts: hierarchical relationships of the different parties.

understandable why these cannot be really effective as a basis for strategies enabling a (female) juvenile to rise above the rank level of its family. Many of these concern, often concerted, actions against the subadult males (cf. Colvin, 1983), who tend to peripheralise. Consequently juveniles dare to take aggressive initiatives against these. By contrast, aggressive initiatives against HF females are rare; these must have been discouraged already very early in life.

The dominant male plays an exceptional role as a helper of infants victimised by HF opponents. In doing so he may support his only (close) relatives in the group. This is in agreement to what has been found in the Japanese macaque (Watanabe, 1979) but in contrast to what has been found in the yellow baboon (Walters, 1980).

Acknowledgements

We thank Jeroen Camping and Hans Meijs for their contribution to the data collection and analysis. We thank Han de Vries for writing the computer programs for data collection and analysis, Anita van Vliet for typing the manuscript, and Mr H. Rypkema for the preparation of the figures.

References

Altmann, J. (1980) *Baboon Mothers and Infants*. Cambridge, Mass.: Harvard University Press

Berman, C. M. (1980) Early agonistic experience and rank acquisition among free-ranging infant rhesus monkeys. *Int. J. Primatol.*, **1**, 153–69

Cheney, D. L. (1977) The acquisition of rank and the development of reciprocal alliances among free-ranging immature baboons. *Behav. Ecol. Sociobiol.*, **2**, 303–18

Colvin, J. (1983) Influences of the social situation on male emigration. In *Primate Social Relationships*, ed. R. A. Hinde, pp. 160–71. Oxford: Blackwell

Datta, S. B. (1983) Relative power and the acquisition of rank. Relative power and the maintenance of dominance. In *Primate Social Relationships*, ed. R. A. Hinde, pp. 93–112. Oxford: Blackwell

Gouzoules, H. (1975) Maternal rank and early social interactions of infant stumptail macaques, *Macaca arctoides*. *Primates*, **16**, 405–18

Horrocks, J. & Hunte, W. (1983) Maternal rank and offspring rank in vervet monkeys: an appraisal of the mechanisms of rank acquisition. *Anim. Behav.*, **31**, 772–82

Kaplan, J. R. (1977) Patterns of fight interference in free-ranging rhesus monkeys. *Am. J. Phys. Anthropol.*, **47**, 279–88

Kaplan, J. R. (1978) Fight interference and altruism in rhesus monkeys. *Am. J. Phys. Anthropol.*, **49**, 241–9

Kawai, M. (1958) On the system of social ranks in a natural troop of Japanese monkeys I, II. *Primates*, **1**, 111–48

Kawai, M. (1965) On the system of social ranks in a natural troop of Japanese monkeys. In *Japanese Monkeys*, ed. S. A. Altmann, pp. 66–86. Chicago: University of Chicago Press

Kurland, J. A. (1977) *Kin Selection in the Japanese Monkey*. (*Contrib. Primatol.*, **12**.) Basel & New York: Karger

Massey, A. (1977) Agonistic aids and kinship in a group of pigtail macaques. *Behav. Ecol. Sociobiol.*, **2**, 31-40

Noordwijk, M. A. van & Schaik, C. P. van (1985) Male migration and rank acquisition in wild long-tailed macaques (*Macaca fascicularis*). *Anim. Behav.*, **33**, 849–61

Sade, D. S. (1967) Determinants of dominance in a group of free-ranging rhesus monkeys. In *Social Communication among Primates*, ed. S. A. Altmann, pp. 99–114. Chicago: University of Chicago Press

Sade, D. S. (1978) Population biology of free-ranging rhesus monkeys on Cayo Santiago, Puerto Rico. In *Biosocial Mechanisms of Population Regulation*, ed. M. N. Cohen, R. S. Malpass & H. G. Klein, pp. 171–87. New Haven: Yale University Press

Waal, F. B. M. de (1977) The organisation of agonistic relations within two captive troops of Java monkeys (*Macaca fascicularis*). *Z. Tierpsychol.*, **44**, 225–82

Walters, J. (1980) Interventions and the development of dominance relationships in female baboons. *Folia Primatol.*, **34**, 61–89

Watanabe, K. (1979) Alliance formation in a free-ranging troop of Japanese macaques. *Primates*, **20**, 459–747.

VI.6

Dominance rank and related interactions in a captive group of female pigtail macaques

P. MESSERI AND C. GIACOMA

Introduction

The concept of dominance, although widely debated, has recently proved useful in describing primate societies (Bernstein, 1981). Its utility, however, depends on the ability to define and measure it. If dominance is defined as an intervening variable which accounts for imbalances in rates of behaviours within dyads (Hinde & Datta, 1981), then directionality is measured first and the role played by rank in non-directional behaviours largely remains to be investigated.

Moreover, relatively less attention has been paid to female dominance, though females of all species of macaques studied to date form linear hierarchies. In particular, female ranks of pigtail macaques are stable (Bernstein, 1969) and independent of other members of the group (Giacoma & Messeri, 1980; Gouzoules, 1980).

For the above reasons traditional measures of dominance were recorded in a captive group of female pigtail macaques and the role played by dominance both in directional and in non-directional interactions was analyzed.

Methods

The group under study was composed of six unrelated female pigtail macaques (*Macaca nemestrina*) housed together from the age of 1 year in an outdoor enclosure of 300 m^2. In 1979 the macaques were moved into a cage 14 m × 14 m × 3 m high. Observations began in the Spring of 1979, when the monkeys were 4 years old, and continued until 1984. In the Summer of 1979 an adult male joined the group for a period of 6 months and infants were born. The infants were withdrawn when 2.5 years old and reintroduced into the group at 4 years. The h of observation were divided, according to the composition of the group, into seven periods.

The behaviours considered were: displacements, attack, threat, present (received), mount, and grooming, defined according to Kaufman & Rosenblum (1966). Because of the good observational conditions and the small size of the group, we were able to keep all the animals under continuous observation and to record all occurrences of these behaviours. This 'all occurrences of some behaviours' sampling method (Altmann, 1974) provides good information about rates of occurrences, and, on its basis, sociometric matrices for each behaviour in each period were obtained. Concordances were tested by Kendall's W concordance coefficient test or by Spearman's rank correlation coefficient (Siegel, 1956).

Results and discussion

A total of 1106 displacements, 277 attacks, 419 threats, 259 presents, 166 mounts and 947 groomings were recorded.

Dominance matrices were constructed so as to minimize the number of reversals in sociometric matrices, but behaviours which occur relatively infrequently sometimes fail to rank all members in a significantly linear hierarchy (Giacoma & Messeri, unpublished). However, ranks for each behaviours, both linear and not, were stable in the seven periods. Kendall's Ws were $W = 0.992$ ($P < 0.001$) for displacement, $W = 0.848$ ($P < 0.001$) for attack, $W = 0.818$ ($P < 0.001$) for threat, $W = 0.501$ ($P < 0.01$) for present, $W = 0.523$ ($P < 0.01$) for mount, and $W = 0.407$ ($P \leq 0.05$) for grooming, indicating that the degree of stability depends not only on frequency, but also on the behaviour considered.

In view of this stability over time, the matrices of the seven periods were summed and then the linearity of the matrices was estimated. All behaviours, except for grooming, were significantly linear (displacement: $\chi^2 = 509.0$, $P < 0.001$; attack: $\chi^2 = 76.2$, $P < 0.001$; threat: $\chi^2 = 116.0$, $P < 0.001$; present: $\chi^2 = 13.4$, $P < 0.001$; mount: $\chi^2 = 8.3$, $P < 0.01$; grooming: $\chi^2 = 0.16$, N.S.).

Mean ranks of the five linear behaviours were calculated for the seven observation periods (Table 1). These mean ranks were preferred to a new overall ranking because, although not nominal numbers, they provided more information than an ordinal series (Bolles, 1981). A highly concordant rank was shown by the more agonistic behaviours, displacement, attack, and threat ($W = 0.872$, $P < 0.001$), which then gave the best measures of dominance. The ranking obtained for present and mount behaviours differed slightly from the mean ranking of the above three behaviours. Two inversions were seen in the case of present (C/D and D/F) and one (C/D) in the case of mount. However, they were in the same direction and significantly correlated with the mean ranks of the three more agonistic behaviours ($r_s = 0.886$ and $r_s = 0.882$, $P < 0.025$, one-tailed for present and mount, respectively). The mean ranks of these five linear and concordant behaviours are taken as the overall dominance ranks. Grooming ordered animals in a ranking totally unrelated to the overall dominance ranking ($r_s = 0.408$, N.S.).

To analyse directionality, we assigned an ordinal rank to each animal according to its overall dominance rank. The mean number of interactions was plotted against the hierarchical distance (rank of the recipient minus the rank of the actor; Fig. 1) for estimating the degree of symmetry of each behaviour. A decrease in linearity, from displacements through grooming, is evident in the increasing number of reversals (dark columns on the right of Fig. 1). Reversals are more frequent between adjacently ranked animals, and become less frequent as the hierarchical distance between subjects increases. The same is also true for linear interactions (barred columns on the left of Fig. 1), which suggests that the control of social conflicts between distantly ranked subjects is more a question of avoidance than of dominance/subordination. This is also suggested by the relatively higher rate of displacement between far-ranking subjects, since displacement is a behavioural pattern for avoiding physical conflict.

Table 1. *Mean individual ranks over the seven periods*

	Individuals					
	A	B	C	D	E	F
Displace	1	2	3	4.1	4.9	6
Attack	1.2	2.2	2.7	4.2	5	5.6
Threat	1.4	2	2.9	4	5.4	5.4
Present	1.4	2.6	4.2	3.2	4.9	4.6
Mount	1.8	2.2	4.3	3	4.8	4.9
means	1.4	2.2	3.4	3.7	5.0	5.3
Grooming	3.9	1.9	3.8	5.4	2.5	3.6

Conclusions

All the behavioural patterns used in this study, except grooming, ranked all the members of the group in linear and concordant hierarchies. They can then be considered independent measures of dominance as an intervening variable (Richards, 1974).

In agonistic encounters dominance generally resulted in linearity, but linking social dominance and conflict seems too restrictive. In fact, dominance appeared to regulate in some way all the behaviours considered. Several behavioural patterns can be used to avoid or resolve physical conflicts and social dominance can be exercised in

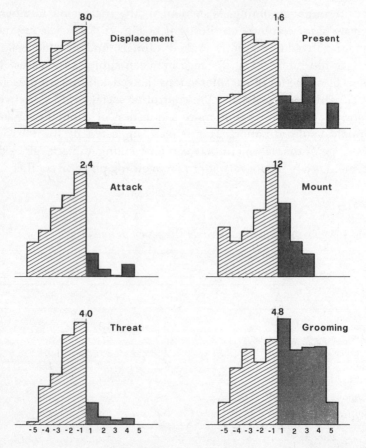

Fig. 1. Mean numbers of interactions plotted against the hierarchical distance, as expressed by the rank of the receiver minus the rank of the actor (vice versa for present). Reversals are shown by the dark columns on the right side of each histogram, while barred columns on the left indicate directional interactions.

different ways. It could result in avoidance (displacement), aggression (attack and threat), appeasement (present and mount) or prevention (grooming). In conclusion, dominance seems better expressed as the ability to control the outcome of interactions at a point of social conflict, no matter what tactic is used (Hand, unpublished).

References

Altmann, J. (1974) Observational study of behavior: sampling methods. *Behaviour*, **49**, 227–67

Bernstein, I. S. (1969) Stability of the status hierarchy in a pigtail monkey group (*Macaca nemestrina*). *Anim. Behav.*, **17**, 452–8

Bernstein, I. S. (1981) Dominance: the baby and the bathwater. *Behav. Brain Sci.*, **4**, 419–29, 449–57

Bolles, R. C. (1981) A parallel to dominance competition. *Behav. Brain Sci.*, **4**, 433-4

Giacoma, C. & Messeri, P. (1980) Social relationships among female pigtail macaques. *Minitore zool. Ital. (N.S.)*, **14**, 107–8

Gouzoules, H. (1980) The alpha female: observations on captive pigtail monkeys. *Folia Primatol.*, **33**, 46–56

Hinde, R. A. & Datta, S. (1981) Dominance: an intervening variable. *Behav. Brain Sci.*, **4**, 442

Kaufman, I. C. & Rosenblum, L. A. (1966) A behavioural taxonomy for *Macaca nemestrina* and *Macaca radiata*: based on longitudinal observation of family groups in the laboratory. *Primates*, **7**, 205–58

Richards, S. M. (1974) The concept of dominance and methods of assessment. *Anim. Behav.*, **22**, 914–30

Siegel, S. (1956) *Nonparametric Statistics for the Behavioral Sciences*. New York: McGraw-Hill

VI.7

A comparative study of aggression and response to aggression in three species of macaque

B. THIERRY

Introduction

The multiple variables of social systems are linked by relations. A change in one variable produces changes in others as a consequence, and relations within systems thus act as constraints in the evolution of species (see von Bertalanffy, 1968; Kummer, 1971). For example, there is growing evidence that maternal protectiveness and development of alloparental care are related to the dominance system particular to the species: the more hierarchically rigid the society, the more exclusive the infant–mother bond (Kaufman & Rosenblum, 1969; McKenna, 1979; Thierry, 1985a).

The aim of the present study is to reveal some systemic relations in the social systems of macaques. The chosen level of analysis is social interactions (cf. Hinde, 1983) and, more specifically, agonistic interactions. Four variables were studied: intensity of aggression; symmetry in aggression; behaviours following aggression; and re-establishment of social relationships after aggression. The study is based on a comparison of the behaviour of three species of macaque living in similar environments: rhesus (*Macaca mulatta*), Java (*M. fascicularis*) and Tonkean macaques (*M. tonkeana*).

Methods

Each species was represented by one group living in conditions of semi-liberty in a wooded area of approximately 0.5 ha. The groups were composed of 12–20 individuals representing various age- and sex-classes. Further information on composition, age and kinship in the groups is available elsewhere (Thierry, 1985b). In this study, 'juveniles' are defined as individuals between 1 and 3 years of age, and 'adults' are individuals over 3 years of age, thus comprising both subadults and true adults.

Data were collected using behaviour-dependent sampling, all occurrences of agonistic interactions being recorded. The behavioural units were based on ethograms from several authors for rhesus (Altmann, 1962; Lindburg, 1971; Rowell & Hinde, 1962), Java (Angst, 1974; de Waal, van Hooff & Netto, 1976) and Tonkean macaques (Thierry, 1984). An agonistic interaction was defined as the display of aggressive behaviour by an individual and an aggressive or non-aggressive response by the aggressee. When more than two individuals became simultaneously involved, the record was broken down into dyadic interactions, and only the two first agonists were taken into account. Only interactions between unrelated individuals are considered here. This resulted in 1049 interactions being analyzed for rhesus, 429 for Java and 522 for Tonkean macaques.

For the study of the re-establishment of social relationships following aggression, a special method was carried out. Agonists were followed during the 20 min following aggression and any contacts between them were recorded. Using this procedure, 234, 145 and 152 cases were observed in rhesus, Java and Tonkean macaques, respectively.

Summed frequencies of interactions were established for each class of dyads. Statistical comparisons between observed and expected frequencies used the chi-squared test (Yates correction).

Results and discussion
First variable: intensity of aggression

Intensity of aggression was measured by the percentage of interactions involving biting (Table 1). The low rate of biting did not permit statistical comparisons, but in rhesus macaques, the observed frequencies of biting were higher than the expected frequencies for every class of dyads; in contrast, in Tonkean macaques, biting was very rare: only one bite was recorded in this species in over 400 agonistic interactions; intermediate results were observed in Java macaques.

Second variable: symmetry in aggression

Aggression may be either unidirectional, when only one of two subjects displays aggressive patterns, or bidirectional (symmetrical), when both subjects display reciprocally (Table 1). Comparisons

Table 1. *Variables of aggression*

		Intensity of aggression (percentage of agonistic interactions with bites)			Symmetry in aggression (percentage of bidirectional agonistic interactions)				Re-establishment of the social relationships (percentage of occurrence of non-agonistic contacts between previous agonists within the 20 min following aggression)			
		rhesus	Java	Tonkean	rhesus	Java	Tonkean	χ^2	rhesus	Java	Tonkean	χ^2
adult male/adult male	OP	5.9	0	0	0	14.9	75.0	14.6 ***	20.0	11.8	62.5	—
	OF	2	0	0	0	13	12		3	2	10	
	EF	1.05	0.43	0.52	13.11	5.36	6.53		6.61	4.10	4.20	
adult male/adult female	OP	4.1	1.1	0.7	4.1	23.4	68.4	71.4 ***	5.3	52.2	72.5	24.0 ***
	OF	8	1	1	22	93	93		2	12	37	
	EF	5.25	2.15	2.62	64.51	26.40	32.10		22.47	13.92	15.00	
adult male/juvenile	OP	6.6	3.3	0	0	27.9	51.6	33.1 ***	8.3	52.2	40.0	6.3 *
	OF	6	2	0	0	17	33		1	12	6	
	EF	4.20	1.72	2.09	26.22	10.73	13.05		8.37	5.19	5.44	
adult female/adult female	OP	3.9	2.0	0	5.4	7.8	59.6	63.5 ***	17.2	66.7	77.5	6.4 *
	OF	15	1	0	21	4	93		16	12	32	
	EF	8.40	3.43	4.18	61.89	25.31	30.80		26.45	16.38	17.17	
adult female/juvenile	OP	2.9	2.3	0	25.6	7.0	50.0	18.9 ***	23.9	13.3	88.0	5.1 *
	OF	9	3	0	80	9	67		10	6	22	
	EF	6.29	2.57	3.13	81.82	33.46	40.72		16.75	10.38	10.88	
juvenile/juvenile	OP	2.1	0	0	3.5	75.0	75.0	8.0 *	25.0	50.0	80.0	—
	OF	6	0	0	1	6	12		1	1	4	
	EF	3.15	1.29	1.57	9.97	4.08	4.96		3.53	2.18	2.29	

OP: observed percentage; OF: observed frequency; EF: expected frequency; Comparisons between observed and expected frequencies used χ^2 test, 2 d.f. (*$P < 0.05$; ***$P < 0.001$)

between groups showed that agonistic interactions were very asymmetric in rhesus macaques; they were slightly less so in Java macaques, while in Tonkean macaques interactions were much more often bidirectional.

An inverse association between intensity of aggression and symmetry in aggression was thus found comparing the three species. A possible explanation is that individuals are more ready to compete as the risk is reduced. In rhesus macaques, where the chances of being bitten are relatively high, the best tactic for the weaker individual is to escape from the aggression. In Tonkean macaques, in contrast, biting is exceptional and the aggressee may therefore counter the aggression.

Third variable: behaviours following aggression

In order to differentiate between behaviour of aggressees and aggressors immediately following aggression, only unidirectional episodes were analyzed (Fig. 1). Rhesus and Java macaques showed similar patterns in the asymmetry of behaviours exhibited by the agonists: the aggressee most often fled and/or submissively vocalized (scream, squeak or gecker). However, mild submissive behaviours (bared teeth, teeth-chattering) were much more frequent in Java macaques. In Tonkean macaques the behaviours displayed by the aggressee were often displayed by the aggressor too: bared teeth, lipsmacking and clasping, which may be called appeasement behaviours in this species (the frequency of lipsmacking is especially noteworthy). These results indicate that behaviours ending aggression are more likely to be initiated by either agonist in the species where the aggression is less intense.

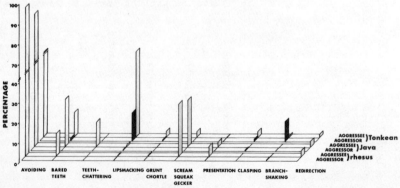

Fig. 1. Behaviours following aggression. Percentages represent means for the six classes of dyads in each species.

In addition, it may be noted that branch-shaking was more frequent in Tonkean macaques, and that redirection of aggression, although of comparable rates in the three species, varied in their focus: in Tonkean macaques, redirection occurred most often towards other groups in neighbouring enclosures while, in rhesus and Java macaques, it occurred towards other individuals in the group (or towards the observer), which resulted in aggression being continued and extended in the group.

Fourth variable: re-establishment of social relationships after aggression

This was measured by the frequency of occurrence of non-agonistic contacts between previous agonists within the 20 min following aggression (Table 1), what has been called 'reconciliation' by de Waal and co-workers (de Waal & van Roosmalen, 1979; de Waal & Yoshihara, 1983). The comparisons showed marked differences between species: contacts after aggression were always less frequent than expected in rhesus macaques, and more frequent in Tonkean macaques; Java macaques showed intermediate scores.

Further studies are needed on this variable: contacts between individuals are generally more frequent in Tonkean macaques than in the other two species (Thierry, unpublished) and the present results might not be specific to aggression. Although it remains to be seen whether and how reconciliation takes place, it can be stated that social relationships are more quickly re-established in Tonkean macaques. This is associated with less intense aggression and frequent initiations of appeasement behaviours in this species.

Conclusion

It should be noted first that only one group of each species was observed in this study. Differences might therefore be due to individual variations, and differences in group composition. The present data require confirmation through the study of other groups. However, comparisons based on interactions between related individuals show the same trends as reported here (Thierry, 1985b) and the variables used in this study appear quite robust with regard to the demographic structure of the groups. In addition, the results are consistent with previous knowledge concerning rhesus (Maxim, 1976; Teas et al., 1982; Bernstein, Williams & Ramsay, 1983; de Waal & Yoshihara, 1983) and Java macaques (de Waal et al., 1976; de Waal, 1977; Zumpe & Michael, 1983). With respect to Tonkean macaques, I have been able to observe a second group of this species (Thierry, 1984) and the patterns were the same as those described here. Interestingly, Bernstein and co-workers (Bernstein, 1971; Bernstein *et al.*, 1983) have also described frequent lipsmacking and a low rate of biting in another species of macaque from Celebes, namely *Macaca nigra*.

Relations have been revealed between intensity and symmetry of aggression, and development of behaviours controlling aggression. It is not yet possible to assume that these variables are causally linked, but they were observed to vary together. The balances between these variables are possibly species-specific, they could be the outcome of evolutionarily stable strategies. It is necessary to consider other variables of the social system in order to describe more completely species-typical strategies; for example, might the degree of symmetry in interactions be correlated with degree of symmetry in relationships? Although strategies are in part an adaptation to particular ecological pressures, it is too early at present to expound the link with the environment (the ecology of Tonkean macaques, for instance, is unknown). However, already existing models such as those developed by Vehrencamp (1983) could be applied to assess the consequences of symmetrical (or egalitarian) interactions and/or relationships.

Acknowledgements

Participation in the congress was supported by grants from the Association for the Study of Animal Behaviour and the French Ministère des Relations Extérieures.

References

Altmann, S. A. (1962) A field study of the sociobiology of rhesus monkeys (*Macaca mulatta*). *Ann. N.Y. Acad. Sci.*, **102**, 338–45

Angst, W. (1974) *Das Ausdruckverhalten des Javaneraffen* Macaca fascicularis Raffles 1821. Berlin: Paul Parey

Bernstein, I. S. (1971) Activity profiles of primate groups. In *Behavior of Non-human Primates*, Vol. 3, ed. A. M. Schrier & F. Stollnitz, pp. 69–106. New York: Academic Press

Bernstein, I., Williams, L. & Ramsay, M. (1983) The expression of aggression in Old World monkeys. *Int. J. Primatol.*, **14**, 113–25

Bertalanffy, L. von (1968) *General Systems Theory*. New York: Braziller

Hinde, R. A. (ed.) (1983) *Primate Social Relationships: An Integrated Approach*. Oxford: Blackwell

Kaufman, I. C. & Rosenblum, L. A. (1969) Effects of separation from mother on the emotional behavior of infant monkeys. *Ann. N.Y. Acad. Sci.*, **159**, 681–95

Kummer, H. (1971) *Primate Societies*. Chicago: Aldine

Lindburg, D. G. (1971) The rhesus monkey in North India: an ecological and behavioral study. In *Primate Behavior*, Vol. 2, ed. L. A. Rosenblum, pp. 1–106. New York: Academic Press

Maxim, P. (1976) An interval scale for studying and quantifying social relations in pairs of rhesus monkeys. *J. Exp. Psychol.*, **105**, 123–47

McKenna, J. J. (1979) The evolution of allomothering behavior among Colobine monkeys: function and opportunism in evolution. *Am. Anthropol.*, **81**, 818–40

Rowell, T. E. & Hinde, R. A. (1962) Vocal communication by the rhesus monkeys (*Macaca mulatta*). *Proc. Zool. Soc. London*, **138**, 279–94

Teas, J., Feldman, H. A., Richie, T. L., Taylor, H. G. & Southwick, C. H. (1982) Aggressive behavior in the free-ranging rhesus monkeys of Kathmandu, Nepal. *Aggress. Behav.*, **8**, 63–77

Thierry, B. (1984) Clasping behaviour in *Macaca tonkeana*. *Behaviour*, **89**, 1–28

Thierry, B. (1985a) Social development in three species of macaque (*Macaca mulatta, M. fascicularis, M. tonkeana*). a preliminary report on the first ten weeks of life. *Behav. Proc.* **11**, 89–85

Thierry, B. (1985b) Patterns of agonistic interactions in three species of macaque (*Macaca mulatta, M. fascicularis, M. tonkeana*). *Aggress. Behav.*, **11**, 223–33

Vehrencamp, S. L. (1983) A model for the evolution of despotic versus egalitarian societies. *Anim. Behav.*, **31**, 667–82

Waal, F. B. M. de (1977) The organization of agonistic relations within two captive groups of Java-monkeys (*Macaca fascicularis*). *Z. Tierpsychol.*, **44**, 225–82

Waal, F. B. M. de, Hooff, J. A. R. A. M. van & Netto, W. J. (1976) An ethological analysis of types of agonistic interaction in a captive group of Java-monkeys (*Macaca fascicularis*). *Primates*, **17**, 257–90

Waal, F. B. M. de & Roosmalen, A. van (1979) Reconciliation and consolation among chimpanzees. *Behav. Ecol. Sociobiol.*, **5**, 55–66

Waal, F. B. M. de & Yoshihara, D. (1983) Reconciliation and redirected affection in rhesus monkeys. *Behav.*, **85**, 224–41

Zumpe, D. & Michael, R. P. (1983) A comparison of the behavior of *Macaca fascicularis* and *Macaca mulatta* in relation to the menstrual cycle. *Am. J. Primatol.*, **4**, 55–72

Part VII

Social and Reproductive Strategies

Editors' introduction

In this concluding section, we present papers on the social behaviour of adults with an emphasis on reproductive physiology and behaviour, and on competition and cooperation between individuals. Many current primate studies are attempting to link together behaviour and its immediate consequences with its function and evolutionary adaptation; examples of this approach follow. Such studies tend to concentrate on reproductive behaviour since variance in the reproductive success of individuals acts to transmit adaptations through successive generations. In addition, since the requirements of males and females for successful reproduction are known to differ, these contributions deal separately with the issues of reproductive competition and cooperation for each sex. An approach offering new insights into these issues is taken by the research presented here. The physiological mechanisms controlling behaviour have been investigated and integrated with observations of reproductive behaviour, with the result that our understanding of how social strategies are implemented has increased greatly. A wide range of primate species is covered, from the 'solitary' lemurs to humans.

Stern & Leiblum (Chapter VII.1) begin by describing the hormonal and behavioural controls on human female sexual activity after giving birth. With the finding that women released from lactation are more willing to engage in sexual contacts, our understanding of processes regulating human reproduction are increased.

Abbott *et al.* (Chapter VII.2) also focus on females and hormones, but in the context of social competition among talapoin monkeys. The presence of dominant females was shown to affect the reproduction of subordinates. Low levels of sexual behaviour among subordinate females were related to hormonal controls of fertility, especially that of ovarian failure. In addition the reproductive behaviour and physiology of males was monitored; subordinate males were inhibited from engaging in sexual interactions by the behaviour alone of the dominant males. Thus, the sexes manipulate different behavioural and endocrine mechanisms to impose social contraception on subordinates.

Wasser and Starling (Chapter VII.3) examined this same problem among female yellow baboons. In this case, the reproductive suppression of subordinates appeared to result from escalated aggressive attacks on subordinates by coalitions of dominant females, especially after the subordinates gave birth. Such attacks resulted in high infant mortality among the subordinates, and thus a reduced reproductive success. Here a behavioural mechanism is proposed for producing reproductive suppression that may have long-term and evolutionary consequences.

Glatston (Chapter VII.4) takes the opposite view in her study of mouse lemurs. She shows that the survivorship of infants is higher when females give birth in communal nests. This is related to the protection afforded to infants when several mothers spread their visits to the nest throughout the foraging period and thus decrease the time that infants are left on their own and vulnerable to predation. This study is particularly interesting in that mouse lemurs forage in a solitary fashion, yet the advantages to cooperative rearing of young lead to the sharing of a communal nest for sleeping.

Manley (Chapter VII.5) discusses a problem that has received little attention. Among territorial langurs, males and females appear to have different reasons for expelling invaders and the reproductive consequences of immigration or territory invasion are different for each sex. Yet males often increase their harem size as a result of the immigration of strange females. The mechanisms by which females initiate immigration and the resolution of the male's conflict between repelling invaders and increasing the number of females are presented.

Manzolillo (Chapter VII.6) also discusses the movement of individuals between groups, here young male baboons. She relates the timing of emigrations to the availability of consort partners in the natal group and immigrations to the opportunities for contacting new groups. Males tended to move into new groups when they were unlikely to face aggressive resistance from resident males. This suggests that male baboons make sophisticated assessment of the opportunities for mating both in their natal and new groups.

Introduction to Part VII

Noë (Chapter VII.7) examines the alliance patterns between male baboons and finds consistent partnerships among cooperating pairs. Again the abilities of males to make complex assessments of rivals for mates are highlighted, and the influence of persisting relationships on male mating tactics is emphasised.

In Chapter VII.8 Blurton-Jones models the complex nature of such assessments by deriving curves of the fitness gained from controlling a resource plotted against the amount of that resource held by an individual. He proposes that such models allow the prediction of asymmetry in contests, and can be useful for interpreting cooperative behaviour such as food-sharing, and competitive behaviour such as dominance and mating tactics. He then discusses how these principles can be applied to understanding conflict in human societies.

VII.1

Postpartum sexual behavior of American women as a function of the absence or frequency of breast feeding: a preliminary communication

J. M. STERN AND S. R. LEIBLUM

Introduction

Compared with the behavioral estrus of the great apes (Table 1), women are often said to be 'continually receptive' (although according to Beach (1974), men who believe that are either too old to remember or too young to know better!). This characteristic is meant to convey the human female's emancipation from sexual desire restricted to the periovulatory period. Nevertheless, in the belief that subtle hormonal/physiological influences may be operating, many

Table 1. *Occurrence of sexual proceptivity/receptivity in great apes as a function of different reproductive phases*[a]

Phase	Orangutan	Gorilla	Chimpanzee
Menstrual cycle	Periovulatory (~10%)	Periovulatory (<10%)	Follicular–ovulatory (~30%)
Pregnancy	Absent	Present[b]	Present[b,c]
Lactation	Absent[d]	Absent[d]	Absent[d,e]
(Birth interval)	(6–7 yr)	(3 yr 10 months)	(5 yr 10 months)

[a] References: Galdikas, 1981; Harcourt *et al.*, 1981; Nadler, 1981; Nadler *et al.*, 1981; Tutin & McGinnis, 1981.
[b] Irregular estrus intervals
[c] + Complete cycles
[d] Until weaning
[e] + Reduced mating frequency

studies of menstrual cycle fluctuations in sexual desire, initiation and activity have been carried out. Although some workers report a modest periovulatory peak in self-initiated sexual activity (e.g. Adams, Burt & Gold, 1978; Mateo & Rissman, 1984), others have failed to find such a peak or have found peaks at other times (mid-follicular; late luteal) (Schreiner-Engel *et al.*, 1981; Sanders & Bancroft, 1982).

An even larger disparity between human and ape female sexuality occurs when menstrual cycles cease altogether for some time, as during pregnancy and lactation. Female apes have little or no sexual activity during these phases until about the time when weaning begins (Table 1). In contrast, most women continue marital sexual relations during these times, though often at a reduced level. Women typically manifest a decline in sexual interest and activity during pregnancy and a period of peripartum abstinence from sexual intercourse; for many women these reductions continue for as long as one year postpartum (Solberg, Butler & Wagner, 1973; Tolor & DiGrazia, 1976; Robson, Brant & Kumar, 1981; Alder & Bancroft, 1983; Grudzinskas & Atkinson, 1984).

Ape mothers feed their babies by nursing them, while human mothers may or may not nurse. The influence of breast feeding, including its duration postpartum and its daily frequency, on the sexual interest and activity of women has not been carefully assessed. Lactating and nonlactating mothers differ markedly in their postpartum hormonal profile. Prolactin levels are very high shortly after birth and return to prepregnancy levels within several weeks in nonlactating women, or decline but remain high for close to 2 years postpartum if frequent daily nursing continues (Delvoye *et al.*, 1978; Stern *et al.*, 1982 and unpublished). Ovarian cyclicity returns within 6–15 weeks postpartum in nonlactating women but may remain suppressed for 6 months or more in lactating women, resulting in postmenopausal levels of ovarian steroids (Howie *et al.*, 1981, 1982; Glasier, McNeilly & Howie, 1983). While the role of ovarian steroids (estrogens and androgens) on the expression of women's sexuality is still controversial (Sanders & Bancroft, 1982), there is increasing evidence from both the animal and clinical literatures that chronic elevation of prolactin is associated with decreased sexual interest and activity (Buvat *et al.*, 1978; Thorner & Besser, 1978; Doherty, Bartke & Smith, 1981; Weizman *et al.*, 1983). Therefore, it is possible that lactational hyperprolactinemia, at least in the early postpartum period when the levels are the highest, may contribute to decreased libido.

The purpose of the present study was to assess the effect of absence or frequency of breast feeding on the resumption and level of sexual activity postpartum, using a prospective design. Because the primary parental adjustment is to the first baby, we sought women having their second (or third) child, so that the possible effects of breast feeding would not be swamped by those of the new maternal role.

Methods
Subjects
Women were recruited around the time of delivery of their second or third child. The sample consisted of young (23–35 yr), married, healthy, non-obese, non-drug-taking (including steroidal contraceptives) Caucasian paid volunteers with no history of infertility problems or irregular menstrual cycles prior to conception, or prepregnancy sexual dysfunction.

Behavioral measures
In the initial interview/questionnaire, information was obtained on the planned feeding mode (bottle or breast; if breast, expected duration postpartum), prepregnancy ('baseline') coital frequency, and sexual behavior changes during pregnancy.

The postpartum longitudinal behavioral measures (from 1 to 9 months postpartum) included a baby-feeding activities form, filled out once per week, and a weekly sexual behavior checklist (daily indices of coitus, sex without coitus, initiations and rejections). This checklist also included three-item scales of physical health (fatigued or ill; neutral; alert and healthy), mood (tense/angry/sad; neutral; happy/contented), and desire for sex (none; minimal; interested). These items subsequently were quantified as 0, 1 and 2 points; thus, for a given scale, a woman's weekly score could range from 0 to 14. For the purposes of this brief communication, daily measures, except coitus, were analyzed only for the first of each 4-week period, beginning week 5 postpartum. The feeding and sexual behavior forms were sent in biweekly. Other written measures included delivery information and onset and establishment of menses.

Physiological measures
In addition to monitoring menses onset, blood samples were obtained at 1, 2, 4, 6 and 9 months postpartum, for radioimmunoassay of serum prolactin, testosterone, androstenedione, and estradiol.

Assays are being carried out in the laboratory of Dr Robert Shelden (Middlesex General University Hospital, New Brunswick, NJ). Subjects were asked to have their blood drawn before noon; nursing mothers were asked to leave at least 1 hour between the last nursing episode and the blood-drawing.

Results

At this writing, data collection and hormone assays are not yet completed. Consequently, a thorough analysis of the many variables collected is not yet possible. Steroid hormone assay results will be reported at a later time.

Behavioral measures

Peripartum abstinence from intercourse did not differ significantly between groups of bottle feeding ($n = 8$) and breast feeding ($n = 16$) women (Fig. 1). Prior to delivery, abstinence averaged (mean ± S.E.; range) 7.9 ± 3.1 weeks (0–26) and 6.4 ± 0.8 weeks (2–12) for bottle- and breast-feeders, respectively. After delivery, bottle-feeders resumed intercourse somewhat sooner (4.9 ± 0.6; 1–7 weeks) than did the nursing mothers (7.0 ± 1.3; 3.5–26 weeks) ($p > 0.1$).

The prepregnancy 'baseline' intercourse rate per week was essentially the same for the bottle-feeders (2.5 ± 0.5; 1–5 per week) and the breast-feeders (2.6 ± 0.3; 0.75–4.0 per week) (Fig. 1). During weeks 5–8

Fig. 1. (Left) Number of weeks bottle-feeding and breast-feeding subjects abstained from sexual intercourse before and after delivery. (Right) Mean sexual intercourse rate per week before pregnancy ('baseline'), 5–8 weeks and 9–12 weeks after delivery for bottle- and breast-feeding subjects.

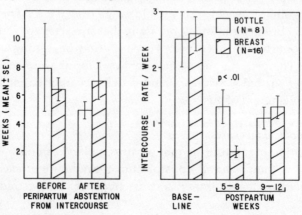

postpartum, the bottle-feeding group was significantly more coitally active than the nursing group (1.3 ± 0.3; 0–2.75 per week vs 0.5 ± 0.1; 0–1.5 per week; $p < 0.01$). By weeks 9–12, the breast-feeders reached the coital activity level of the bottle-feeders (Fig. 1).

In Fig. 2 it can be seen that the average sexual intercourse rate per week remains about one-half of the 'baseline' rate for up to 9 months postpartum for both bottle- and breast-feeding women. Sexual contacts short of intercourse (Fig. 2) were highest for the bottle-feeders in the early postpartum period. At week 5, a similar percentage of subjects in each group engaged in sex without intercourse (53.3 vs 55.5%, breast vs bottle, respectively), but the bottle-feeders did so at twice the rate (2.2 ± 0.6 vs 1.1 ± 0.1 per week; $p < 0.05$). More bottle-feeders than breast-feeders were coitally active at week 5 (55.6 vs 25%), and of those active, the rate was higher (2.0 ± 0.5 vs 1.5 ± 0.3), but these differences were not significant.

Further analysis of the week-5 data revealed that mood was significantly elevated among the breast-feeders (11.4 ± 0.7 vs 8.9 ± 0.9; $p < 0.05$), while sexual desire and physical state, though rated lower by the breast-feeders, did not differ significantly between groups (desire: 5.6 ± 0.9 vs 7.0 ± 0.9; physical state: 7.0 ± 0.9 vs 9.4 ± 1.0). Subsequently, these measures did not differentiate the groups.

The most dramatic finding to date is that in those women continuing to nurse, there is a significant negative correlation between the coital rate per week and the nursing frequency per day in each 4-week period studied between 13 and 24 weeks postpartum. This relation,

Fig. 2. Mean frequency per week of sexual intercourse and sexual contacts without intercourse in bottle- and breast-feeding subjects. Only the first of each 4-week period is illustrated.

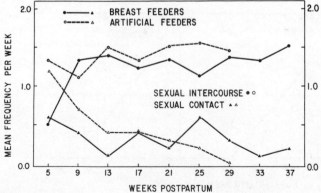

illustrated in Fig. 3, accounts for 33.6% and 44.9% of the variance for weeks 17–20 and 21–24, respectively. By inspection of Fig. 3, suppression of coital frequency was most pronounced among women nursing 5 or more times per day. There was no such relation between the number of artificial feeds and the coital rate among non-lactating women.

One possible basis for the findings illustrated in Fig. 3 may be in the realm of night feeds. More breast-feeders continued to wake up during the night to feed their babies (37.5% at week 13 to 18.7% at weeks 17 and 24 for breast-feeders; 0% during this period for bottle-feeders), but only once per night for the most part. Further, in contrast to the strong negative correlation found between total breast feeds and coital rate, no such relation was found between night feeds only and coital rate.

The percentage of subjects still nursing declined to 75, 68.7 and 25% at 4, 6 and 9 months, respectively. Of those still nursing, the mean ± SE (range) breast feeds per day was 8.6 ± 0.8 (5–16, $n = 16$), 6.4 ± 0.6 (2–12, $n = 16$), 5.6 ± 0.6 (2–9, $n = 13$) and 4.5 ± 0.8 (1–8, $n = 13$) at 1, 2, 4 and 6 months postpartum, respectively.

Physiological measures

As expected, prolactin levels were very high in the nursing women early postpartum and then declined markedly, but remained elevated compared to the non-lactators at 6 months postpartum (Fig. 4). Among the non-lactators, all but three displayed prolactin levels of menstruating, non-lactating women at 1 and 2 months

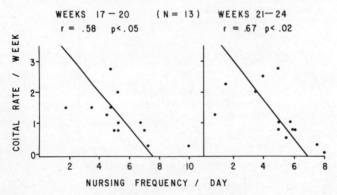

Fig. 3. Correlations between coital rate per week and nursing frequency per day for continuing nursers only during weeks 17–20 and 21–24 postpartum.

postpartum and all eight did so thereafter. (The three exceptions had levels of prolactin (mean = 33 ng/ml) within the lower range of the lactating levels (24.3–204.7 ng/ml).)

Among the continuing breast feeders, all the correlations between prolactin at 4 and 6 months and coital rate (the week closest to the blood sampling as well as the 4-week periods before and after it) were negative, but not statistically significant.

Also predictably, 100% of the non-lactating women resumed menses by 3 months postpartum (median = 8 weeks), whereas 87.5% of the breast-feeders resumed menses by 9 months (median = 26 weeks) (Fig. 4). Typically, menses resumed within 4 weeks after complete weaning or reduction in breast-feeding frequency to 1–2 nursing episodes per day. Menses onset was not associated with any apparent change in coital frequency.

Discussion

In addition to an average peripartum abstinence from sexual intercourse of 6–8 weeks before and 5–7 weeks after delivery, the birth of a second baby (or third in a few cases) also heralds a prolonged period in which sexual activity is reduced to 50% of the prepregnancy levels for most couples, whether or not the wife nurses the new baby. The mean change in intercourse rate from about 2.5 to 1.2 per week for multiparous women in New Jersey is essentially the same as that reported recently by Alder *et al.* (in press) for primiparous women in Scotland. Fatigue and time restrictions probably account for much of this reduction.

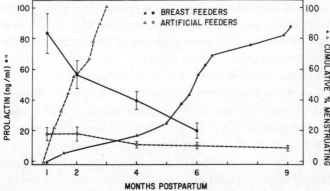

Fig. 4. Prolactin levels (left axis) and cumulative percentage with return of menses (right axis) as a function of time postpartum.

Two substantial effects of breast feeding on the nursing mother's sexual activity are apparent. The first is the lower rate of sexual activity in the early postpartum period among nursing compared to non-nursing mothers, contributed both by a later onset of resuming sexual intercourse and a lower coital rate once resumed. The second effect emerges about 13 weeks postpartum when women continuing to nurse diverge markedly with respect to the number of breast-feeding episodes per day. Among these women, the greater the frequency of nursing, the lower the frequency of coitus. Although the overall differences in postpartum sexual activity between bottle- and breast-feeding women are not significant, such differences may emerge once data collection and analysis are complete.

Are these results due to the experimental or physiological effects of breast feeding, or both? In the puerperium, the mood elevation among the nursing mothers in our sample suggests that they may be deriving satisfaction from their physical involvement with their new baby sufficient to reduce their interest in having sexual intercourse. Later postpartum, the comment from a 'high-intensity' nursing mother interviewed by one of us (JMS) previously, pinpoints the negative relation observed: 'I have a body on me all day long; I don't want another one on me at night'. Unlike the findings of Alder & Bancroft (1983), the presence of night feeds does not seem to be related to the differences in sexual activity.

Physiological differences may also contribute to the effects of breast feeding on sexual behavior. During the early postpartum period, the levels of prolactin in lactating women are very high; as in states of pathological hyperprolactinemia, these high levels of prolactin may directly contribute to reduced libido (see Introduction), and moderate levels subsequently may do so as well. Although the negative correlations between prolactin levels at 4 and 6 months postpartum and coital rate are not significant, it is more likely that daily nursing frequency is more indicative of the 24-hour prolactin mean (Stern *et al.*, 1982 and unpublished) than is a single daily sample, which is inversely related to the prior nursing interval.

The return of menses was not associated with a change in sexual activity, comparable to the findings of Alder *et al.* (in press). However, these workers found lower levels of plasma testosterone and androstenedione in lactating than in non-lactating mothers; further, lactating women who reported a severe loss in sexual interest (5 of 14; 37.5%) had the lowest levels of these hormones. Although a definitive role of androgens in the expression of women's sexuality has yet to be established, these findings are consonant with the view that they are libido-enhancing (Sanders & Bancroft, 1982).

In sum, caring for a new baby is associated with a prolonged reduction in a couple's weekly sexual intercourse rate, and frequent daily nursing may accentuate this reduction. Thus, the human female bears a resemblance to the great apes by reducing the frequency of her 'continual receptivity' whilst escalating the intensity of her nurturance.

References

Adams, D., Burt, A. & Gold, A. R. (1978) Rise in female-initiated sexual activity at ovulation and its suppression by oral contraceptives. *New Engl. J. Med.*, **299**, 1145–50

Alder, E. & Bancroft, J. (1983) Sexual behaviour of lactating women: a preliminary communication. *J. Reprod. Infant Psychol.*, **1**, 47–52

Alder, E., Cook, A., Davidson, D., West, C. & Bancroft, J. (in press) Hormones, mood and sexuality in lactating women. *Br. J. Psychiatry*

Beach, F. A. (1974) Human sexuality and evolution. In *Reproductive Behavior*, ed. W. Montagna & W. Sadler, pp. 333–65. New York: Plenum Press

Buvat, J., Asfour, M., Bavst-Herbant, M. & Fossat, P. (1978) Prolactin and human sexual behavior. In *Progress in Prolactin Physiology and Pathology*, ed. C. Robyn & M. Harter. Amsterdam: Elsevier/North Holland Biomedical Press

Delvoye, P., Demaegd, M., Uwayitu-Nyampeta & Robyn, C. (1978) Serum prolactin, gonadotropins and estradiol in menstruating and amenorrheic mothers during two years lactation. *Am. J. Obstet. Gynecol.*, **130**, 635

Doherty, P. C., Bartke, A. & Smith, M. S. (1981) Differential effects of bromocriptine treatment on LH release and copulatory behavior in hyperprolactinemic male rats. *Horm. Behav.* **15**, 436–50

Falicov, C. J. (1973) Sexual adjustment during first pregnancy and postpartum. *Am. J. Obstet. Gynecol.*, **117**, 991–1000

Galdikas, B. M. F. (1981) Orangutan reproduction in the wild. In *Reproductive Biology of the Great Apes*, ed. A. Harcourt, pp. 281–300. New York: Academic Press

Glasier, A., McNeilly, A. S. & Howie, P. W. (1983) Fertility after childbirth and changes in serum gonadotrophin levels in bottle and breast feeding women. *Clin. Endocrinol.*, **19**, 493–501

Grudzinskas, J. G. & Atkinson, L. (1984) Sexual function during the puerperium. *Arch. Sex. Behav.*, **13**, 85–91

Harcourt, A. H., Stewart, K. J. & Fossey, D. (1981) Gorilla reproduction in the wild. In *Reproductive Biology of the Great Apes*, ed. A. Harcourt, pp. 265–79. New York: Academic Press

Howie, P. W., McNeilly, A. S., Houston, M. J., Cook, A. & Boyle, H. (1981) Effect of supplementary food on suckling patterns and ovarian activity during lactation. *Br. Med. J.*, **283**, 757–9

Howie, P. W., McNeilly, A. S., Houston, M. J., Cook, A. & Boyle, H. (1982) Fertility after childbirth: postpartum ovulation and menstruation in bottle and breast feeding mothers. *Clin. Endocrinol.*, **17**, 323–32

Matteo, S. & Rissman, E. (1984) Increased sexual activity during the mid-cycle portion of the human menstrual cycle. *Horm. Behav.*, **18**, 249–55

Nadler, R. D. (1981) Laboratory research on sexual behavior of the great apes. In *Reproductive Biology of the Great Apes*, ed. A. Harcourt, pp. 191–238. New York: Academic Press

Nadler, R. D., Graham, C. E., Collins, D. C. & Kling, O. R. (1981) Postpartum amenorrhea and behavior of apes. In *Reproductive Biology of the Great Apes*, ed. A. Harcourt, pp. 69–81. New York: Academic Press

Robson, K. M., Brant, H. A. & Kumar, R. (1981) Maternal sexuality during first pregnancy and after childbirth. *Br. J. Obstet. Gynaecol.*, **88**, 882–9

Sanders, D. & Bancroft, J. (1982) Hormones and the sexuality of women – the menstrual cycle. *Clin. Endocrinol. and Metab.*, **11**, 639–59

Schreiner-Engel, P., Shiavi, R. C., Smith, H. & White, D. (1981) Sexual arousability and the menstrual cycle. *Psychosom. Med.*, **43**, 199–214

Solberg, D. A., Butler, J. & Wagner, N. W. (1973) Sexual behavior in pregnancy. *New Engl. J. Med.*, **288**, 1098–103

Stern, J. M., Konner, M., Herman, T. & Reichlin, S. (1982) Postpartum amenorrhea, nursing behavior and prolactin during prolonged lactation in Caucasian and !Kung mothers. *Endocrine Society Abstracts*

Thorner, M. O. & Besser, G. M. (1978) Hyperprolactinemia and gonadal function: results of bromocriptine treatment. In *Prolactin and Human Reproduction*, ed. P. G. Crossignoni & C. Robyn. New York: Academic Press

Tolor, A. & DiGrazia, P. V. (1976) Sexual attitudes and behavior patterns during and following pregnancy. *Arch. Sex. Behav.*, **5**, 539–51

Tutin, C. E. G. & McGinnis, P. R. (1981) Chimpanzee reproduction in the wild. In *Reproductive Biology of the Great Apes*, ed. A. Harcourt, pp. 239–64. New York: Academic Press

Weizman, R., Weizman, A., Levi, J. L., Gura, V., Zevin, D., Maoz, B., Wijsenbeek, H. & Ben David, M. (1983) Sexual dysfunction associated with hyperprolactinemia in males and females undergoing hemodialysis. *Psychosom. Med.*, **45**, 259–69

VII.2

Social suppression of reproduction in subordinate talapoin monkeys, *Miopithecus talapoin*

D. H. ABBOTT, E. B. KEVERNE, G. F. MOORE
AND U. YODYINGYUAD

Introduction

Social suppression of reproduction is prevalent among many primate societies (Rowell, 1972; Drickamer, 1974; Epple, 1975; Sade et al., 1977; Dunbar, 1980; Abbott, 1984; French, Abbott & Snowdon, 1984; Keverne et al., 1984). In laboratory studies of groups of talapoin monkeys, low-ranking or socially subordinate males (Scruton & Herbert, 1970; Dixson & Herbert, 1977a; Eberhart, Keverne & Meller, 1980b) and females (Bowman, Dilley & Keverne, 1978; Keverne, 1979) were shown to participate less frequently in sexual behaviour than their high-ranking counterparts. Low-ranking female talapoin monkeys also have impaired pituitary LH responses to an oestrogen challenge (Bowman et al., 1978; Keverne, 1979) suggesting a reduced capacity to undergo normal ovarian cycles.

However, these previous studies on talapoin monkeys did not examine: (1) the nature of the behavioural constraints imposed on low-ranking males and females by high-ranking males and females, respectively; or (2) whether the endocrinological impairment found in low-ranking, ovariectomised females could be directly applicable to intact, low-ranking females. This paper will therefore present data from a preliminary study designed to examine the behavioural nature of reproductive inhibition in intact, low-ranking talapoin monkeys and the effect of low social rank on the fertility of intact females.

Methods
Animals

Eleven intact adult male, 11 ovariectomised adult female, and 9 intact adult female talapoin monkeys were used. Males weighed between 1.2 and 2.2 kg and females between 0.7 and 1.2 kg. All through this study, the ovariectomised females were implanted with a subcutaneous silastic capsule of oestradiol-17β which promoted the maximal development of the perineal sex skin swelling and maintained serum oestradiol concentrations at approximately 200 pg/ml (within the normal ovarian cycle range (Keverne, 1979)). Each of the four observed groups (see Table 1) of talapoin monkeys occupied a single large cage (3.5 × 1.5 × 1.7 m). Each cage was in a separate room without direct contact with the rest of the colony. All rooms were maintained at 21–27°C and 50–70% relative humidity, and natural light was supplemented with fluorescent illumination between 08.00 and 20.00 h. Once daily, monkeys received Mazuri Primate Diet (British Petroleum Ltd), mixed seeds and fresh fruit (08.00 h).

Behavioural observation

Males were partitioned from females, except during periods of observation (100 min/day, 3–5 days/week for 8 weeks: 50 min between 09.00 and 11.00 h and 50 min between 16.00 and 18.00 h). Observation was carried out behind one-way glass and selected aggressive and socio-sexual behaviours (see relevant figures below) exhibited by any animal were continuously recorded on a computerised keyboard. The behaviours recorded are described in detail elsewhere (Dixson, Scruton & Herbert, 1975).

Blood collection and progesterone assay

Twice weekly at 11.00 h, immediately following the morning observation, a 2 ml blood sample was taken from each intact female by femoral venipuncture, under ketamine (Vetalar, Parke-Davis, 0.15 ml) anaesthesia. The females were well habituated to the procedure. Blood samples were kept at room temperature for 1 h and at 4°C overnight before the serum was separated and stored at −20°C until assayed. Progesterone was measured in serum (duplicate aliquots of 100 μl) by radioimmunoassay (Clarke, Scaramuzzi & Short, 1977). The antiserum to progesterone (1:4000 dilution) showed little cross-reactivity (<1%) with other steroids apart from dexoxycorticosterone (21.2%), 11-ketoprogesterone (17.1%), corticosterone (9.7%), 5α-hydroxyprogesterone (8.3%), 17α-hydroxyprogesterone (7.1%), 11β-hydroxyprogesterone

(6.6%), 11α-hydroxyprogesterone (4.6%) and 20α-dihydroprogesterone (1.5%). Inter- and intra-assay coefficients of variation were 16.9% and 8.1%, respectively, assay sensitivity was 31 pg/tube, and recovery of tritiated progesterone averaged 63%. There was a parallel change in displacement of antiserum binding to tritiated progesterone produced by serial dilutions of the reference standard (7.8–2000 pg) and female talapoin serum.

Detection of ovarian cycles and pregnancy

Ovarian cycles were determined from a combined measure of sex skin swelling and serum progesterone concentrations. Sex skin swelling scores ranged from 0 (the minimum distension of the perineal sex skin observed for an individual female), with increments of 0.5, to 5 (the maximum distension of the sex skin observed for an individual female). Sex skin swelling scores were made at the time of blood collection. An ovulatory cycle was deemed to have occurred when the sex skin swelling score of a female rose and then fell as serum progesterone concentrations increased to over 1.0 ng/ml. The latter progesterone rise alone was also taken to indicate an ovulatory cycle if sex skin swelling was not scored. If the sex skin swelling score fluctuated without a concomitant rise in serum progesterone concentrations as the sex skin deflated, an ovulatory cycle was not recorded.

Pregnancy was confirmed from a prolonged (more than 30 days) elevation of serum progesterone concentrations over 1.0 ng/ml and abdominal palpation of the enlarging uterus. Conception was estimated to have occurred one day before either the last rise in serum progesterone over 1.0 ng/ml or the last deflation of the sex skin.

Behavioural analysis

Social dominance was assessed from the direction of aggression observed between animals in each group (e.g. Yodyingyuad, Eberhart & Keverne, 1982): aggression was essentially unidirectional and demonstrated a linear hierarchy within each sex in each group, as summarised in Table 1.

The behaviour sequence analysis was based on an initial behavioural transition matrix constructed from the recorded observations (over 50 h) of group 3 (Table 1). During the construction of the matrix, dyads of behaviours which occurred together (less than 1 min apart) more frequently than expected by chance (chi-squared test, $p < 0.0026$; Fagen & Mankovich, 1980) were combined and re-named as a single behavioural unit, e.g. 'attack' followed by 'threat' became 'Aggression' (see Dawkins, 1976). The transition matrix was then re-constructed and similar combination and re-naming processes occurred until four behavioural clusters of talapoin behaviour were identified (see Fig. 1). These behavioural clusters were arbitrarily named: (1) Affiliation (or grooming behaviour); (2) Investigation; (3) Sex; and (4) Aggression (which included submissive behaviour). The subsequent behavioural analyses (described below) examine the

Table 1. *Sex, social rank and reproductive condition of the talapoin monkeys in the four social groups*

Group	Males	Male social rank	Reproductive condition	Females	Female social rank	Reproductive condition
1	M3	1	Intact	FE6	1	Intact
	M15	2	Intact	F2	2	Ovex + E_2[a]
	IGN	3	Intact	F10	3	Ovex + E_2
				FE1	4	Intact
				F2054	5	Ovex + E_2
				WEL	6[b]	Intact
1A[c]	M3	1	Intact	A13	1	Intact
	M15	2	Intact	A16	2	Intact
	IGN	3	Intact	A23	3	Intact
				F10	4	Ovex + E_2
				F2054	5	Ovex + E_2
2	M2	1	Intact	F12	1[d]	Ovex + E_2
	M17	2	Intact	F2056	2	Ovex + E_2
	M13	3	Intact	F318	3	Ovex + E_2
	M21	4	Intact	FE22	4[d]	Intact
				FE18	5	Intact
				FE5	6	Intact
3	STA	1	Intact	F1	1	Ovex + E_2
	DUM	2	Intact	F2	2	Ovex + E_2
	IGN	3	Intact	F3	3	Ovex + E_2
				F7	4	Ovex + E_2

[a] Ovariectomised, plus oestradiol implant
[b] Removed from the group after 32 days because of the severity of injuries from attacks
[c] Group 1A was formed by the addition of three intact females to Group 1 following the removal of all the previous intact females and an ovariectomised female
[d] F12 was removed because of illness from group 2, 114 days after group formation and, within 2 weeks, FE22 had taken over the now vacant Rank 1 position. The rank order of the other females remained unchanged in relation to one another

transitions made by individual animals between these four behavioural clusters. At this stage of the analysis, a fifth behavioural state was identified and named 'Idle'. An individual animal entered 'Idle' when there was a gap of more than 1 min in between the occurrence of any of the four behavioural clusters. 'Idle' will not form part of this preliminary analysis.

Whether or not individual animals showed preferential transitions from one behavioural cluster to another, for example, from 'Investigation' to 'Sex', was determined using the chi-squared test ($p < 0.05$).

Reproductive suppression in subordinate male talapoin monkeys

Whilst the expression of male sexual behaviour was directly related to social rank, with males ranking 3 or below displaying virtually no mounts or ejaculations with females (e.g. Eberhart et al., 1980b), such low-ranking males still showed normal testicular spermatogenesis and sperm production (E. B. Keverne, unpublished). Furthermore, administration of exogenous testosterone to subordinate male talapoins failed to increase their sexual activity (Dixson & Herbert, 1977b) and only succeeded in increasing the aggression they

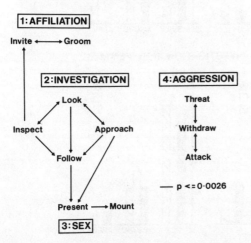

Fig. 1. Four behavioural clusters identified from the construction of a transition matrix based on the behavioural data obtained from group 3 (Table 1). The direction of the arrows indicate those behaviours which followed others more often than expected ($p < 0.0026$). A double-headed arrow indicates that either behaviour could follow the other. 'Aggression' was not significantly linked to any other behavioural cluster and could equally well erupt before or after any of the other behavioural 'states'.

received from high-ranking males. Apparently, only behavioural constraints, imposed by high-ranking males, prevented subordinate male talapoins from being reproductively successful. How were these constraints imposed?

The transition matrix analysis, using the four behavioural clusters identified above (see Fig. 1) provided at least part of the answer. In this context, the outcome of 'Investigation' with females for males of different social rank is of interest in determining what happens when a male and female talapoin first start to interact together. Using this transition analysis, the outcome or consequence of 'Investigation' could only be 'Sex', 'Aggression' or 'Affiliation' (Fig. 1). In the two groups examined (Groups 1 and 3 in Table 1), 'Sex' with a female was the most likely outcome of 'Investigation' of a female for both Rank 1 males (Fig. 2a,b) and not 'Aggression' or 'Affiliation'. In contrast, for

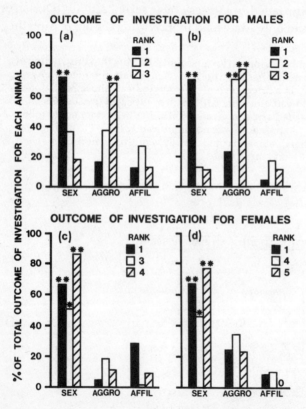

Fig. 2. Outcome of 'Investigation' for (a) group 3 males, (b) group 1 males, (c) three females in group 3 and (d) three females in group 1. * $p < 0.05$, ** $p < 0.01$: chi-squared test (see Methods).

males ranking 2 and 3 in both groups, 'Sex' with a female was not the most likely outcome of 'Investigation' (Fig. 2a,b). Indeed, for the male ranking 3 in group 3 and for the males ranking 2 and 3 in group 1, 'Aggression', mostly with a higher-ranking male, was the most likely outcome of 'Investigation' with a female. There was no significantly biased outcome of 'Investigation' for the male ranking 2 in group 3.

Consequently, the particularly restricted sexual behaviour of captive subordinate male talapoins is apparently solely due to the aggressive intervention of the dominant or higher-ranking male when the subordinate males attempt to interact with females.

Reproductive suppression in subordinate female talapoin monkeys

In the three groups containing intact females (Table 1), 3 of the 5 females who held Ranks 1–3 at some time during the study became pregnant, whereas only 1 of the 7 females who held Ranks 4–6 became pregnant. Female A13, ranking 1 in group 1A, never underwent any ovarian cycles before or after group formation.

There were two possible causes of the infertility among subordinate females. The first was ovarian failure. Both female Rank 6 (WEL) in group 1 and female Rank 4 (FE22) in group 2 stopped cycling after they were placed in their respective groups (the latter is illustrated in Fig. 3). The subordinate female in group 1 only started cycling again after her removal from the group to a single cage (D. H. Abbott, unpublished), whilst the subordinate female in group 2 recommenced ovarian cyclicity only after the original Rank 1 female was removed and the subordinate had taken over the vacant Rank 1 position (Fig. 3). As Rank 1 female, FE22 conceived on her second cycle. However, ovarian failure did not fully explain the poor reproductive performance of subordinate female talapoins. For example, the lowest ranking females in group 2, FE18 and FE5 (Table 1), underwent ovarian cycles throughout the study (e.g. Fig. 3). Nevertheless, they received very few mounts from males (Fig. 3), and no ejaculatory mounts, similar to the situation of female FE22 before she attained the Rank 1 position. Was there a comparable behavioural inhibition of sexual behaviour for subordinate female talapoins similar to that found for subordinate males? The behavioural transition matrix analysis suggested otherwise.

In the two groups examined (groups 1 and 3, Table 1), 'Sex' with a male was the most likely outcome of 'Investigation' not only for the Rank 1 females, but also for the two lowest-ranking females (Fig. 2c,d), despite the rare occurrences of 'Sex' in the latter (e.g. Fig. 3; also found by Bowman *et al.* (1978) and Keverne (1983)). 'Aggression' and 'Affiliation' were not significantly likely outcomes of 'Investigation' with males. In contrast, the intermediate ranking female(s) in both groups (group 3: F2; group 1: F2 and F10; Table 1) reacted to 'Investigation' by any males with 'Aggression' towards the investigating males. Seemingly, these females ensured their own celibacy. In this primate, females can outrank all males in a group (Dixson *et al.*, 1975; Dixson & Herbert, 1977a).

The behavioural problems of the lowest-ranking female talapoins actually just began when they entered into 'Sex' with a male. Courtship

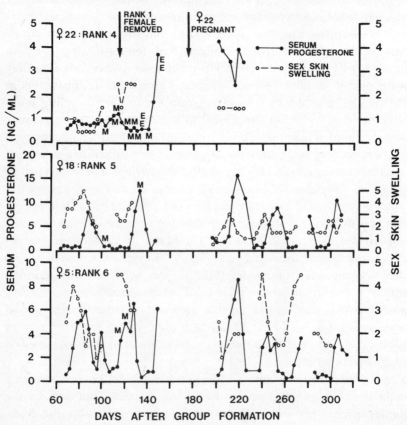

Fig. 3. Serum progesterone concentrations and sex skin swelling scores (see Methods) for the three lowest-ranking females in group 2. 'M' denotes one or more mounts of the female by one or more males on a particular day and 'E' denotes each ejaculatory mount received by the female on a particular day.

and sexual behaviour in talapoin monkeys is an elaborate affair and involves 'Sex', 'Investigation' and 'Aggression', all of which can be initiated by either the male or the female. Among the lowest-ranking females, 'Aggression' with males, displayed after 'Sex', disrupted the subsequent sexual interaction, unlike the situation with Rank 1 females (Fig. 4). Following 'Aggression', Rank 1 females were equally as likely as their male partner to re-initiate further 'Sex' or 'Investigation'. However, with lower-ranking females, it was predominantly the male which had to re-initiate 'Sex' or 'Investigation' (Fig. 4). This failure of subordinate females to re-initiate 'Sex' or 'Investigation' with males following 'Aggression' with the same males may well have contributed to the infrequent mounts they received and the lack of ejaculatory mounts.

There is therefore no apparent direct behavioural penalty imposed by the high-ranking females on low-ranking females interacting with males during 'Investigation' or 'Sex'. Nevertheless, ovarian failure in some subordinate females and insufficient sexual interactions with males in others combined to suppress reproduction among subordinate female talapoins.

Discussion

Social suppression of reproduction in low-ranking talapoins by high-ranking animals limits reproduction to only the high-ranking individuals in captive groups of talapoin monkeys.

Fig. 4. Re-initiation of 'Sex' or 'Investigation' with a male partner following 'Aggression' for three individual females in (a) group 3 and (b) group 1. * $p < 0.05$, ** $p < 0.01$: chi-squared test (see Methods).

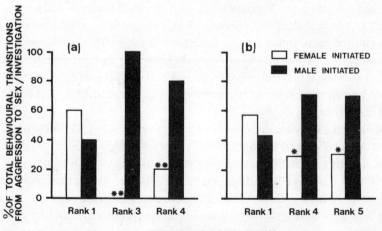

In males, reproductive suppression appears to be behaviourally mediated and particularly restricts ejaculatory mounts with females to the Rank 1 males or, in some instances, to the two top-ranking males (Eberhart et al., 1980b). This specific behavioural block to reproduction for subordinate males may explain why, in the absence of any higher-ranking males, subordinate males require several weeks to show ejaculatory mounts with females, despite the increased sexual solicitations from females (Keverne, 1983). Such persistence of inadequate sexual behaviour from these 'anxious' males, which continue to show 'high levels of stress hormones' (Keverne, 1983), could well be due to the previous specific behavioural deterrent imposed by high-ranking males.

In female talapoins, reproductive suppression in subordinates was linked with endocrinological as well as behavioural factors. In two subordinate females, ovarian cyclicity ceased. Whilst one of the females was the lowest ranking in her group (WEL, Rank 6; Table 1), the other female ranked above two cycling subordinates (FE22, Rank 4; Table 1). These two acyclic females may have become anovulatory because they were receiving more aggression from higher-ranking females than were other low-ranking intact females. The fact that one of them (WEL) had to be removed from her group after 32 days because of the severity of attacks received provides some evidence for this notion. The ovarian failure in subordinates may well be due to inhibited gonadotrophin secretion, because low-ranking, ovariectomised female talapoin monkeys have previously been shown to be unable to respond with an LH surge to an oestrogen challenge (Bowman et al., 1978; Keverne, 1979). Gonadotrophin insufficiency is certainly the cause of ovarian acyclicity in subordinate female marmosets (Abbott et al., 1981). Nevertheless, ovarian failure did not occur in all subordinate, intact female talapoin monkeys. This was perhaps not surprising, given the multi-female breeding system of talapoin monkeys in the wild (Rowell & Dixson, 1975) as opposed to the single female breeding system of the marmoset (Stevenson, 1978; Rylands, 1981; Hubrecht, 1984).

Subordinate female talapoins also failed to interact sufficiently with males to receive ejaculatory mounts because they either attacked males showing sexual interest or were unable to cope with aggression which was displayed during courtship. Further transition matrix analysis is required to elucidate the cause of this intimidation, but there was apparently no direct intervention of high-ranking females, unlike the situation with the males. However, high-ranking females may play a less specific role than high-ranking males in behaviourally deterring subordinates from copulating. When oestrogen was selectively given to the lowest-ranking female in a group, where all the females were ovariectomised, aggression received by the treated female from other females increased considerably, as did the males' interest in the treated female (looks and inspection of her sex skin), but no mounts or ejaculations occurred with this female (Eberhart et al., 1980a). Such females also avoided males and did not solicit their attention with sexual invitations (Eberhart et al., 1980a; Keverne, 1983). Subordinate female talapoins therefore, while still appearing attractive to males, shun sexual interactions with them. The origins of this celibacy are less straightforward than is the case with subordinate males.

In the wild, talapoin monkeys live in large social troops (Rowell & Dixson, 1975). During the breeding season there is an increase in male intra-sexual aggression and males seem to copulate 'at a distance' from other males. Courtship between a male and female can be disrupted at the approach of other males, whilst on one occasion, a male was observed by another courting and copulating without any reaction on the part of the observing male (Rowell & Dixson, 1975). The results of the present study lend weight to the interpretation by Eberhart et al. (1980b) of the above observations and would predict that the highest-ranking males in large social troops would copulate without restraints, unlike the more numerous subordinate males who might well seek to copulate away from the latter. Comparison of reproductive suppression in subordinate females in captive and free-living groups is more difficult because of the difficulty in obtaining field data (Rowell & Dixson, 1975). It is not known whether or not all adult females ovulate during the breeding season, but certainly only a maximum of 70% of females in a troop showed sex skin swellings (Rowell & Dixson, 1975). The degree of female copulatory success in the wild is also difficult to assess because mounts (with or without thrusts) accompanied by copulatory calls are not necessarily indicative of intromitted, ejaculatory mounts (D. H. Abbott, unpublished). To what extent socially induced reproductive suppression, as seen in the laboratory, operates in free-living groups of talapoin monkeys remains to be seen.

Conclusions

In captive groups of talapoin monkeys, high-ranking males specifically deterred otherwise fertile subordinate males from copulating. In contrast, such specific behavioural constraints were not placed on subordinate female talapoins. However, subordinate female talapoins attacked males showing sexual interest or were intimidated by aggression which occurred during courtship and copulation and failed to receive ejaculatory mounts. Ovarian failure was also observed in two subordinate females. In talapoin monkeys, the sexes seem to employ different tactics to suppress reproduction of subordinates.

Acknowledgements

This work was supported by an MRC Programme Grant to Drs Joe Herbert and E. B. Keverne. We thank Dr N. D. Martensz for the progesterone antiserum, H. M. Shiers and K. Batty for excellent technical assistance with the progesterone assay, B. Tuite for assistance with the observations and animal handling, T. J. Dennett and C. J. Sharp for preparing the figures, and D. J. Morris for typing the manuscript. U. Y. was supported by a Girton College Scholarship.

References

Abbott, D. H. (1984) Behavioural and physiological suppression of fertility in subordinate marmoset monkeys. *Am. J. Primatol.*, **6**, 169–86

Abbott, D. H., McNeilly, A. S., Lunn, S. F., Hulme, M. J. & Burden, F. J. (1981) Inhibition of ovarian function in subordinate female marmoset monkeys (*Callithrix jacchus jacchus*). *J. Reprod. Fertil.*, **63**, 335–45

Bowman, L. A., Dilley, S. R. & Keverne, E. B. (1978) Suppression of oestrogen-induced LH surges by social subordination in talapoin monkeys. *Nature (Lond.)*, **275**, 56–8

Clarke, I. J., Scaramuzzi, R. J. & Short, R. V. (1977) Ovulation in prenatally androgenised sheep. *J. Endocrinol.*, **73**, 385–9

Dawkins, R. (1976) Hierarchical organisation: a candidate principle for ethology. In *Growing Points in Ethology*, ed. P. P. G. Bateson & R. A. Hinde, pp. 7–54. Cambridge: Cambridge University Press

Dixson, A. F. & Herbert, J. (1977a) Testosterone, aggressive behaviour and dominance rank in captive adult male talapoin monkeys (*Miopithecus talapoin*). *Physiol. Behav.*, **18**, 539–43

Dixson, A. F. & Herbert, J. (1977b) Gonadal hormones and sexual behaviour in groups of adult talapoin monkeys (*Miopithecus talapoin*). *Hormones Behav.*, **8**, 141–54

Dixson, A. F., Scruton, D. M. & Herbert, J. (1975) Behaviour of the talapoin monkey (*Miopithecus talapoin*) studied in groups in the laboratory. *J. Zool., Lond.*, **176**, 177–210

Drickamer, L. C. (1974) A ten-year summary of reproductive data for free-ranging *Macaca mulatta*. *Folia primatol.*, **21**, 61–80

Dunbar, R. I. M. (1980) Determinants and evolutionary consequences of dominance among female gelada baboons. *Behav. Ecol. Sociobiol.*, **7**, 253–65

Eberhart, J. A., Herbert, J., Keverne, E. B. & Meller, R. E. (1980a) Some hormonal aspects of primate social behaviour. In *Endocrinology*, ed. R. Short, pp. 622–5. Melbourne: Australian Academy of Sciences

Eberhart, J. A., Keverne, E. B. & Meller, R. E. (1980b) Social influences on plasma testosterone levels in male talapoin monkeys. *Hormones Behav.*, **14**, 247–66

Epple, G. (1975) The behaviour of marmoset monkeys (Callithricidae). In

Primate Behaviour, Vol. 4, ed. L. A. Rosenblum, pp. 195–239. New York: Academic Press

Fagen, R. M. & Mankovich, N. J. (1980) Two-act transitions, partitional contingency tables and the 'significant cells' problem. *Anim. Behav.*, **28**, 1017–23

French, J. A., Abbott, D. H. & Snowdon, C. T. (1984) The effect of social environment on oestrogen excretion, scent marking and sociosexual behaviour in tamarins (*Sanguinus oedipus*). *Am. J. Primatol.*, **6**, 155–67

Hubrecht, R. (1984) Field observations on group size and composition of the common marmoset (*Callithrix jacchus jacchus*) at Tacapura, Brazil. *Primates*, **25**, 13–21

Keverne, E. B. (1979) Sexual and aggressive behaviour in social groups of talapoin monkeys. *Ciba Found. Symp.* (N.S.), **62**, 271–97

Keverne, E. B. (1983) Endocrine determinants and constraints on sexual behaviour in monkeys. In *Mate Choice*, ed. P. P. G. Bateson, pp. 407–20. Cambridge: Cambridge University Press

Keverne, E. B., Eberhart, J. A., Yodyingyuad, U. & Abbott, D. H. (1984) Social influences on sex differences in the behaviour and endocrine state of talapoin monkeys. *Prog. Brain Res.*, **61**, 325–41

Rowell, T. E. (1970) Baboon menstrual cycles affected by social environment. *J. Reprod. Fertil.*, **21**, 133–41

Rowell, T. E. (1972) Female reproductive cycles and social behaviour in primates. In *Adv. Stud. Behav.*, **4**, 69–105

Rowell, T. E. & Dixson, A. F. (1975) Changes in social organisation during the breeding season of wild talapoin monkeys. *J. Reprod. Fertil.*, **43**, 419–34

Rylands, A. B. (1981) Preliminary field observations on the marmoset, *Callithrix humeralifer intermedius* (Hershkovitz, 1977), at Dardanelos, Rio Aripuna, Mato Grosso. *Primates*, **22**, 46–59

Sade, D. S., Cushing, P., Dunaif, J., Figueros, A., Kaplan, J. R., Lauer, G., Rhodes, D. & Schneider, J. (1977) Population dynamics in relation to social structure on Cayo Santiago. *Yearbook Phys. Anthropol.*, **20**, 253–62

Scruton, D. M. & Herbert, J. (1970) The menstrual cycle and its effect on behaviour in the talapoin monkey (*Miopithecus talapoin*). *J. Zool., Lond.*, **162**, 419–36

Stevenson, M. F. (1978) The behaviour and ecology of the common marmoset (*Callithrix jacchus jacchus*) in its natural environment. In *Biology and Behaviour of Marmosets*, ed. M. Rothe, H.-J. Wolters & J. P. Hearn, p. 298. Göttingen: Eigenverlag Rothe

Yodyingyuad, U., Eberhart, J. A. & Keverne, E. B. (1982) Effects of rank and novel females on behaviour and hormones in male talapoin monkeys. *Physiol. Behav.*, **28**, 995–1005

VII.3

Reproductive competition among female yellow baboons

S. K. WASSER AND A. K. STARLING

Introduction

Altmann (1980) found infant mortality among yellow baboons to be approximately 30% during each of the infants' first 2 years. She speculated that mortality was dependent on the availability of easily obtained, eaten, and digested 'weaning foods' in relation to the number and relative ages of other infants simultaneously present in their group. Resource availability has been shown to limit infant survivorship at Mikumi National Park; baboons born around the same time as many others experience decreased survivorship and this is particularly the case for female infants (Rhine *et al.*, unpublished; see Dittus, 1979, for toque macaques). Individuals born late relative to others in their birth clump should also have an age disadvantage in competition with their peers. Under such conditions of density-dependent mortality, females may be able to improve the survivorship of their offspring by suppressing the reproduction of their competitors; and females of a variety of species appear to do just that (Wasser & Barash, 1983).

Previous work at Mikumi suggests that socially mediated reproductive suppression is important among female baboons. Such suppression is mediated by attack coalition behavior, defined as when two or more females simultaneously threat, chase, hit, or bite another, and by infant-handling behavior where females handle infants other than their own (Wasser, 1983). Several predictions were generated from this previous research. We subsequently returned to Mikumi for 15 months to evaluate these predictions, testing the hypothesis that attack coalition behavior serves to suppress the reproduction of the recipients and thereby reduce competition between young. The predictions and results of this study are presented below.

Methods

We observed three troops of yellow baboons. Troops 1 and 2 had widely overlapping home ranges; the home range of troop 3 was much smaller, fully within the ranges of troops 1 and 2. Troops 1 and 2 contained approximately 65 animals each, both with 21 adult females. Troop 3 contained 18 animals, 4 of whom were adult females. The reproductive states of all females were recorded each morning as they left their sleeping trees. Females were marked visibly pregnant as soon as their pericollosal skin began to turn visibly pink (approximately 2 months into pregnancy). Generally, each troop was followed for three to four 5-day periods within every month.

A maximum of three observers moved continuously through the troop being followed, collecting social data on focal animals. Whenever a coalition was seen or heard (they were loud, fast moving, and easily distinguishable), the nearest observer(s) would stop any present activities, go to the scene, and record the identities of all attackers, recipients, and bystanders. With this system, we were able to record a reasonably accurate measure of actual coalition rates; over 1400 coalitions were recorded during the 15-month study period.

Predictions

1. The rates of attack coalition behavior per hour per adult female should be greater in the two larger troops, since they have the highest annual production of newborns and presumably the greatest amount of resource competition between these newborns.
2. Given increased reproductive competition with the number of reproductive females present in one's group, females should become increasingly involved in attack coalition behavior as they approach sexual maturity. Moreover, natal females should be far more involved in attack coalitions between females than are similar-aged natal males; a comparable age-related change in coalition behavior should not take place amongst natal males, as they will disperse at sexual maturity. If resource availability alone were driving coalitions, natal males should be involved in coalitions to a degree comparable with female age-mates.
3. Similarly, coalition rates should change with season and in consistent manners between troops if resource availability is driving them. If reproductive competition is driving coalitions, coalition rates should track the number of females

per reproductive state more closely than seasons. Thus, coalition rates should, alternatively, increase with the extent of projected birth clumping and hence the projected number of infants present at weaning. The concentration of females in a birth clump, as well as its overall size, can be predicted by the number of females present in the troop who are simultaneously in the visible third trimester of pregnancy (P3) (but see also prediction 8). Thus, coalition rates should increase with the number of P3 females in the troop. The first offspring born in a birth clump would have an age advantage over other offspring in that clump; thus coalition rates should peak as the first pregnant females in this clump give birth; these first females should also be frequent recipients of aggression at this time (providing they are not the highest-ranking females in the troop).

4. Visibly pregnant females should be among the most common attackers of the newly lactating females as the pregnant females try to offset this age advantage over their own offspring.
5. Rates of aggression should decrease in each troop once the majority of females have given birth and need to ensure the safety of their vulnerable infants.
6. This decrease should be greatest in the troop having the greatest number of newly lactating females.
7. Since females in the follicular phase of their menstrual cycle are in the physiologically most suppressable states (Wasser, 1983), and are 'jockeying' for their times of conception, these females should be both recipients and attackers in significantly more coalitions than expected based on their availabilities.
8. Coalition rates should increase with the number of females simultaneously in estrus since this too would reflect the projected size of the forthcoming birth clump and hence the degree of reproductive competition.

Results

1. Of the 1400 coalitions observed, rates of coalition attacks per individual per hour were 0.051 for troop 1, 0.038 for troop 2, and 0.006 for the smaller troop 3 – a ratio of 1.34:1.00:0.16 for troops 1–3, respectively. Coalition rates recorded over the year following this study showed comparable troop differences,

although the coalition rates per individual per hour doubled in troop 3 (see below; all subsequent results focus on the larger two troops unless stated otherwise).
2. Females became significantly more involved in attack coalition behavior as they approached recruitment into the breeding pool in all three troops. In fact, the doubling in overall coalition rates per individual per hour for troop 3 corresponded with the sexual maturation of 2 of 3 large juvenile females in that troop – a 50% increase in their number of sexually mature females. Overall, the involvement of natal females in attack coalition behavior was nearly sixfold greater than that of the similar-aged natal males. These natal males did become increasingly involved in coalition behavior as they went from the small juvenile to the large juvenile stage. But their involvement decreased below that of small juveniles as they approached subadulthood. It should be noted, however, that female–female coalitions occasionally started in response to harassment of a female by a subadult or large juvenile male. But these coalitions served to distract the male's harassment as the target female (of the male's harassment), or her relative, redirected aggression onto another female by starting a coalition against her; the males generally did not participate in the subsequent coalition.
3. Neither troop 1 nor troop 2 showed any seasonal trend in rates of coalitions; nor did these two troops show any intertroop consistencies within any given month. By contrast, coalition rates were very much a function of the number of females in the various reproductive states. Thus, troops 1 and 2 both showed a significant increase in their rates of coalitions with an increase in the number of females simultaneously in the third trimester of pregnancy, and coalition rates peaked as the first couple of pregnant females in the troop gave birth (Table 1, see also result 8). In fact, coalition rates in troop 2 corresponding with the onset of its birth clump (2.05/h) exceeded all other rates of coalitions for any month, for either troop. The projected size of the birth clump in this troop at its onset in April (as suggested by the number of females in the visible second and third trimesters of pregnancy) was nearly twice as large as that of troop 1 at its onset in May–June. Thus, these troop differences provide further support for prediction 3. Finally, the earliest of the newly lactating females were

among the most frequent recipients of attacks in both troops 1 and 2 at these times, and, in fact, two infanticides by adult females were recorded. (The first infanticide was observed; the second one was inferred by intensive attacks on a female who had just given birth within the previous couple of hours; the female had numerous bite wounds, and her infant was already missing, when the observed attacks took place.)
4. The most frequent attackers of the newly lactating females were visibly pregnant females, most especially females in the second trimester of pregnancy. However, the observed infanticide was committed by a newly lactating female and a weanling daughter of the most dominant female in the troop. The infant killed was the next oldest infant (4 months older),

Table 1. *The observed minus expected frequencies of attack coalitions as a function of the number of third-trimester pregnant females (P 3) and newly lactating females (Lact 1) simultaneously present in the troop*

Troop 1

		Lact 1					
		0	1	2	3	4	5
P 3	0	–	–	–	–	101.0 / 79.9 / 21.1**	–
	1	77.0 / 61.6 / 15.4*	138.0 / 211.9 / −73.9***	1 / 0 / 1	11.0 / 15.6 / −4.6	65.0 / 50.1 / 14.9**	46.0 / 53.5 / −7.5
	2	20.0 / 11.5 / 8.5**	63.0 / 77.9 / −14.9*	–	–	23.0 / 16.2 / 6.8	–
	3	–	–	–	11.0 / 18.3 / −7.3*	–	–
	4	–	23.0 / 26.4 / −3.4	67.0 / 45.4 / 21.6***	–	–	–
	5	–	31.0 / 9.5 / 21.5***	–	–	–	–

TOTAL 677

continued

same sex (male), as the infant of the newly lactating female who committed the infanticide. Thus, this observation is still consistent with predictions 3 and 4.
5. There was a significant decrease in rates of aggression in each troop once the majority of pregnant females had given birth (Fig. 1).
6. This decrease was significantly greater in troop 2 (which had the greater number of newly lactating females) than in troop 1.
7. Menstruating females and females in the swelling phase of estrus received the highest rates of coalitions, followed by estrus females in full swell. Reproductive states of attackers,

Table 1. Continued
Troop 2

		Lact 1						
		0	1	2	3	4	5	6
P3	0	30.0 27.1 2.9	33.0 50.0 −17.0**	3.0 29.2 −26.2***	40.0 28.8 11.2*	—	1.0 4.6 −3.6	26.0 25.0 1.0
	1	55.0 65.5 −10.5	—	22.0 22.0 0.0	—	—	11.0 20.0 −9.0*	—
	2	23.0 19.2 3.8	—	—	—	—	18.0 13.8 −4.2	—
	3	1.0 5.8 −4.8*	—	49.0 12.9 36.9***	5.0 7.9 −2.9	9.0 7.1 1.9	—	—
	4	14.0 20.4 −6.4	48.0 31.3 16.3***	8.0 5.8 2.2	10.0 5.8 4.2	—	—	—
	5	11.0 20.0 −9.0*	—	—	—	—	—	—

TOTAL 417

The top value per cell is the observed frequency under that condition; the middle value per cell is the expected frequency, calculated by multiplying the total number of attacks for that troop by the proportion of total study time under the conditions of that cell; the lower cell value is the observed − expected frequency. Asterisks correspond to p values from the binomial statistic: $0.05 \geqslant * < 0.01$, $0.01 \geqslant ** < 0.001$, and $*** \leqslant 0.001$

however, deviated somewhat from predictions. Menstruating females did attack others significantly more than expected, but only in troop 1 did estrous females in full swell attack others somewhat more than expected based on their availabilities, and this trend was not statistically significant. The lack of this pattern appeared to be due, in part, to the relatively high proportion of attacks by visibly pregnant females. Analyses of cycling females alone did reveal the predicted patterns.

0. Only in troop 1 did coalition rates increase with the number of females simultaneously in estrus, while troop 2 showed the reverse trend (Fig. 2). This difference appeared to result, in part, from differences in the times in which high levels of estrous synchrony occurred in each troop. Troop 2 did not show high levels of estrous synchrony until August, 1983, when a small number of visibly pregnant females and a large number of newly lactating females were present. Troop 1, by contrast, showed high levels of estrous synchrony on and off over the entire study period. Moreover, when troop 1 did have its birth surge, the surge was half the size of that in troop 2.

Fig. 1. Differences in observed (O) and expected (E) frequencies of attack coalitions in troops 1 (open bars) and 2 (hatched bars), based on the number of simultaneously available females 3–6 months into lactation, in each troop. A separate expected value is calculated for each number of simultaneously lactating females, based on the proportion of total study time in which that number of simultaneously lactating females were simultaneously present in the troop. Asterisks correspond to p values from the binomial statistic: $0.05 \geq * < 0.01$, $0.01 \geq ** < 0.001$, and $*** \leq 0.001$.

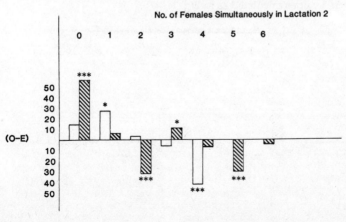

Unlike the case in troop 1, aggressive effects associated with high estrous synchrony appear to have been offset in troop 2 by their restricted estrous synchrony until a large number of relatively non-aggressive newly lactating females were present (see results 5 and 6 above). These troop-specific reproductive patterns appear to provide the most parsimonious explanation for overall differences in rates of aggression observed between these two troops. One clump of lactating females in troop 2 appears to be setting the stage for conception by the next; the high concentration of lactating females results in low aggression and correspondingly high rates of conception by females cycling during that time. The pattern then repeats itself. By contrast, the relatively lower degrees of birth synchrony in troop 1, coupled with relatively high rates of estrous synchrony during most months of the year, probably allowed for much higher rates of aggression in that troop and correspondingly lower rates of conception; also a repeating pattern. These differences appear to be resulting in the aggressive patterns seen today, as further suggested by the year-to-year stability in relative degrees of aggressiveness between these three troops (see result 1 above). In fact, the large observed values in cells (0,4) and (1,4) of Table 1 for troop 1, corresponded with such a situation. There was an average of eight females simultaneously in estrus in troop 1 during that time; but the overall number of newly lactating females in that troop never came close to that in troop 2 (see also Fig. 1). Hence the high rates of aggression in troop 1 could not be buffered by a large number of newly lactating females as in troop 2 (see also Fig. 2).

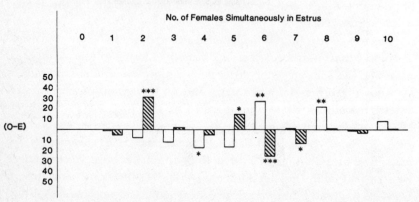

Fig. 2. Differences in observed and expected frequencies of attack coalitions in troops 1 (open bars) and 2 (hatched bars), based on the number of simultaneously available females in estrus, in each troop. Expected values calculated as in Fig. 1.

Rank and relatedness

The social ranks of attackers and recipients varied somewhat between the troops. In troop 1, high-ranking females (1–11) rarely received attacks; mid-ranking females (12–19) received relatively moderate rates of attack, and the lowest-ranking females (20–23) received excessively high rates of attack. In troop 2, high-ranking females (1–7) also rarely received attacks. Unlike troop 1, however, high- to middle-ranking females (8–14) in troop 2 received moderate rates of attack and these exceeded those of middle- to low-ranking females (15–18). Low-ranking females (19–24) received the highest rates of attack in troop 2, as per troop 1.

The tendency of individuals to attack others in coalitions based on social rank follows somewhat of a bell-shaped curve in troop 1. The most frequent attackers were mid-ranking females (11–17), followed by high-ranking (1–8) and low-ranking females, although the two lowest-ranking females in that troop rarely attacked others in coalitions. Troop 2, on the other hand, shows a more bi-modal trend, with middle–high-, and middle–low-ranking females being the most frequent attackers in coalitions. High-ranking females and mid-ranking females attacked others second most frequently, and the lowest three ranking females in troop 2 rarely attacked others in coalitions.

In comparing the ranks of attack recipients relative to those of the attackers, the differences between troops 1 and 2 become particularly apparent. Attackers in troop 1 showed a strong tendency to form coalitions against females much lower in rank than themselves, whereas attackers in troop 2 showed a significant tendency to attack most frequently those females in ranks just below that of themselves (Wasser & Starling, unpublished). Given the positive association between rank and relatedness among baboons, differences in degrees of relatedness in the two troops may, in part, account for the above disparate tendencies.

The sizes of the adult female matrilines appears to be much greater in troop 2 than in troop 1. Troop 2 has nine known mother–daughter pairs and five known sister–sister pairs among the adults. Troop 1 has six known mother–daughter pairs and no known sister–sister pairs among the adults. As the adult matriline size increases, so does its effectiveness as a support network against other matrilines (Quiatt, 1966; Kurland, 1977; Silk, Samuels & Rodman, 1981). Hence, large matrilines present more of a threat than do smaller ones. Accordingly, we suspect that the greater adult matriline sizes in troop 2 *versus* troop 1 has led to greater intermatrilineal strife among troop 2, as suggested by the above differences in their patterns of attack. This suggestion is further supported by the following: whereas in both troops the very highest-ranking females showed some tendency to support known kin in coalitions, adult females in troop 2 showed a much stronger tendency to be attackers against known kin than did females in troop 1. In all cases of 'kin-directed' attacks in troop 1, mothers were acting as collaborators (i.e. attackers other than the initiators) in these coalition attacks, and appeared to be attempting to divert the coalition from their daughters onto other individuals during these times. In the kin-directed attacks in troop 2, higher-ranking females appeared to instigate target-females of the next-lowest matriline to attack the target's own kin by deliberately initiating such attacks on targets when the target female was in close proximity with kin. Some targets would then go after their mothers and sisters, effectively diverting the aggression from themselves onto these kin who would rarely retaliate. Using this strategy, the high-ranking females may actually be weakening the cohesiveness of the target's matriline in troop 2.

Conclusions

Wrangham (1980) argued that the primary factor driving female–female coalitions was the availability of food. While this may be important over the long run, and was perhaps supported by the relatively low levels of coalition behavior in the small troop 3, food availability does not seem to be the primary variable driving coalitions in the immediate sense among our study troops. Thus, neither troop 1 nor troop 2 showed any seasonal trend in rates of coalitions; nor did these two troops show any intertroop consistencies within any given month (see result 3). Yet, these two troops had widely overlapping home ranges, were virtually identical in size, and neither troop was consistently dominant over the other. Moreover, if resource availability was the primary factor driving these coalitions, we would also expect natal males to be involved in those coalitions to comparable degrees as their same-aged female peers. This was not the case either (result 2).

Attack coalition behavior has also been described as a means to establish and maintain social rank among baboons (Cheney, 1977; Walters, 1980). This hypothesis for attack coalition behavior may also be somewhat important among our baboons over the long run. Thus, if attack coalitions function to maintain social rank, one would also expect females to attack others close in rank to themselves. But this was only the case in troop 2, which had substantially larger adult matrilines than did troop 1. Our more aggressive troop 1 with its smaller matrilines showed the opposite pattern, with the majority of females focusing their attacks on the lowest-ranking females in the troop. These observations suggest that the role of coalitions in rank maintenance becomes particularly important as the size of the adult matrilines grows. Larger matrilines have potentially more influence. Thus, as matrilines grow in size, there becomes progressively more reason to restrain them. Accordingly, females in the more aggressive troop 1, with its smaller matrilines, appear to direct their suppression attempts at those females who are easiest to suppress – the lowest-ranking females in the troop. The larger matrilines in troop 2, on the other hand, called for greater selectivity – suppression of females producing infants in the matrilines just below them.

But the primary role of coalitions as a reproductive suppressor, *versus* simply a means of maintaining social rank, is most strongly suggested by the following: social rank has been shown to be highly stable, at least over the short term, among female baboons (Walters, 1980), and this was the case in all of our troops over the study period as well. Yet, coalition rates in all troops varied considerably over time. Moreover, changes in these coalition rates both within individuals and within troops appear to be very much a function of female reproductive state.

Based on the above, we conclude that reproductive competition among females is the primary variable driving these coalition attacks. Food availability and social rank are important determinants of offspring survivorship over the long run, but a much more sensitive and manipulatable factor at any point in time is the number of competitors these offspring will face in the future. Hence, female–female attack coalitions seem to function foremost as a reproductive suppressor.

Acknowledgements

We wish to thank Ramon Rhine and Guy Norton for their hospitality at Mikumi National Park, and for use of their long-term data on the Mikumi baboon population. We also wish to thank the Senior Park Warden, Mr E. Kishe, for his tremendous generosity and assistance in helping our project run smoothly, and the Tanzanian National Parks, the Tanzanian National Scientific Research Council, and the Serengeti Research Institute for permission to conduct our research at Mikumi. Charles Kidungho helped with all aspects of data collection and protection against the many hazards one encounters in the bush. Diana Blane assisted in all aspects of data entry, preparation of figures and the final manuscript. Finally, we would like to thank the Harry Frank Guggenheim Foundation who funded this project in its entirety, and will be continuing to fund our work for some time in the future.

References

Altmann, J. (1980) *Baboon Mothers and Infants*. Cambridge, Mass.: Harvard University Press

Cheney, D. L. (1977) The acquisition of rank and the development of social alliances among free-ranging immature baboons. *Behav. Ecol. Sociobiol.*, **2**, 303–18

Dittus, W. P. J. (1979) The evolution of behaviors regulating density and age-specific sex ratios in a primate population. *Behaviour*, **49**, 265–302

Kurland, J. A. (1977) Kin selection in the Japanese monkey. *Contrib. Primatol.*, **12**. Basel: Karger

Quiatt, D. D. (1966) Social dynamics of rhesus monkey groups. Unpublished doctoral dissertation, University of Colorado, Boulder

Silk, J. B., Samuels, A. & Rodman, P. S. (1981) Hierarchical organization of female *Macaca radiata*. *Primates*, **22**, 84–95

Walters, J. (1980) Interventions and the development of dominance relationships in female baboons. *Folia Primatol.*, **34**. 61–89

Wasser, S. K. (1983) Reproductive competition and cooperation among female yellow baboons. In *Social Behavior of Female Vertebrates*, ed. S. K. Wasser. New York: Academic Press

Wasser, S. K. and Barash. D. P. (1983) Reproductive suppression among female mammals: implications for biomedicine and sexual selection theory. *Q. Rev. Biol.*, **58**, 513–38

Wrangham, R. W. (1980) An ecological model of female-bonded primate groups. *Behaviour*, **75**, 262–300

VII.4

The influence of other females on maternal behaviour and breeding success in the lesser mouse lemur (*Microcebus murinus*)

A. R. GLATSTON

Introduction

The grey lesser mouse lemur, *Microcebus murinus,* is a small bodied nocturnal prosimian. It is generally described as a solitary species (Wilson, 1975) due to the fact that individuals are usually observed in the field foraging and travelling alone. Eisenberg (1981) awards the mouse lemur a social complexity rating of 2.5, indicating its virtually solitary habits. However, reports from the field also indicate that female mouse lemurs often sleep together in nests during the daylight hours (Martin, 1972). These nesting groups usually number between two and five individuals and they are reported to be stable. Furthermore, it has been observed that females rear their infants in these nests in the presence of other females.

In this study, the behaviour and reproductive success of females housed alone are compared with those of females housed together in groups, in order to ascertain whether there are any advantages in terms of reproductive success accruing to those females who nest together.

Methods

A small colony of lesser mouse lemurs numbering between 16 and 20 individuals has been maintained in an off-display facility at the Rotterdam Zoo for the last 6 years. All these animals are captive-born and are the descendants of a colony founded in London in 1970.

The housing and management of the animals in the Rotterdam colony has been discussed elsewhere (Glatston, 1981). It suffices here to say that each mouse lemur was housed in a 75 cm^3 cage module.

Where groups of mouse lemurs were housed together, sliding doors between adjacent cage modules were opened to provide adequate housing for the number of animals in the group.

For the purposes of this study females were either housed alone or in pairs; usually mother/daughter or sister pairs. However, the data on fertility also include information from two larger groups, one of six animals (two males, four females) and one of five animals (two males, three females). In both these groups the males were not present during the infant rearing period.

The fertility and infant mortality data were collected from 12 different females over a 5-year period and the behavioural data were collected from eight females over 3 years.

All early maternal care takes place within the nest-box. Therefore in order to obtain behavioural data the rear wall of the nest-box was replaced by an infra-red filter supported between two perspex sheets. An infra-red-sensitive video camera connected to a time-lapse video recorder was used to record data. The time-lapse recorder was used at one-tenth normal speed. Recordings were 50 min in duration and eight of these recordings were made at set times throughout the day (03.00, 07.00, 11.00, 13.00, 15.00, 19.00, 22.00, 24.00). Observations were continued for the first 24 days of the infants' lives.

The recordings were analysed and data were collected on the following behaviours: the length of time a female spent in the nest-box with infants; the length of time spent suckling the infants; the length of time spent looking out of the nest-box (vigilance); the length of time spent grooming the infants; and the frequency of visiting behaviour (a visit is defined as a return to the nest-box of less than one minute in duration; in reality a visit is only a few seconds long and during a visit a female may sniff, groom or manipulate infants).

Results
Reproductive success

Examination of the fertility records of females housed alone and of those housed in groups indicates that the rate of fertility is the same in both situations (73% of females housed alone became pregnant and 75% of females housed in groups did so). The pregnancy rate in the Rotterdam colony compares favourably with that in the wild (Martin (1972) reported that 75% of the wild females he observed were pregnant during the breeding season).

In addition to the similar pregnancy rates recorded in the group and single-housed situations, the rate of foetal loss (abortion or miscarriage) was the same in both groups (10–15% of pregnant females lost their foetuses in both cases). However, a marked difference is found when infant mortality is compared between the two groups: infant mortality in the group-housed females at 3% is significantly lower than the 27% found in the single-housed females.

Behaviour

Although behavioural data were collected throughout the 24-hour cycle, for the purposes of this discussion only those collected during the artificial night will be referred to. During the 'day' both single- and group-housed females remained asleep in the nest-box with their infants.

From the data presented in Fig. 1 it can be seen that the percentage of time that a mother spent in the nest-box with her offspring was the same in both single- and group-housed females. However the amount of time that the infants were left on their own was substantially less in the group-housed situation. This was due to the fact that when the mother was away from the nest-box another female was usually present.

The amount of time spent suckling the infants did not differ between the single- and group-housed females. Whenever the mother was in the nest-box at least one of her infants was suckling. Thus the percentage of time spent suckling was essentially the same as that which the mother spent in the nest-box (see Fig. 1).

The amount of grooming received by infants born to group-housed mothers was higher than that received by those born to single-housed mothers. Infants born to group-housed mothers received on average 16 mins/observation of grooming in their first week of life, 13 min in their second and 19 min in their third. Infants born to single-housed mothers received 14, 9 and 8 mins/observation, respectively. The differences in the second and third weeks are significant (Mann–Whitney U-test, $p = 0.05$). The differences are due to the fact that females other than the mother frequently groomed infants.

One notable difference between the behaviour of single- and group-housed mothers is to be found in the amount of time the mothers stayed away from the nest per sortie. The average length of these sorties was much longer in group-housed females than in their single-housed counterparts (Fig. 2). For single-housed females more than 75% of sorties were of less than one minute in duration during the first 3 weeks after birth. Indeed, in the first week more than 90% of sorties were of this length. In contrast, in the group-housed situation sorties were longer and, even in the first week, 50% of sorties were of more than 2 min in length.

The main reason for the reduced sortie length in single-housed females was the high level of visiting behaviour in these females. The number of visits recorded in each observation period was positively correlated to the total amount of time the female was away from the nest in that period. In contrast, in the group-housed females, not only was visiting behaviour a very rarely recorded occurrence but on those

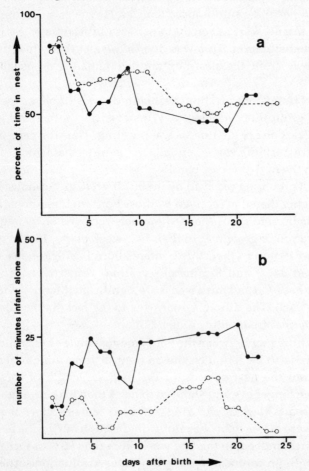

Fig. 1. (a) The percentage of time that the mother spent in the nest-box. (b) The amount of time that the infants were left alone per observation. The open circles and dotted lines represent group-housed females; solid dots and solid lines represent single-housed females.

occasions when it was recorded there was no correlation between it and the amount of time the mother was away from the nest (see Table 1).

Discussions and conclusions

From the data presented here it would seem that, in captivity at least, the group nesting behaviour of female mouse lemurs appears to promote the survival of infants. One possible behavioural mechanism by which infant survival could be promoted is to be found in the time which females other than the mother spend in the nest-box with infants. Arguably the more regular presence of an adult in the

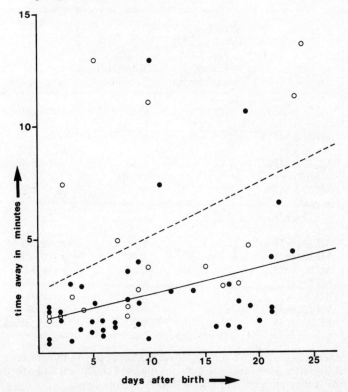

Fig. 2. The number of minutes that the mother was away from the nest per sortie. Dots represent the mean sortie length per observation. The lines represent the regression of sortie length on the number of days after birth, $n = 2100$ for single-house females and $n = 256$ for group-housed females. Solid dots and solid line represent single-housed females; open circles and dotted line represent group-housed females.

nest-box would result in the infants being maintained at an optimum temperature for a longer period. Observations on hand-reared mouse lemurs have indicated that young infants need to be maintained at a temperature of 30°C in order for them to grow normally (Glatston, 1979, 1981). As the temperature in the colony room was lower than this (22–27°C) the body warmth of the mother or another female was needed to keep the infants adequately warm. Infants born to group-housed mothers were therefore maintained at a warmer temperature for a greater proportion of the time than their counterparts born to single-housed mothers. The maintenance of infants at an adequate temperature may also be the reason why the sorties of single-housed mothers are so short. It might be suggested that the visiting behaviour observed in the single-housed mothers is a mechanism for assessing whether the infant is warm enough.

In addition to the advantages accruing to group-housed mothers in terms of infant survival further advantages may be gained in terms of

Table 1. *The correlation between visiting behaviour and time away from the nest-box in single- and group-housed mothers*

	Female I.D.	Number of observations when 'visiting' was observed	Correlation coefficient	Significance
Groups	MF22 & MF11 (yr 1)	8	0.5	not significant
	MF22 & MF11 (yr 2)	2	–	–
	MF20 & MF32 (yr 3)	7	0.3	not significant
	MF31 & MF26 (yr 3)	2	–	–
	MF12 & MF30 (yr 3)	0	–	–
Single	MF20 (yr 1)	22	0.7	$p = 0.001$
	MF12 (yr 1)	15	0.84	$p = 0.001$
	MF20 (yr 2)	7	0.94	$p = 0.001$
	MF12 (yr 2)	9	0.83	$p = 0.01$
	MF26 (yr 2)	15	0.7	$p = 0.01$
	MF12 (yr 3)	9	0.7	$p = 0.05$
	MF22 (yr 3)	7	0.4	not significant

MF12 appears twice in year 3 because she reared her young in the presence of MF30 until the latter gave birth. MF30 then killed MF12's infant and deserted her own. MF12 then adopted and reared MF30's infants alone.

improved foraging possibilities. The longer sorties of the group-housed mothers would lead to more time being available for the collection of food. This may be of limited importance in captivity, where the food supply is never far away, but to the wild female mouse lemur, the longer periods of foraging would be invaluable: she would be able to cover greater distances searching for food and have more time available to find it.

Finally, it is to be noted that in the wild, female nesting groups are larger than those observed in this study. The aggregation of more females could result in the infants being left alone even less frequently, perhaps not at all. This, in addition to the provision of continuous body warmth, could also act as a deterrent to predators, improving still further the chances of survival of young born to females in groups.

Acknowledgements

Thanks are due to the Rotterdam Zoo for providing the facilities for this study and especially to the mouse lemur keepers who so carefully avoided moving the video camera. Further thanks are due to Mr J. van den Bosch who prepared the line drawings and Dr R. Latter who corrected the manuscript.

References

Eisenberg, J. F. (1981) *The Mammalian Radiations.* Chicago: University of Chicago Press

Glatston, A. R. (1979) Reproduction and behaviour of the Lesser Mouse Lemur (*Microcebus murinus*, Miller 1777) in captivity. Unpublished Ph.D. thesis, University of London

Glatston, A. R. (1981) The husbandry, breeding and hand-rearing of the Lesser Mouse Lemur (*Microcebus murinus*) in Rotterdam Zoo. *Int. Zoo Yearbook*, 21, 131–7

Martin, R. D. (1972) A preliminary field-study of the Lesser Mouse Lemur (*Microcebus murinus*, J. F. Miller 1777). *Adv. Ethol.*, 9, 43–89

Wilson, E. O. (1975) *Sociobiology.* Cambridge, Mass.: Belknap

VII.5

Through the territorial barrier: harem accretion in *Presbytis senex*

G. H. MANLEY

Introduction

The purple-faced leaf monkey *Presbytis senex*, endemic in Sri Lanka, characteristically forms reproductive units of the harem type, in which a single adult male is associated with as many as seven adult females. Equally characteristic of the species is a pronounced territoriality, informed by ceaseless male vigilance, signalled by a spectacular male display, and enforced by violent male aggression.

It is in the reproductive interests of a male to acquire as many breeding females as can be maintained within a suitable area which he can defend. It is in the reproductive interests of the female to associate herself with a male who reserves for her and her offspring a food supply and other necessities of life.

The questions examined here are: If, as is the case, resident males commonly oust female intruders as readily and violently as they do males, how are they to acquire harem members? How does a female who wishes to join a harem penetrate the barrier of territorial hostility? And, indeed, where does she come from in the first place?

The study population

Two subspecies of the purple-faced langur were studied in Sri Lanka during 1968 and 1969. These were the typical form *P. s. senex* at Polonnaruwa, NCP, and the montane form *P. s. monticola* on the Horton Plains, CP (Manley, in Roonwal and Mohnot, 1977).

As stated above, the reproductive unit of the species is a one-male troop or Harem composed of one (or very occasionally two) adult males, from 1 to 7 adult females and a number of younger individuals. In all, troops range in size between 3 and 16 members ($n = 47$; troops of both subspecies combined). In addition, and principally accounting for the 'excess' males excluded from the breeding population, there are Male Troops of between 2 and 14 members (Fig. 1). A third, small and mobile social unit of 1–3 members, the Wanderers, completes the picture of the main group types of the species.

As has been argued elsewhere (Manley, 1978), wanderers are primarily maturing males leaving their natal harems, but such animals may be accompanied by other immature individuals of either sex. Wanderers are the chief building blocks of the male troops. Harem-owning adult males who are thrown out at take-overs also gravitate towards male troops and they too may be accompanied by one or more faithful adult females and surviving younger members of their former troops, again of both sexes. As a consequence of such activities, the male troops, whilst predominantly comprising adult and subadult males, not uncommonly include the odd adult and subadult female.

Results
New females in harems

During the course of the study it was observed that established harems both gained and lost adult and subadult female members.

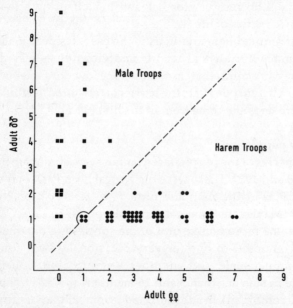

Fig. 1. Adult male and female representation in troops of *Presbytis senex* ($n = 66$; *P. s. senex* and *P. s. monticola* combined).

Of 10 harem troops of *P. s. monticola* studied for 6–21 months (average 14.4 months), two troops acquired four additional adult females in three episodes (+1, +1 and +2) but no females were lost.

Of 17 harem troops of *P. s. senex* studied for 7–15 months (average 11.5 months), no fewer than 10 acquired additional adult or subadult females, 12 in all, and in this population six losses were also recorded (apparently recent acquisitions moving on rather than long-term residents leaving).*

What are the origins of those females who add themselves to existing harems? In one case on the Horton Plains, two adult females, who together joined the three females of Troop D, were known to have belonged to the neighbouring Troop HO (comprising originally the adult male and the two females, a juvenile female and an infant). Rather unusually for this population, Troop HO lost both the juvenile and infant within a short period of time and, either upon the further death of the male or deserting him, the two adult females transferred.

Otherwise, the available evidence suggests that the two principal immediate sources of new females are: (1) the odd adult and subadult females found in male troops; and (2) the subadult females who are companion wanderers.

Of nine male troops of *P. s. monticola*, two possessed three adult females; and of nine male troops of *P. s. senex*, five had five adult and one subadult females. Juvenile females, not considered here, were present in the male troops of both subspecies and represent potential additions to the breeding harem units. In four of the 21 wanderer groups of *P. s. monticola*, a subadult female was present (Manley, 1978).

The territorial barrier

Presbytis senex is a markedly territorial langur. This is very clear from home-range maps based on location data of troops studied over many months, but even more so from actual behaviour. The adult male is far and away the chief actor in this but certain adult females may contribute (see below).

Three features of territorial behaviour may be considered briefly here. Each harem male spends a considerable part of every day 'on

* It should be noted that the study site at Polonnaruwa, an archaeological reserve known as the Sacred Area, was for *Presbytis senex* essentially an island of forest and parkland, and the effects of crowding were conspicuous. In contrast to pristine conditions on the Horton Plains (population density 116 individuals/km^2), at Polonnaruwa density was 327/km^2, social change of every kind was more frequent, many fewer immature animals survived, fewer males survived, and injuries and illness were more widespread.

watch', seated at vantage points which are usually elevated positions such as the crowns of emergents. This has little to do with predation but is the constant monitoring of the integrity of his territory's boundaries.

Adult males have a territorial display, the Great Call Sequence, in which powerful calls are linked to vigorous bounding runs and spectacular dropping leaps. This is given spontaneously (without apparent external stimulus), in response to the similar calling, sight and proximity of neighbours, and as a victory call after ousting intruders.

Finally, harem males engage in very active aggressive defence of the territory. Often galloping at great speed through the canopy or, where appropriate, over the ground, they evict trespassing individuals or whole troops by the very appearance of their approach, by chasing, and by the trauma of actual attack. A territory-owning male, on catching up with an intruder, sometimes mounts a series of 'hit and run' attacks, a canine slashing movement coming at each hit point. More commonly, grappling occurs with the two animals locked together often falling to the ground, accompanied by pulling, scratching, biting and canine slashing. The attack may be so violent that the intruder is killed outright.

Fig. 2. Juvenile male (Troop DRAM), post mortem, showing wounds inflicted in territorial encounter.

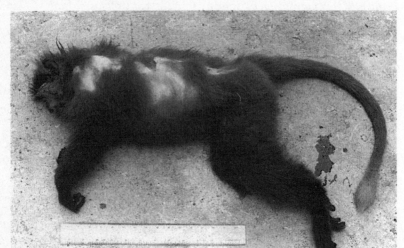

'Bulbtail', a large juvenile male and a member of the Horton Plains male Troop DRAM, was killed in such circumstances by the adult male of the adjoining Troop SY. Post-mortem examination showed extensive denudation, many nail scratch marks, at least eight single or double canine puncture wounds, torn skin and muscle tissue. Death was apparently caused by profound canine puncture damage in the post-otic squamous temporal and occipital region, although it is doubtful if the animal would have survived such severe mauling anyway (Fig. 2).

What is of direct concern here is that *female* trespassers are not exempted: numerous records of pursuits and attacks exist for both subspecies.

Moreover, older/larger adult females of harems occasionally produce a reduced version of the territory-occupancy display and more pertinently, in confrontations between harem troops at least, specifically repel other females by aggressive behaviour which may extend to pursuits, grappling and biting.

Harem accretion

How does an outsider female get through this barrier of territorial aggression? To attempt to answer this for females 'marking time'* in male troops, the general behaviour (indeed strategy) of male troops needs to be outlined.

Located ordinarily within defined home ranges, male troops, in contingents or all together, are frequent trespassers into the areas of adjoining troops. At first sight such excursions seem to have a straightforward foraging function, the animals exploiting (sometimes particular) food trees in the territories of neighbours. When, as often happens, invasion coincides with the absence of the owners who may be at the other end of their range, and when male troop animals flee at the slightest indication of the owners' return, this first explanation would seem to be adequate.

However, a close scrutiny of many trespass episodes, and an awareness that take-overs are a real reproductive strategy of outsider males, lead one to believe that male troop trespassing may serve another precise function: that of continually testing the fitness and resolve of surrounding harem-owners.

---

* Females living in male troops have not been known to exhibit sexual behaviour nor have infants. The only exception to this was sexual presentation and head-shaking to a male who successfully 'distilled' a large, heterogeneous male troop down to a harem-type structure of an adult female, three subadult females and a subadult male.

In the instance of the take-over of Troop ST1 (at Polonnaruwa where all five witnessed take-overs occurred), the beginning of the sustained series of challenges by the take-over male which were to lead to the overthrow of the resident happened in mid-June 1969, but were anticipated by testing incursions by the male Troop SD1M dating back to February if not earlier. Then, ST1 male was seen to be driven back and pursued within his own territory and it may well be that even a single such episode of revealed weakness was sufficient to seal his fate.

If certain males in male troops keep themselves so informed of the state of adjoining harem troops, there is no reason to suppose that an adult female in a male troop is not similarly informed and well aware, for example, of the occurrence of take-overs.

For such a female wishing to join a harem, the whole period during and following a take-over surely would be auspicious. The territorial barrier hitherto so assiduously maintained has been breached and the original male overthrown; to the new male she would be no more alien than the resident harem females; the latter, at this time of much disruption and confusion might well be less inclined to harass a newcomer. Moreover, should the incoming female derive from the same male troop as the take-over male, the process would be further facilitated: she would be accompanying or joining a male who knows her and who might even protect her against resident harem antagonism.

Of the five recorded take-overs of harem troops at Polonnaruwa, three saw a new female added to the harem at, or not long after, the time when the male changed. In two of these additions, when the male came from a known male troop, this troop appears to have provided the additional female as well. Thus, Troop ST1 was taken over by a male from Troop SD1M and an adult or subadult female was seen in spasmodic association with Troop SD1M in the months preceding; Troop POT was taken over by a male from Troop NGM and an adult female was an established member of that troop and disappeared from it at about the same time.

Harem-joining behaviour

Over and beyond the situation just described, it is quite clear that harem accretion takes place in other circumstances as well. Indeed, in the majority of known cases there is nothing to suggest the occurrence of take-overs; the new female was simply discovered within the stable troop structure. How is the territorial barrier penetrated in this case?

The answer apparently lies in the adoption of very particular behaviour by both male and female – and in neither is the behaviour of the kind associated with oestrus and sexuality (in the female: presentation postures and positioning, and head-shaking; in the male: interest in the female's genitalia, etc.). Witnessed episodes on which a harem-joining interpretation has been placed are few but very distinctive; ironically, the most clear-cut involved not a harem troop but a wanderer female joining an established two-male male troop.

To point up the contrast between this behaviour and other forms of active male–female interaction (sexual behaviour has already been mentioned), it is worth noting that: (1) in the territorial expulsion context, the resident male's approach is fast, direct, and variously overtly aggressive as outlined earlier; the female as trespasser usually responds with precipitate flight, often squealing; and (2) in herding, which harem males occasionally employ to control the positions of their own adult females, the male again makes fast and direct pursuits but these tend to be short, include heading-off elements, and no attacks are pressed home; the females flee directly, never very far.

Situations in which females appear to be attempting to join harems are characterised by a number of distinctive elements, and even though the recorded episodes are not identical, in all of these episodes the shared peculiar quality of the whole is evident.

In both the subspecies studied an alien female has been seen to approach a harem male in his own territory and then lead *slowly* away, deliberately keeping a set and rather short distance between herself and the (slowly) following male, *not looking at him* and *always keeping her back towards him* in movement and when seated. Unfortunately in none of these cases was the troop sufficiently well known at the time to be sure of the outcome.

The case of wanderer female 'Miss Myb' joining 'Mr Mash and Mr Hobbs' (Horton Plains population) is particularly striking. She appeared in their territory on a forested hill slope above them and only 15 m distant. On catching sight of the female, staring intently initially, each male, in turn but in an identical manner, moved gradually up the slope towards her, feeding. Both males took all of 20 min to cover this short distance, and frequently managed to keep their (seated) backs towards her despite the general direction of the approach. The female meanwhile maintained station, watching the approach of the males whilst appearing not to do so. When each male reached the tree in which the female was seated, virtually identical rituals were again enacted: on arrival and for 5 min following, each pair made a number of position changes close together in the tree, always with one or both having the seated back turned squarely towards the other. With gaze always directed out and away, the air of studied indifference was pervasive. (In this instance the female remained a close member of the troop and was still with the two males when they were last seen 2.5 months later.)

Whilst a harem entrant may disarm the territorial male in the manner described, the important issue of her reception by the established harem females is unresolved. On theoretical grounds additional females in a troop would seem to have both positive and negative aspects. Brief aggressive interactions between harem females are seen from time to time but the present study can provide no evidence of established females attempting to repel or drive out a newcomer.

Summary

The problem in *Presbytis senex* of harem accretion in the face of strong and indiscriminate male territorial defence has been resolved in two ways. The first is opportunistic and relies upon the occurrence of take-overs and the circumstances accompanying them: the new female moves in when the erstwhile guardian of the territory is being dispossessed and when antagonism to her arrival is likely to be minimal. But the addition of females to harems cannot wait upon take-overs. Entry into, and acceptance by, stable one-male troops is achieved by the performance by female and male protagonists of a distinct ritual, a complex of harem-joining behaviours.

Acknowledgements

This paper stems from a study carried out as a contribution to the Smithsonian Institution's Biological Program in Sri Lanka (Primate Survey), supported by NIMH grant Ro1MH15673-01 and Foreign Currency Program grant SFC-7004.

References

Manley, G. H. (1978) 'Wanderers' in *Presbytis senex*. In *Recent Advances in Primatology*, Vol. 1, *Behaviour*, ed. D. J. Chivers & J. Herbert, pp. 193–5. London: Academic Press

Roonwal, M. L. & Mohnot, S. M. (1977) *Primates of South Asia*. Cambridge, Mass.: Harvard University Press

VII.6

Factors affecting intertroop transfer by adult male *Papio anubis*

D. L. MANZOLILLO

Introduction

Intergroup transfer has been observed in almost all species of primates studied in the wild. This behavior, often remarked upon in earlier studies as a rare event, is now considered a major factor in the reproductive activity of individuals, and an important process in the dynamic changes characteristic of social groups.

In both early and subsequent studies reporting regular occurrence of transfer, it was assumed that transfers functioned to reduce inbreeding and to increase reproductive success (Carpenter, 1964; Lindburg, 1969; Altmann & Altmann, 1970; Itani, 1972; Harcourt, 1978; Packer, 1979; Greenwood, 1980; Cheney, 1983). Although ultimate functions of intertroop transfer have been proposed on the basis of such studies, identification of the proximate causes of transfer behavior has proved difficult. Studies of macaques documented one proximate cause of intergroup migration by males: the availability of fertile females within a discrete mating season. When migration was observed in species which did not have a mating season, further questions arose: (1) What events occur in the group prior to migration that would encourage, or force individuals to move? (2) What characteristics in neighboring groups will attract migrating individuals? and (3) How do individuals become integrated into the new group, and what is the response of the residents? In testing hypotheses derived from these questions, investigators have shown that there is a great deal of variability underlying transfer patterns both within and between different primate species.

Several studies have provided evidence that migrating males tend to choose troops in which opportunities to mate are increased. Packer (1979) found that baboon males chose troops based not just on the number of females, but on the number of cycling females, or even more specifically, the number of females who were actually in consort in a given period. If males are attracted to new troops by the availability of females, we might expect to see evidence of competition between newcomer males and residents for these females. Different studies on the response of resident baboon males to immigrants yield very different results. Whereas Altmann & Altmann (1970), Hausfater (1973), Buskirk, Buskirk & Hamilton (1974) and Packer (1979) report aggression to newcomers by residents, Rowell (1966, 1969), Anderson (1981) and Ransom (1981) found little or no opposition to immigration.

In this paper the results of a study of the proximate factors affecting intertroop transfer by adult male baboons are presented. The two principal, interrelated proximate factors are the behavioral responses of resident males to migrant males ('newcomers') and the availability of potentially fertile females at or near the time of transfer.

Methods

A 32-month study was undertaken on a previously habituated troop of olive baboons (*Papio anubis*) at Gilgil, Kenya. When the study began in 1976, the troop (PHG) consisted of 88 individuals including 9 adult males. It increased in size to 105 individuals in March 1978 with 10 adult males, and then decreased to 88 in August 1979, with 7 adult males. Focal samples (Altmann, 1974) were collected on all adult males in the troop, as well as immigrating adult and subadult males. When possible, emigrating males were followed into their new troop and sampled. In addition to the focal samples, *ad libitum* data were taken on all aggressive, grooming, male–infant and consort interactions in the troop. A daily record of the sexual states of all adult females was kept.

Behaviors used by baboons to communicate with one another have been described in detail elsewhere (Hall & DeVore, 1965; Hausfater, 1975; Ransom, 1981). Some of these behaviors are generally recognized as being aggressive, whereas others are considered to be friendly, cohesive or reassuring (Hall & DeVore, 1965; Kummer, 1971; Ransom, 1981; Pelaez, 1982). Those behaviors identified as aggressive include slapping, biting, chasing, pant-grunting, and certain facial threats. Non-aggressive, friendly or reassuring behaviors include certain vocalizations (e.g. grunting), facial expressions such as narrowing of the eyes, and flattening of the ears, lipsmacking, genital pulling, presenting the hindquarters or grasping the hindquarters of another individual. In certain situations, friendly behaviors are combined into a 'greeting'. Pelaez (1982) describes greeting behavior as 'ritualized behaviors between two individuals characterized in a general way by an approach, an exchange of cohesive signals, and a posterior separation'. In this paper, such greetings are used as a measure of friendly behavior between adult males.

Rates of aggressive interactions and greetings between adult males were used to evaluate the response of residents to newcomers. The data were used to test the hypothesis that resident males will attempt to resist immigration of other males in order to minimize competition for resources, and that this resistance will show up in the form of aggression towards newcomers (Packer, 1979; Wrangham, 1980; Cheney, 1983).

Results

Male membership was relatively stable during the first half of the study (November 1976–December 1977). All immigrations occurred during the second half (January 1978–August 1979). For this reason, and because I have data for two full years in the middle of the study, particular emphasis is placed on comparison of interaction rates between 1977 and 1978. For analysis of some of the data, the study period was divided into nine time blocks of approximately 4 months each, which will be referred to as T1, T2, T3, etc. (Fig. 4). Immigrations occurred in time blocks five through nine.

From the focal samples collected on each male, hourly rates of aggression for 1977 and 1978 were calculated, and compared for each individual from one year to the next. The rates were then combined across individuals to show the overall hourly rates of aggression for all adult males in the troop in each year of the study. Because all immigration occurred in 1978, it was expected that individual and overall aggressive interaction rates would be higher in 1978 than in 1977 (Fig. 1). This was not supported by the data (for individual rates, Wilcoxon matched-pairs signed-ranks test $\alpha = 0.01$). Overall hourly rates of aggression were 0.374 interactions/hour in 1977 and 0.310 interactions/hour in 1978.

Further analysis examined whether the *focus* of aggression might have shifted towards immigrants and whether residents were directing whatever aggression did occur towards newcomers. The proportion of aggression occurring among different dyads: resident–resident, resident–immigrant and immigrant–immigrant was calculated for T5, T6, T7 and T9 (there were too few recorded bouts of aggression in T8 for analysis). The frequencies of interactions that could be expected by chance were determined from the number of dyads in the troop during each time period. For T5–T7, differences were not significant (for T5, $\chi^2 = 0.32, p > 0.05$; T6: $\chi^2 = 1.812, p > 0.05$; T7: $\chi^2 = 1.66; p > 0.05$) indicating that aggression within dyads was not based on residency. Results for T9 were significant (binomial test, $p = 0.002$). There was only one immigrant (AT) in T9. Although he was involved in more dyadic aggressive bouts than expected, all of these occurred during consorts, and all took place after he had been in the troop for several weeks.

I will show that AT's participation in a higher frequency of aggressive interactions than expected was the result of an increase in consort activity in the troop, not simply a result of his status as an immigrant.

Three subject males emigrated in 1977 and 1978 (BR, DR, ST). Although I do not have hourly aggression rates for BR for 1978, both other emigrants showed the lowest hourly rates for 1978. This is worth noting, in that it was reflected in their friendly interactions (see below).

Because immigration patterns did not affect aggressive interactions as expected, I examined friendly behavior between males to see if the arrival of newcomers had any effect. An analysis of greeting interactions shows that overall greeting rates in focal samples increased from 1977 to 1978 (0.408 greetings/hour to 0.724 greetings/hour). If the

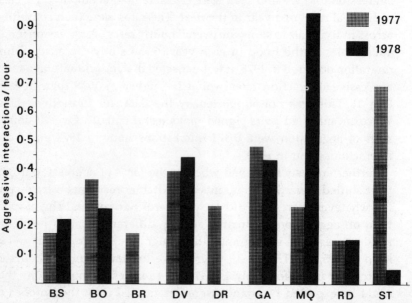

Fig. 1. The hourly rate of aggressive interactions for resident males 1977–1978. (Note that BR was not present in 1978.)

emigrants (DR, ST) are excluded from analysis, a significant increase in greeting rates of all other residents occurred (Wilcoxon matched-pairs signed-ranks test, $\alpha = 0.01$). Resident males showed a significant rise in greeting rates, whereas emigrants showed a drop in hourly interaction rates (greetings and aggression), in the 6–8 months before leaving the troop (Fig. 2). Analysis of greeting bouts by residence category for T5–T9 showed that greetings between residents and immigrants occurred significantly more than expected in each time block, with immigrants initiating most of the greetings (T5: $\chi^2 = 34.25$, $p < 0.01$; T6: $\chi^2 = 13.2$, $p < 0.01$; T7: $\chi^2 = 18.31$, $p < 0.01$; T8: $\chi^2 = 47.05$, $p < 0.01$; T9: $\chi^2 = 108.61$, $p < 0.01$). Immigrants were also greeting each other more than expected.

These data show that PHG males did not direct an increased amount of aggression towards immigrants. In fact, it appears that resident males increased the frequency of friendly behavior towards newcomers. This friendly behavior was initiated by newcomers, but residents responded. Emigrants showed a low level of interaction rates with all other males prior to departure.

Although there was a decrease in aggressive interaction rates for almost all males in the troop, one male (MQ) is notable for the high rate of aggression in his focal samples (Fig. 1). Most of the aggression in

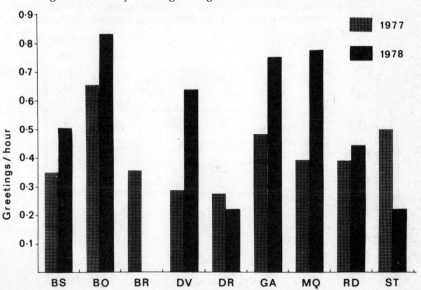

Fig. 2. The hourly rate of greetings for resident males in 1977–1978.

which MQ took part occurred in the context of consort behavior. MQ was involved in more consorts than any other male for most of the study. During the last third of 1978, he was involved in more aggressive interactions ($N = 19$), than any other male. Eighteen out of the 19 aggressive bouts were fights over consorts. Fifteen of the bouts were polyadic, and most involved residents as well as immigrants. Because of the apparent connection between aggression and consort behavior in PHG, overall hourly rates of aggression and availability of consort females were compared throughout the study. Availability of females was determined from the number of days females were seen in consort during the study. Only those consorts occurring around the maximum likelihood of conception, between the seventh day before deflation and the second day after deflation of the females' sexual swelling, were used (see Hausfater, 1975; Packer, 1979). All consort days available for each time block were added together, and divided by the number of observer days during that time period (see Packer, 1979).

Most aggression during the study period was linked to consort activity in the troop. Fig. 3 shows the relation between hourly rates of aggression, female consort days available, and the percentage of aggression involving consorts. It indicates that there are times when conditions in the troop contribute to a low overall level of aggression. If males immigrate when conditions are right, they encounter less

Fig. 3. Variation in rates of aggression, the percent of consort-related aggression and the availability of consorts during the study. Time blocks represent periods of approximately 4 months each (see also Fig. 4).

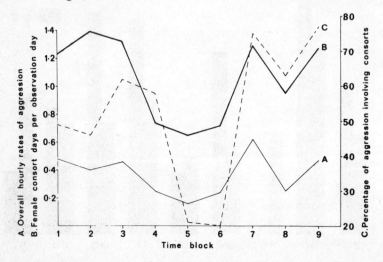

aggression from resident adult males than they would by attempting to immigrate during periods of much consort activity. The only immigrant in T9 (AT), entered the troop at a time when consorts were relatively infrequent and overall aggression rates were low. Consort activity and, therefore, consort-related aggression in the troop increased after his arrival. This is reflected in the higher than expected number ($N = 6$) of dyadic aggressive bouts (all fights over consorts) in which he participated. Although AT met with very little aggression when he first entered the troop, he later became a focus of consort-related aggression. From his first appearance, however, he was also a focus of greeting behavior, participating in a far higher number of greetings ($N = 43$) than aggressive bouts.

It has been proposed that the availability of females in a troop is an important variable affecting transfer into and out of the troop (Harcourt, 1978; Packer, 1979). Two measures of female availability have been mentioned most often: female consort days available and the absolute number of cycling females available per adult male. (It is important to note that cycling females were not always seen in consort during the swollen phase of their cycle, especially those females who had recently begun cycling again after lactation. Therefore, the 'absolute number of cycling females' included those who did not form consortships, as well as those who did. 'Female consort days' are calculated only on the basis of females who formed consort associations with males.) In order to determine which of these factors might be affecting transfer of males, the two measures were compared for the entire study period.

Fig. 4 shows the availability of females in the troop throughout the study period. Migration into and out of the troop by adult males was significantly related to the ratio of cycling females to males, whereas it showed a poor relation to the number of female consort days available (Fisher Exact Probability Test, $\alpha = 0.025$). All immigrations occurred at a time of high availability of cycling females, and low female consort days.

Conclusion

The prediction that resident males will resist immigration of newcomers by directing increased amounts of aggression towards them was not supported by the data. On the contrary, the relationship between residents and immigrants was characterized by an increase in friendly behavior. If aggression is linked to consort activity, then males who transfer into a troop when consort activity is low might encounter low levels of aggression. If intertroop transfer is costly for males because it involves the risk of injury, an individual can reduce that risk by transferring at a time when aggression is at a low level in the troop. A male who will not begin consorting as soon as he enters the troop (the immigrants in this study were slow to attain the same level of consort activity as residents) can optimize his use of females by immigrating at a time when most females are cycling, but not yet consorting. These females will most likely be ready to conceive when the immigrant becomes socially integrated.

What advantages are gained by resident males from friendly behavior exhibited as a response to the immigrants' greetings? All interactions between the newcomer and residents serve to place the newcomer within the social network of the troop. One may expect the newcomer to initiate these interactions. In the absence of some immediate stimulus for competition (e.g. consort females), the cost of an aggressive response by a resident may be greater than any benefit that could be derived from excluding potential immigrants. Whereas the newcomer has the advantage of being able to watch residents interact within the social network, residents probably have little information for assessing the newcomer, who is not yet part of the network. For residents to evaluate the newcomer, they must interact

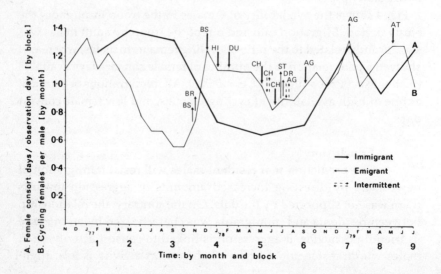

Fig. 4. The relations between the availability of estrous females, cycling females and male migration into and out of Troop PHG. Time blocks 1–9 are shown.

with him. Greeting is less costly than aggression, especially if the current conditions do not present an immediate source of competition. If access to resources, or potential mates is at stake, then immigrant males might become the target for aggression in an attempt to exclude them as potential competitors.

The conditions in the troop at the time of migration are important in helping us to understand the proximate factors affecting intertroop transfer. The response of resident individuals may not be determined only by expectations of future competition, but by factors based in the dynamics of the troop life cycle.

Acknowledgements

I would like to thank the Office of the President of the Republic of Kenya for permission to conduct research. I would also like to thank Mr Richard Leakey of the National Museums of Kenya, and Dr James G. Else of the Institute of Primate Research for sponsoring my fieldwork. Thanks also to Gema Holdings Ltd, for permission to work on Kekopey Ranch, and to R. S. O. Harding and S. C. Strum for the opportunity to work on The Pumphouse Gang. Funds to conduct this research were provided, in part, by Grant Number BMS77-01065 from the National Science Foundation. J. L. Manzolillo provided extra funds to enable me to extend my field time. Thanks to Tony Collins, Phyllis Lee, Fiona Marshall, Ronald Noe, and especially Dana Olson and Tom Pilgram for their comments and help on earlier drafts of this manuscript.

References

Altmann, J. (1974) Observational study of behavior: sampling methods. *Behaviour*, **49**, 227–67

Altmann, S. A. & Altmann, J. (1970) *Baboon Ecology: African Field Research*. Chicago: Chicago University Press

Anderson, C. M. (1981) Intertroop relations of Chacma baboons (*Papio ursinus*). *Int. J. Primatol.*, **2**, 285–310

Buskirk, W. H., Buskirk, R. E. & Hamilton III, W. J. (1974) Troop mobilizing behavior of adult male Chacma baboons. *Folia primatol.*, **22**, 9–18

Carpenter, C. R. (1964) *Naturalistic Behavior of Nonhuman Primates*. University Park, Pa.: Pennsylvania University Press

Cheney, D. L. (1983) Proximate and ultimate factors related to the distribution of male migration. In *Primate Social Relationships: An Integrated Approach*, ed. R. A. Hinde. Cambridge: Cambridge University Press

Greenwood, P. J. (1980) Mating systems, philopatry and dispersal in birds and mammals. *Anim. Behav.*, **28**, 1140–62

Hall, K. R. L. & DeVore, I. (1965) Baboon Social Behavior. In *Primate Behavior, Field Studies of Monkeys and Apes*, ed. I. DeVore, pp. 53–110. New York: Holt, Rinehart & Winston

Harcourt, A. H. (1978) Strategies of emigration and transfer by primates, with particular reference to gorillas. *Z. Tierpsychol.*, **48**, 401–20

Hausfater, G. (1973) Aggressive dominance and reproductive success in yellow baboon (*Papio cynocephalus*). *Am. Zoologist*, **13**, 1261

Hausfater, G. (1975) Dominance and reproduction in baboons (*Papio cynocephalus*). *Contrib. Primatol.*, **7**, 1–150

Itani, J. (1972) A preliminary essay on the relationship between social organization and incest avoidance in non-human primates. In *Primate Socialization*, ed. F. E. Poirier, pp. 165–71. New York: Random House

Kummer, H. (1971) *Primate Societies: Group Techniques of Ecological Adaptation.* Chicago: Aldine–Atherton Inc.

Lindburg, D. G. (1969) Rhesus monkeys: mating season mobility of adult males. *Science*, **166**, 1176–8

Packer, C. (1979) Intertroop transfer and inbreeding avoidance in *Papio anubis*. *Anim. Behav.*, **27**, 1–36

Pelaez, F. (1982) Greeting movements among adult males in a colony of baboons, *Papio hamadryas, Papio cynocephalus*, and their hybrids. *Primates*, **23**(2), 233–44

Ransom, T. W. (1981) *Beach Troop of the Gombe*. London: Bucknell University Press

Rowell, T. E. (1966) Forest living baboons in Uganda. *J. Zool., London*, **149**, 344–64

Rowell, T. E. (1969) Long term changes in a population of Ugandan baboons. *Folia primatol.*, **11**, 241–54

Wrangham, R. W. (1980) An ecological model of female-bonded primate groups. *Behav.*, **75**, 262–99

VII.7

Lasting alliances among adult male savannah baboons

R. NOË

Introduction

The phenomenon of two individuals acting in concert against one opponent is described under labels such as 'coalition', 'alliance', 'support', 'aid-giving', 'interference' and 'intervention' for several primate species (e.g. macaques: Kaplan, 1977; de Waal, 1977; Watanabe, 1979; Datta, 1983; baboons: Hall & DeVore, 1965; Stoltz & Saayman, 1979; Walters, 1980; geladas: Bramblett, 1970; vervets: Cheney, 1983). Packer (1977, 1979) reported that some pairs of non-related male baboons repeatedly joined each other in conflicts. de Waal (1978, 1982) described long-lasting alliances between adult male chimpanzees, the composition of which was dependent on the relative power of each individual male.

I investigated such lasting alliances in adult male savannah baboons. My first goal was to establish whether alliances existed, that is, whether some males joined forces exceptionally often, while they did not join others in conflicts against their partner in the same period. Once alliances are found, a number of characteristics are interesting: the balance between cost and benefit to each ally; the social status of the allies and their opponents; the mechanism of alliance formation; the duration; the circumstances in which alliances are terminated; and the goals reached by means of alliances.

In order to distinguish between idiosyncrasies and generalities I gathered data of three groups of free-ranging baboons. In this paper I present the data from one group. Several alliances were observed in this group which provides the opportunity to illustrate some of the general characteristics of alliances in a nutshell.

Savannah baboons follow the more common pattern of male dispersal (Harcourt, 1978; Packer, 1979). Most adult males present in a baboon group will have been born in another group, although males do sometimes return to their natal group (Altmann, Altmann & Hausfater, 1981) or stay part of their early adult life in their natal group. In Amboseli, Kenya (Hausfater, personal communication; own observation), and Gombe, Tanzania (Packer, 1979), males were rarely observed to take part in mating activities in their natal group, but in a group ranging near Gilgil, Kenya, this was not unusual (own observation).

There are some reasons to believe that kin-selection (Hamilton, 1964) did not play a major part in the selection for behaviour leading to the formation of long-term alliances among adult male baboons. First, many male baboons will spend considerable periods of their adult life in groups where no closely related adult males are present. Even if related males are present, it will not always be possible for the immigrant to differentiate between kin and non-kin. When familiarity provides the proximate mechanism of kin recognition, this mechanism will only be reliable when males have been simultaneously present in the natal group of at least the younger of them. Secondly, when the choice of the ally depends on the circumstances, as in de Waal's chimpanzees (de Waal, 1982), there is another reason to doubt the explanatory value of kin-selection theory for the existence of alliances. Kin-selection theory would predict that in the presence of relatives a male does not change partners. He is expected to prefer consistently his closest relative over more distant relatives and non-relatives (S. A. Altmann, 1979). Therefore Trivers' (1971) theory of 'reciprocal altruism' is considered to give the basic explanation for the existence of lasting alliances.

Among adult male baboons agonistic support is most likely reciprocated with agonistic support. They do not exchange other clearly recognizable (for observer as well as recipient) 'altruistic' acts, like grooming (cf. Seyfarth, 1977; Seyfarth & Cheney, 1984) and food-sharing. Adult male baboons only groom each other in rare cases of extreme tension between actor and recipient (own observations).

At least one of the participants in each coalition is likely to have a direct gain, like the acquisition of a resource, or an indirect gain, like the improvement of status. (See for the distinction between the terms 'coalition' and 'alliance' the section 'Definitions used' below.) In this paper I will present some data on the acquisition of consorts with females in oestrus, as an example of direct gains. The control over females in oestrus is one of the most obvious sources of conflict among the males.

Methods

Site and subjects

The study was conducted in Amboseli National Park, Kenya. The area consists of arid savannah and woodland with one larger swamp and several permanent waterholes (see Altmann & Altmann, 1970).

Alto's Group has been under observation almost continuously since 1971 (see Hausfater, 1975, and Altmann *et al.*, 1981, for site history). During the present study the number of group members ranged between 51 and 57. Six adult non-natal males were present over the whole study period. These males are the subjects in this paper.

Observational methods and observation time

Data were gathered by two observers (B. Sluijter and the author) simultaneously while walking among the well-habituated animals. The data used in this paper stem from data-sets gathered according to the *'ad libitum'* sampling method (J. Altmann, 1974). Agonistic interactions of all group members were described in detail with the help of cassette recorders. All males above 3 years of age had priority in *'ad libitum'* observations. These males were also systematically followed during 20-min 'focal animal samples' with 20-min pauses in between. Most of the time one observer was engaged with these focal samples. The size of the group and the flat and open habitat permitted almost continuous observation of the adult males.

Data for this study were gathered from November 1981 until November 1982. The total observation time was 786 hours during 83 days. The observations of Alto's Group were alternated with observations of another study group.

Dominance criteria

The dominance relationship between two individuals was established on the basis of dyadic agonistic interactions in which only one of the two opponents showed submission. Eight submissive elements were used. In each possible pair of these elements the main direction was the same in 95% of all dyads of individuals in which both elements were observed (Noë, unpublished). Therefore the occurrence of one of these behaviours in a dyadic conflict was considered to be sufficient to establish the dominance relationship of two animals at that moment.

Definitions used

In this paper the following terms will be used.

A COALITION is any event when individuals combine forces, either one comes to the aid of the other(s), or two or more individuals direct aggression on the same target at the same time. An individual is said to INTERFERE in a conflict when he aims aggressive behaviour on one (or both) opponents during the conflict or within 5 seconds after the last agonistic behaviour by one of the opponents. Participants in a conflict in which A comes to the aid of B, who is in conflict with C, will be termed INTERFERER (A), BENEFICIARY (B) and TARGET (C) after Kaplan (1977). Two individuals starting to aggress a third simultaneously will be called PARALLEL AGGRESSORS and their opponent again TARGET. Some parallel interactions observed may in fact have been interferences of which we missed the start. Most of them are, however, true parallel attacks, often synchronized after an exchange of 'sexual greetings' between the two attackers (Noë, unpublished).

The term ALLIANCE will be used to describe the relationship between two individuals, the ALLIES, who combine forces several times over some period of time.

ADULTHOOD is not defined by age, but is an intervening variable for a complex of physical (weight, mantle-length, scrotum size, etc.; see Altmann et al., 1981) and behavioural (dominance relations, sexual

Table 1. *Adult males in Alto's group in 1982*

Rank Jan '82	Rank Nov '82	Name	Date of immigration	Date of emigration	Origin
–	1	(Nol)	Oct '82	–	unknown
1	2	Kojak	Oct '80	–	unknown
2	5	Teta	May '78	–	Stud's Group
3	3	Leroy	Sep '80	–	Stud's Group
4	9	Aly	Dec '78	–	Hook's Group
5	6	Omen	Dec '80	–	unknown
6	7	Kush	Dec '78	–	Stud's Group
–	8	(Harvey)	Jun '82	Nov '82	Hook's Group
7	–	(Ozzie)	–	Mar '82	natal (born Dec '74)
8	4	(Fred)	–	–	natal (born Jan '75)

(NAME): data of this individual not used

behaviour, migration) parameters. The two groups of parameters show temporal relations and are likely to have causal relations in both ways. One typical pattern for a natal male reaching adulthood is a rather sudden rise in rank among the adult immigrants present, followed by emigration. Numerous variations on this theme are possible, however (Noë, unpublished).

Results

During the study period the subgroup of adult males in Alto's Group was rather stable in composition as well as in dominance relationships. The adult male members of the group are given in Table 1. At any time a clear linear rank-order of the six subjects of this study could be constructed.

Shortly before we started regular observations in November 1981, Omen and Aly, then rank 4 and 5, reversed rank in favour of Aly. There were two rank changes during the observation period: a rank reversal between Teta and Leroy (rank 2 and 3) in the beginning and a drop in rank of Aly (was rank 4) below the males Omen and Kush (were rank 5 and 6) half-way through the study. The latter reversal is pivotal in this paper. The reversal itself took place in the period from 25 to 28 August when the group was not under observation. Aly had some fresh wounds on our return on 28 August.

In this paper absolute numbers of coalitions, consorts, etc., observed during 'ad libitum sampling' are used. The only way to control for observer bias in these data would be a comparison with the frequencies in the focal samples. For these relatively rare events the probability of occurrence in the samples is too low, however. Table 2 gives the total number of conflicts in which each male was observed and the total

Table 2. *Bias in* ad lib *observations of coalitions*

Male	Total number of conflicts observed	Total number of coalitions observed
Kojak	541	16
Leroy	454	13
Teta	549	65
Omen	500	62
Aly	527	47
Kush	418	29
Total:	2989	232

number of coalitions in which he was observed to take part. Two conclusions can be drawn from this table. First, our coverage of each individual male was reasonably high. The lower scores for the total number of conflicts observed have plausible explanations other than observer bias. Kush was likely to be the oldest male of this group; he was relatively inactive. Leroy was the most peripheral male of the group. Secondly, a high involvement in conflicts is not correlated with a high involvement in coalitions (Kendall's tau $= 0.33$, $p = 0.235$). If males randomly joined in conflicts when other males were involved, one would expect the two parameters to be more strongly correlated.

Fig. 1 shows that the use of coalitions was not equally divided over the male dyads. This is significant in both matrices ($p < 0.05$ before 27 August and $p < 0.01$ after 27 August, Kolmogorov–Smirnov one sample test, two-tailed). Of three pairs which formed coalitions frequently it was the dominant individual who came to the aid of his partner more frequently than the other way around.

Table 3 gives the number of coalitions formed against each male. It is clear that the two top-ranking males, Kojak and Leroy, were the target of coalitions disproportionally often. These two males were the least involved in the formation of coalitions (see Table 2). The Spearman rank correlation coefficient between the frequencies of being participant in a coalition and being the target of a coalition is negative (-0.77), but not significant. When the type of coalitions used against

Fig. 1. Frequency of coalitions for each pair of males. Coalitions are divided into interferences by the dominant (in the right margin), interferences by the subordinate (at the top) and simultaneous attacks (parallel coalitions). The study period is divided into the periods before and after some crucial rank reversals (see text).

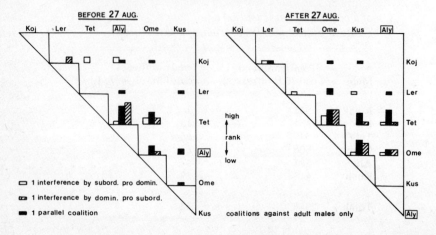

the two top males is compared with the type of coalitions used against the four lower-ranking males it shows that coalitions against the top-ranking males started relatively more often as parallel aggression and against the lower ranking ones as an interference (chi-square = 7.71, $p < 0.01$, two-tailed test).

The drop in rank of Aly at the end of August coincided with changes in the alliances among the adult males. Teta changed his attitude towards Aly and Omen: after the rank reversal his choice in Aly–Omen conflicts was always against Aly where previously he was always aggressive towards Omen when he interfered. Before, as well as after, this change Teta sided with both Aly and Omen when they were involved in conflicts with one of the other males (13 times with Aly, 12 times with Omen; directly after Aly's drop in rank Teta once supported another low-ranking male, Harvey, against Aly). Omen and Kush formed several coalitions after 27 August, mainly (6 out of 12) against Aly. Coalitions were often formed during attempts to take oestrus females from consorting higher-ranking males. These attempts were almost always successful in Alto's Group. In Table 4, consorts with oestrus females are categorized according to the way in which they were obtained. For each male the number of consorts he obtained from each category is given. Although the males Kojak, Teta and Omen all obtained a considerable number of consorts, they differed considerably in the way they obtained them and when. Alpha-male Kojak obtained the lion's share of the first consorts of the day. There are reasons to believe that the chances of fertilization are on average better for these consorts than for consorts obtained from another male later

Table 3. *The number of times a male was the target of a coalition of at least two adult males*

Male	Target of interference	Target of parallel aggression	Total
Kojak	6	15	21
Leroy	7	11	18
Teta	1	0	1
Omen	5	4	9
Aly	10	3	13
Kush	5	2	7
Total:	34	35	69

in the day (van Noordwijk, in press; Noë, unpublished). Middle-ranking males Teta and Omen obtained most of their consorts during the day directly from another male. For them, coalitions, which are hard to form in the sleeping trees, were an important instrument to achieve this.

Discussion

A strict definition of what is an 'alliance' based on frequencies of coalitions is meaningless, because the observed frequencies depend on parameters such as the number of oestrus females and total frequency of conflicts in the group and on the quality of observations (visibility, percentage of time spent in observation). The typical alliances found in Amboseli showed at least five coalitions per 200 hours of observation (20 field-days).

The consistency of the mutual preference of the partners for each other over certain classes of opponents forms an important distinction between the relationships I call alliances and the casual relationships of males who form coalitions incidentally. I use the term 'certain classes of opponents' to leave room for special cases in which a male defends a female or infant against his ally. Alliances between males are mainly used against other males, but occasionally against members of other classes, possibly to show a willingness to support the ally without too much risk. I consider both an interference to the advantage

Table 4. *How the males obtained their consorts*

	Male						
	Koj	Ler	Tet	Ome	Aly	Kus	Total
Not direct from other male							
First consort of the day	25	5	3	3	2	6	44
After female was abandoned	1	0	4	3	3	2	13
Direct from other male							
Without conflict	1	2	5	4	1	0	13
Dyadic conflict	1	2	1	2	0	1	7
Polyadic conflict	0	2	7	8	0	1	18
Total:	28	11	20	20	6	10	95

1 unit = 1 consort > 30 min.
Consorts during the period of the menstrual cycle of 10 days to 1 day before the deturgescence of the sex-skin

of somebody and an act of parallel aggression with somebody to be a choice in favour of that individual. The consistency of an individual's preference for a partner is statistically ascertained according to the binomial test when he is observed to join this partner six times against a certain class of opponents but never observed to support one of those opponents against the partner in the same period. Under these restrictions Teta and Aly had an alliance before the Omen–Aly reversal and Omen and Teta afterwards. Omen and Teta had a support relationship of a weaker kind in the period before the reversal, and after the reversal the pairs Teta–Aly and Omen–Kush had such semi-alliances. The change of partners Teta showed half-way through the study makes the existence of unknown kinship-ties a less likely explanation for his partner choice.

What did the four males involved in alliances have in common? All four were past-prime males. Aly, Omen and Kush had worn teeth; Teta was likely to be younger, but was also past his prime. He was thought to be a young adult when he immigrated in 1978 and was alpha-male until Kojak came in 1980. Teta, Aly and Kush were immigrants of long residence: 3 years at the start of the study. Omen was a resident for 20 months at the time of his rank reversal with Aly. After Teta's drop in rank at the beginning of the study they were the four lower-ranking of the six males. The real alliances were found between males with adjacent ranks in the middle of the adult male rank-order. A similar picture was found for a second study group (Noë, unpublished). Thus the dominance structure in the group seems to be the most important factor, but a minimum period of shared residence is likely to be necessary for a solid alliance.

Alliances are not only expected to bring the subordinate ally closer in rank to the more dominant one, but also to be formed between individuals who are already close in rank based on their individual strength. Each male is likely to prefer as an ally the highest-ranked male who is willing to reciprocate. The willingness of a male to reciprocate support will depend on the ability of the solicitor to contribute, in his turn, substantially to the combined power. Seyfarth (1977) and Colvin (1983) give models explaining high frequencies of grooming between individuals of adjacent rank. Packer (1977) states that alliances between adult males and other age-sex classes are not very likely, because of the difference in fighting abilities. There is no reason not to accept the validity of this argument within the adult male class, since fighting abilities of males apparently differ considerably too (own observations).

Even if the allies are of adjacent rank, the difference in rank between them will cause asymmetry in their alliance. First, in most coalitions the risk of taking part is not the same for the participants, since the difference in strength with the target is not the same. A coalition is often cheaper for the dominant than for the subordinate partner. This makes it likely that alliances are out of balance as far as the number of acts is counted. Secondly, if some kind of resource is taken from a third individual through combining forces, the dominant could potentially take hold of it in all cases. This suggests that it is the dominant ally who determines how benefits of this nature are shared out, which means that there are more possibilities for the dominant than for the subordinate ally to use the alliance. A third asymmetry may result from the difference in the goals the two participants try to reach through the alliance.

Although the alliances observed consisted of irregular series of cooperative and altruistic acts, one can ask whether they are the result of two individuals using a strategy resembling the TIT FOR TAT strategy in an iterated Prisoners' Dilemma game as described by Axelrod & Hamilton (1981).

The alliances found were clearly reciprocal in a sense that both partners invested energy to the advantage of the other. The asymmetries mentioned above do not obstruct the formation of alliances under the assumptions of the model, as long as both partners have a net gain and there is uncertainty for both over the length of the period in which the alliance is useful. But did all allies reach recognizable goals? In one alliance (Omen–Teta) both partners obtained consorts with help of the partner. In another alliance (Teta–Aly) the gain for the dominant partner was of the same kind. Kush and Aly seemed to be able to improve their dominance status with the help of their partner, but direct improvements of their mating success could not be recognized.

In some coalitions the low-ranking participants seemed only to join higher-ranking males against males holding a consort for the very small chance of obtaining the consort themselves. In the field it is often hard to make a distinction between an interference on behalf of another male or an interference that failed to give the actor the immediate reward he was after. Males may take part in alliances for a very marginal bettering of their position. Older, low-ranking males that have nothing to lose are likely to follow such strategies. A male that is effectively excluded from reproduction may take high risks for small chances.

To be able to follow even such a simple strategy like TIT FOR TAT, a male who wants to form an alliance with an unrelated male has: (1) to choose the right partner, while taking into account the relationships between the other group members; (2) to control for cheating or defection by the partner; (3) to pick the right moments to show his own willingness to invest; and (4) to evaluate constantly the cost–benefit ratio of the alliance. This is likely to require simultaneously a number of skills not needed in a society without cooperation. It is possible that the threshold this requirement poses was passed when support behaviour among immature paternal siblings was selected for. Among immature natal males the cost of altruism resulting from imperfect control on the reciprocation may be balanced by an increase in inclusive fitness if recipients are likely to be related (J. Altmann, 1979). This model requires that the lifetime reproductive success of a male is affected by costs or benefits of support given or received during his stay in the natal group. The final stability, however, of the strategy of alliance formation is more likely to depend on the reciprocal character of the alliances.

Acknowledgements

I gratefully acknowledge the permission to carry out research in the Amboseli National Park from the Office of the President, the National Council for Science and Technology, the Ministry of Environment and Natural Resources, and the Warden. I thank J. Altmann, S. A. Altmann and G. Hausfater for their permission to work in the Amboseli Baboon Research Project. The following persons gave support to my study: M. Buteo, J. Else, J. van Hooff, R. Mututua, M. Kirega, M. Pereira, A. Samuels, S. Sloane. B. Sluijter did a large part of the field-work and the data-processing. J. Altmann, S. A. Altmann, G. Hausfater, J. van Hooff, P. Lee, M. van Noordwijk and C. van Schaik gave useful comments on earlier drafts. Financial support: WOTRO to the author and grants by U.S.P.H.S. (MH 19617) and N.S.F. (BNS 76-0659) to the long-term project.

References

Altmann, J. (1974) Observational study of behavior: sampling methods. *Behavior*, **49**, 227–67

Altmann, J. (1979) Age-cohorts as paternal sibships. *Behav. Ecol. Sociobiol.*, **6**, 161–9

Altmann, J., Altmann, S. A. & Hausfater, G. (1981) Physical maturation and age estimates of yellow baboons, *Papio cynocephalus*, in Amboseli National Park, Kenya. *Am. J. Primatol.*, **1**, 389–99

Altmann, S. A. (1979) Altruistic behaviour: the fallacy of kin deployment. *Anim. Behav.*, **27**, 958–9

Altmann, S. A. & Altmann, J. (1970) *Baboon Ecology. African Field Research.* Chicago: University of Chicago Press

Axelrod, R. & Hamilton, W. D. (1981) The evolution of cooperation. *Science*, **211**, 1390–6

Bramblett, C. (1970) Coalitions among Gelada baboons. *Primates*, **11**, 327–33

Cheney, D. L. (1983) Extrafamilial alliances among vervet monkeys. In *Primate Social Relationships*, ed. R. A. Hinde. Oxford: Blackwell

Colvin, J. (1983) Familiarity, rank and the structure of rhesus male peer networks. In *Primate Social Relationships*, ed. R. A. Hinde. Oxford: Blackwell

Datta, S. B. (1983) Patterns of agonistic interference. In *Primate Social Relationships*, ed. R. A. Hinde. Oxford: Blackwell

Hall, K. & DeVore, I. (1965) Baboon social behavior. In *Primate Behavior*, ed. I. DeVore. New York: Holt

Hamilton, W. D. (1964) The genetical evolution of social behaviour, I and II. *J. Theor. Biol.*, **7**, 1–52

Harcourt, A. H. (1978) Strategies of emigration and transfer by primates, with particular reference to gorillas. *Z. Tierpsychol.*, **48**, 401–20

Hausfater, G. (1975) Dominance and reproduction in baboons (*Papio cynocephalus*). *Contrib. Primatol.*, **7**. Basel: Karger

Kaplan, J. (1977) Patterns of fight interference in free-ranging rhesus monkeys. *Am. J. Phys. Anthropol.*, **47**, 279–88

Noordwijk, M. A. van (in press) Sexual behaviour of Sumatran long-tailed macaques. *Z. Tierpsychol.*

Packer, C. (1977) Reciprocal altruism in *Papio anubis*. *Nature*, **265**, 441–3

Packer, C. (1979) Inter-troop transfer and inbreeding avoidance in *Papio anubis*. *Anim. Behav.*, **27**, 1–36

Seyfarth, R. M. (1977) A model of social grooming among adult female monkeys. *J. Theoret. Biol.*, **65**, 671–98

Seyfarth, R. M. & Cheney, D. L. (1984) Grooming, alliances and reciprocal altruism in vervet monkeys. *Nature*, **308**, 541–3

Stolz, L. P. & Saayman, G. S. (1970) Ecology and behaviour of baboons in Northern Transvaal. *Ann. Transv. Mus.*, **26**, 99–143

Trivers, R. (1971) The evolution of reciprocal altruism. *Q. Rev. Biol.*, **46**, 225–85

Waal, F. M. B. de (1977) The organisation of agonistic relations within two captive groups of Java monkeys (*Macaca fascicularis*). *Z. Tierpsychol.*, **44**, 225–82

Waal, F. M. B. de (1979) Exploitative and familiarity dependent support strategies in a colony of semi-free living chimpanzees. *Behaviour*, **66**, 268–312

Waal, F. M. B. de (1982) *Chimpanzee Politics*. London: Jonathan Cape

Walters, J. (1980) Interventions and the development of dominance relationships in female baboons. *Folia primatol.*, **34**, 61–89

Watanabe, K. (1979) Alliance formation in a free-ranging troop of Japanese macaques. *Primates*, **20**, 459–74

VII.8

Fitness returns from resources and the outcome of contests: some implications for primatology and anthropology

N. G. BLURTON JONES

Introduction

Parker (1974) and several other investigators, e.g. Hammerstein & Parker (1982), argue that contests over resources will usually be settled by asymmetries in the contest – either inequality in value of the resources to the contestants or inequality in strength or fighting skill (resource-holding potential). The individual for which the resource represents the highest benefit will be expected to be able to expend most costs to gain the resource and thus to prevail in the contest. So far, apart from important but neglected indications in Parker (1974) and some comments by Clutton-Brock & Harvey (1976) and Popp & DeVore (1979), the theoretical studies have not been directed towards the ecological circumstances of contests. Thus they have yet to yield systematic predictions about the contest behavior to be expected under various ecological circumstances. Indeed it may be too early in the development of the models of contests over resources for such an attempt to be wise. However, a very simple addition to the theory may help us to begin to see a way to apply the contest models to the issue of why particular creatures show their particular forms of allocation of resources, e.g. sharing, rank order, territory, hoarding, and various forms of male competition over females.

What circumstances give rise to asymmetric contests, and in particular to contests asymmetric with respect to resource value? One such set of circumstances are those that determine the shape of the curve of fitness gained from a resource plotted against amount of resource currently held (Fig. 1). The main purpose of this paper is to suggest consequences of such curves for the outcome of contests over resources, and to suggest ways in which we might extend this way of thinking to explain the variety of behaviour over resources. In doing this, my own main interest is in explaining origins of human behaviour and variation in behaviour with circumstances. But the principles should be applicable to any animal and are thus appropriately debated in a biological forum.

Four possible curves of fitness benefit against amount of resource held are illustrated in Fig. 1. They are (A) a straight line, (B) a diminishing-returns curve, (C) an increasing-returns curve, and (D) an S-shaped curve. These curves differ in the extent of fitness gains to contestants at different positions on the x axis. These are shown on Fig. 1 as y_1, y_2, and y_3. x is the size of the contested unit of resource. y is the fitness gain from the resource of size x to an individual low on the x axis (y_1), high on the x axis (y_2), and in an intermediate position (y_3).

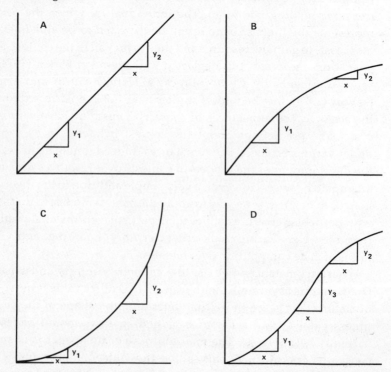

Fig. 1. Curves modelling fitness benefit as different functions of the amount of resources held by an individual: (a) straight line; (b) diminishing-returns curve; (c) increasing-returns curve; (d) S-shaped curve. x = the size of the contested unit of resource. y = fitness gain.

Assumptions and argument

1. Contests over resources will be settled by asymmetries in the contest: (a) inequality in the fitness value of the resources to the contestants, or (b) inequality in strength or fighting skill (resource-holding potential, RHP).

    That such behaviour is an Evolutionarily Stable Strategy (ESS) under a wide range of circumstances has recently been confirmed by Hammerstein & Parker (1982).
2. We will consider only cases in which the RHP of the contestants is equal, or near enough equal to be considered so.
3. We will ignore differences in relatedness between individuals.
4. We will consider only contests between two individuals, not exploring potentially interesting questions about alliances.
5. We will ignore individual differences in steepness of slope (not shape) of the curves (see Discussion).
6. The y axis of the curves illustrated is a measure of the fitness benefit that accrues from the amount of resource held.
7. The x axis of these curves is the amount of resource held, measured in normal units, e.g. grams of food, number of calories, number of females in a harem, square metres of territory.
8. Individuals often occupy different positions on the x axis at any moment, i.e. they often hold different amounts of resource (or differ in nutritional status). (Individuals may often be at different positions on the x axis by chance alone. But in addition some kinds of resource may increase the likelihood of individuals being at different positions, e.g. one being well fed, the other being hungry.)
9. If the position of each contestant on the x axis can be approximately determined, the potential fitness gain to each individual from the next contested item of resource can be estimated and compared.
10. The contestant that will gain least from the added resource will relinquish it to the other contestant.

Thus if we know the shape of the curve for a particular resource for the species concerned, we can predict the outcome of contests. Predicting the outcome of repeated contests may allow us to predict features of distribution and accumulation of resources.

Results of the argument: outcome of particular contests

1. On curve A in Fig. 1 (a straight-line relation of fitness to amount of resource), no matter where an individual starts on the x axis, a unit of resource acquired brings the same gain in fitness ($y_1 = y_2$ in Fig. 1A). There will be no asymmetry in resource value. Under these conditions of symmetry all contestants have the same prospective fitness gain, no matter what resources they hold. Contests will not be resolved by asymmetry in resource value.
2. On curve B (a diminishing-returns curve), a unit of resource acquired brings a greater gain in fitness to the individual that is lower on the x axis ($y_1 > y_2$ in Fig. 1B). There will be an asymmetry such that an individual that has more should give way to another individual that has less.
3. On curve C (an increasing-returns curve), a unit of resource acquired brings a greater gain in fitness to the individual that is higher on the x axis ($y_2 > y_1$ in Fig. 1C). Thus asymmetry will arise but in this case the individual that has least will give way to the one that has most, yet another portion of the resource is defendable.
4. Curve D (the S-shaped curve) is a combination of curves C and B. If contestants are both in the C portion, the outcome will be as in C, ($y_3 > y_1$ in Fig. 1D), individuals that have more will win contests. If the contestants have both reached the B portion, the outcomes will resemble those for B ($y_3 > y_2$ in Fig. 1D). Some contests between an individual low on the x axis and an individual high on the x axis will result in symmetry ($y_1 = y_2$ in Fig. 1D), and others will result in asymmetries in either direction. The ultimate outcome of the operation of the S-shaped curve in a population may differ from either C or B (see below).

Outcome of repeated contests

The argument so far seems clear and secure. But we may be able to take the argument beyond the outcome of particular contests to consider the outcome of repeated contests in a population. Some intuitive conclusions are worth reporting because they show the directions in which the model may develop.

For example, the diminishing-returns argument, that an individual that has more will give way to one that has less, would result in equal shares among all individuals. Any time an individual has more, it is vulnerable to individuals that have less. Any time an individual has less, it is likely to be able to win resources from an individual that has more.

Conversely, the increasing-returns curve should result in one individual obtaining all the resources (clearly only a local phenomenon, even in human societies!).

The S-shaped curve may then lead to a situation where part of the population accumulates substantial resources and part remains with very little indeed. However, this conclusion probably needs particularly careful checking, for there will be points low on the curve where the slope is the same as at a point high on the curve. At these points a 'poor' individual will be equally matched with a 'rich' individual. But it does seem safe to conclude that the diminishing returns at the top of the curve ensure a different result from the increasing-returns curve; there is no pay-off for accumulating all of the available resources.

The straight-line relation of fitness to amount of resource held allows no determination of outcome by asymmetry in resource value. This must imply either (1) less ready resolution of contests and more fighting, or (2) determination of outcome by asymmetry in resource-holding potential. Thus inequalities in size and fighting ability will be more frequent factors determining outcome, and perhaps will be more obviously selected for. Since size and weaponry do not change from moment to moment, in contrast with the amount of resource held (and therefore the asymmetry in resource value), the outcomes of contests will be predictable from characteristics of the individual. In other words individuals can then be ranked in order of their chance of winning contests against each other. This is usually known as a dominance hierarchy. So we must conclude that dominance rank will be an outcome of the straight-line relation of fitness against amount of resource.

Some real examples?

For these suggestions to be of any use we must be able to attach real-life resources to these theoretical curves.

In a previous paper (Blurton-Jones, 1983), I suggested that food is a resource that belongs on a diminishing-returns curve (B). I then argued that large prey, taken by different individuals at unpredictable long intervals, might be a resource that frequently puts individuals on markedly different positions on curve B. I proposed the resulting frequent reversals of asymmetry as a possible origin of human food-sharing. Parker (1974) made a similar suggestion about the decay in value of a large food item over time as an individual devours it. This individual soon relinquishes the object to a latecomer. The series of events that Parker describes foreshadows the descriptions of food-sharing by baboons (Hausfater, 1975) and by chimpanzees (Teleki, 1973). However, the existence of begging gestures in chimpanzees and the extensive food-sharing described for pygmy chimpanzees (IPS, 1985), suggest that food-sharing in primates can go well beyond the process that Parker describes.

My food-sharing (tolerated theft) argument depends on the added argument that some kinds of food (large items, found at unpredictable intervals and places) tend to result in individuals often being at different points on the x axis, and that these differences are often reversed on subsequent occasions. The converse might be expected for small food items evenly spread or predictably located. Also the gains represented by each portion of the food item would be more nearly equal than if it was a large item extending over a substantial length of the x axis on curve B. This, and the even spread of food, might result in individuals being more often close to each other on the x axis, thus giving rise to more nearly symmetrical contests. Then contests would be resolved by differences in RHP. In a group of individuals that live together the outcome might show itself as a ranking of contestants. Thus dominance orders that concern food might be expected in species that exploit smaller, more evenly spread foods, and to disappear in those that exploit large unpredictable resources. Since a dominance hierarchy is almost by definition a series of contests that are settled by resource-holding potential (consistent features of individuals), it must be more likely to occur when differences in the value of the resources to the contestants are small. This might indeed be likely when resources are scarce and no individual fills its 'dietary requirement' early in the day. But scarcity cannot be the key issue. Scarcity varies among many kinds of social system and has even been invoked to explain sharing as well as hierarchies. Clutton-Brock & Harvey (1976) and Parker & Rubenstein (1981) discuss these same issues and point out that several experiments have shown that individuals will temporarily overturn their rank if sufficiently hungry (Boelkins, 1967; Castell & Heinrich, 1971; Hazlett, Rubenstein & Rittscheff, 1975; Rubenstein, 1977). The implication is that the nearly starved individual stands to gain more in a contest over food than do the dominant individuals. It is important to remember that my argument concerns the relative value of the resource to the contestants, not its overall scarcity or absolute value.

If a rank order may be expected when contests occur over resources that give the straight-line fitness relation, then females as a resource contested by males might provide particularly interesting examples. For many species, one can assume that the fitness gain from winning access to any additional female is as great as the gain from each previous female – curve A, constant returns. In such a case more females will always be worth fighting for. In this case since all females are of equal value to all males, access to females can only be determined by resource-holding potential, and selection will favour larger, stronger, more dangerous males. The pinnipeds and deer come to mind as obvious examples.

But this constant return only applies when each additional female has the same likelihood of raising her offspring, as does the first female acquired. In many mammals, when the female is solely responsible for parental care, this will be the case. But in many birds, and in some mammals, when paternal care facilitates the survivorship of offspring, females will belong on curve B. As the male's care is divided among more and more offspring so their survivorship declines, and the male's fitness gain declines (e.g. Silverin, 1980). These diminishing returns imply that the added females will not be defendable against a male that has no females. Thus the cost of defense of added females will limit male gains from polygyny. This limit will be well below the point at which no additional offspring could be raised from added females in the absence of rivals. Males will be expected to retain exclusive access to only one or at most a few females.

This argument applies to competition over, and defence of sexual access to, females. It does not reduce the gains from surreptitious copulations with additional females, whose offspring may be raised by some other male and the females that he defends. The argument implies that monogamy (in the sense of an attempt to retain exclusive access to one female and not more than one female) may evolve more easily than hitherto seemed likely but leaves unmodified the view that male faithfulness may be difficult to evolve. Nor does this argument replace the threshold theory of polygyny, which concerns the costs and benefits for the female, on the assumption that a male will always gain from polygyny. I have argued that he may not, and we thus need to examine the situation of both male and female in deciding whether we expect to see polygyny or not.

Resources that give the increasing returns curve, C, are harder to imagine. Females whose cooperation increased the number of offspring that they raise would be an example. Curve C may be particularly interesting in the case of human cultures. In such a case the individual that has more stands to gain more by acquiring another unit than does an individual that has less. Such economies of scale may exist in some kinds of agriculture, herding, and technology.

It is tempting to suggest that an animal territory may arise from a resource that follows the S-shaped curve, D. In so far as the sigmoid curve gives a population of resource-holders and a population with no resources it may parallel the existence in territorial species of a population of territory holders and a population of 'floaters'. Contests between neighbours and the limits to territory size seem to be well accounted for by diminishing returns from added area as suggested by Tullock (1983). The economic defensibility of territory (Brown, 1964) does seem to be an issue in which we should take a closer look at the interests of BOTH contestants. Parker (1974) and Hammerstein & Parker (1982) imply that we should more energetically seek differences in resource value of territory to holder and intruder, rather than look for 'conventional' solutions which will apply to Hawk–Dove games and not the more realistic war of attrition contests. Davies & Houston's (1981) classic paper on wagtail territories is a successful example of examining differences in resource value to owner and intruder. There are at least two steps needed in a theory of territory. One concerns whether land is a resource. This has to do with patchiness and predictability. The next concerns whether an individual can defend such an area. A good patch will be attractive to all individuals. Why do good patches sometimes end up occupied exclusively by one individual and other times by many?

Discussion

My propositions have some antecedents in papers by Parker (1974) and Popp & DeVore (1979). Parker included some consideration of changes in value of a resource with time. Examples were the value of an item of food decreasing as the holder consumed the resource, value of a territory increasing as the owner learns how best to exploit it. To this extent Parker's original theory of contests over resources did attend to implications of the nature of resources for the outcome of contests. Since 1974 this aspect has been overshadowed by the necessary exercise of establishing the conditions under which differences in resource value will be a criterion that determines the outcome of contests. In this paper I have tried to set out some features of resources that may influence the relative value of resources to contestants and hence the outcomes, and the patterns of distribution of resources.

Thus I am suggesting that we direct attention to this neglected aspect of Parker's original work, and I argue that it gives the potential for putting the theory to work in a unified account of resource allocation that may explain many features of social organisation, and may predict associations between resources and behaviour.

Of particular interest in connection with rank orders is Popp & DeVore's (1979) statement that we should not expect rank orders to generalize across resources. They emphasize that some resources are of intrinsically greater fitness value to one individual than to another, comparing as an example the value to a low-ranking monkey of a piece of food and of its baby. The low-ranking animal will relinquish a piece of food but will defend its baby, and probably win against a bigger stronger animal who will stand to gain little from harming the baby. However, Popp & DeVore gave no explicit attention to the effect of different curves of fitness against amount of resource held upon the value of a particular contested item of resource to different individuals.

An important implication of both Popp & DeVore (1979) and of my model is that difference in the value of a resource to the contestants says more about expected behaviour than does scarcity or abundance of resource for the population as a whole. We should anyway have been suspicious of interpretations and predictions based on differences in scarcity. If populations are density-dependent, they should increase up to a limiting value of resources. Our basic assumption should be that a resource will be scarce at all times, unless we can show that some other resource is the limiting resource. We should be particularly wary of assuming that food is not a limiting resource and that its scarcity differs between populations. Explanations that depend on postulated differences in scarcity are widespread in accounts of cultural adaptation. More surprisingly there are instances implied in accounts of animal behaviour, particularly primate social organization.

Subsequent work should be directed towards further developing the theory and then predicting the behaviour that is to be expected in association with various resources. There are at least two steps involved: (1) testing my suggestions about the systems of resource tenure or social relations that arise from repeated contests; and (2) deciding which resources, or which circumstances, give a particular curve for a particular creature. Then we can look to see if the predicted association between resource and behaviour occurs in nature. To predict about contests between particular individuals on particular occasions we would need to be able to tell whether individuals are at different points on the x axis, and if so which individuals are higher and which lower on the x axis.

The decision to ignore individual differences in steepness of the curves needs some discussion. The slope of the curves could be subject to individual variation. For instance, a fertile male gains more fitness from each female added to his harem than does an infertile male (giving the counter-intuitive prediction that the less-fertile males will not fight hard enough to compensate for their low fertility by gathering more mates). This individual variation would lead to unexpected outcomes of particular contests. However, I have assumed that such variation will be normally distributed and unrelated to the other parameters considered, and that it may therefore be ignored. It will introduce variation in outcome but not change the overall conclusions.

The assumption that differences in resource-holding potential (RHP) can be ignored is also untested. RHP comes back into my argument as the only realistic solution to the symmetric contests produced by the straight-line relation. But we do not know how to expect moderate differences in RHP to be weighted against moderate differences in resource value, since there has been little explicit investigation of the way asymmetries in RHP and resource value can be expected to interact other than Parker (1974) and Hammerstein (1981).

The model of the origin of food-sharing offered in 1983 and summarised here draws attention to some features of human food-sharing that have often been neglected. An undercurrent of hostility associated with sharing has often been noted in the literature (e.g. Lee, 1969, 1979), yet not linked to possible reasons for the existence of sharing. The difficulty with which hunter-gatherers in sharing societies change to habits of accumulating possessions or wealth has also often been reported. The explanations that are offered usually centre on the reliability of the environment, the function of sharing networks as an insurance policy (presumably the environment is not all that reliable after all), a feckless personality, or social inertia. The tolerated theft model of sharing implies that wealth would not be defendable in the circumstances in which many hunter-gatherers live. Only if the conditions for tolerated theft disappear will we expect to see accumulation of wealth or stores of food. The conditions for tolerated theft (food found unpredictably and in large packages) are unusual, seen in the perspective of the animal kingdom as a whole but are rather typical of many of the contemporary hunter-gatherer societies that have been studied. A further implication of this view is that agriculture may have been difficult to begin, unless conditions for tolerated theft had disappeared. The harvest of the first farmer in a group would not be defendable. Harvests are defendable if all the members of a group produce a harvest at the same time. If 'catches' occur at the same time as each other, the condition for tolerated theft (catches made at different times by different individuals) disappears. This could explain the occurrence of food storage in some hunter-gatherer societies, where there are sharply synchronized gluts like salmon runs or caribou migration. Societies using such resources did store food. But there are more hunter-gatherer societies that also stored food and it is not clear whether they fit the model. The effect of a hard winter on the fitness curve for food during the summer is not obvious. The model does suggest that the occurrence of storage is not determined by the knowledge of storage technology but by ecological circumstances. Nonetheless there is a conspicuous difference between hoarding in man and in other animals. Most human hoarding involves accumulation of a large store in one place. The commonest form of hoarding in animals appears to be scatter-hoarding, and hoarding seems to occur mostly in territorial species.

One may wonder whether tolerated theft penalizes good hunters to the extent that they should attempt to leave the group. Kaplan & Hill (1985) and Vehrencamp (1983) consider this issue more extensively, using other models with many interesting consequences. Scroungers are presumably always at liberty to follow the successful hunter. In the tolerated theft model, although good hunters will not receive the advantage that may appear due to them, they will nonetheless gain from their efforts. They are likely to gain more from their efforts than from the efforts of less good hunters – an equal share of their own greater returns will be greater than an equal share of a poor hunter's lesser returns. Further elaboration of the tolerated theft theory suggests that under most conditions poor hunters will be the ones to opt first for a pure scrounging strategy, or a less active mixed hunter–'resting' strategy. Good hunters should persist despite the drain of 'spongers'. 'Some people like to hunt, other people just like to eat', as a !kung informant said to Konner and Blurton Jones.

The increasing-returns curve raises interesting questions. It is very hard to find clear examples. The most promising places to look would seem to be in ownership of land or livestock in simple societies, or capital (almost by definition) in complex societies. Kaplan (personal communication) suggested that food might follow such a curve in an animal in which added food was readily converted into increased body size, which in turn led to increased resource-holding potential. This seems very likely for species in which body size varies greatly between individuals, and in which growth rates vary. Thus in a young male primate, food may lead to attainment of greater adult body size and competitive ability, and thus to a longer tenure of a single male troop. This suggestion about increasing returns from food may also allow us to link my suggested framework to Chapais & Schulman's (1980) treatment of primate dominance. They treated rank as a resource, and went from that premise to explain some features of primate behaviour. One of my aims has been to understand why some species show rank orders and others do not. Clearly in that context we cannot assume that rank is a resource without showing when it will act as a close approximation to real resources and when it will not. The time would seem ripe for primatologists (and child ethologists who study rank order) to try to organise their interest in rank around some clearly stated and basic throughts about contests over resources. It might be fruitful to combine the thoughts behind Vehrencamp's (1983) model and the thoughts outlined in this paper and replace traditional ideas about scarcity and abundance as explanatory principles.

One last point needs to be made about hierarchies. Garfinkel (1981) argued that economies of scale (which are implied by the increasing-returns curve) are the source of specialisation and stratification of human societies. If this is correct (there are many other good theories of the ecological causes of stratification), then the difference between stratification of human societies and the rank orders of monkey societies would be profound. Human stratification is suggested to depend on the increasing-returns curve. I have suggested that animal rank orders depend on the constant returns or straight-line relation.

The propositions outlined in this paper were originally developed with a view to accounting for variation among human societies, where additional issues like central site sharing, hoarding, and scrounging arise. But if they are valid propositions about the fitness pay-offs for fighting over resources they should apply to any animal. These propositions do promise to expand theories of conflict resolution in the direction of a unified account of theories and phenomena that are currently thought about in isolation from each other, e.g. economic defensibility of territory, dominance hierarchies, polygyny, food-sharing, hoarding. They might also be helpful in making us attend to the role of conflict over resources in an even wider range of phenomena such as fat deposition, hibernation, and speed of eating.

Acknowledgements

I wish to thank Eric Charnov for advice and for curve C, Richard M. Sibly for comments, advice and curve D; Geoffrey Parker for comments on an earlier draft, particularly for clarifying and updating my understanding of his own leading work in this field; and Ronald M. Weigel for unearthing some of the hidden assumptions. I am also grateful to the following for helpful discussions, critiques and advice: Kristen Hawkes, Kim Hill, Hillard Kaplan, also for access to their valuable work on food-sharing, and for discussion of the anthropological implications of the models.

References

Blurton Jones, N. G. (1983) A selfish origin for human food sharing: tolerated theft. *Ethol. Sociobiol.*, **5**, 1–3

Boelkins, R. C. (1967) Determination of dominance hierarchies in monkeys. *Psychoanalyt. Sci.*, **7**, 317–18

Brown, J. L. (1964) The evolution of diversity in avian territorial systems. *Wilson Bull.*, **76**, 160–9

Castell, R. & Heinrich, B. (1971) Rank order in a captive female squirrel monkey colony. *Folia Primatol.*, **14**, 182–9

Chapais, B. & Schulman, S. R. (1980) An evolutionary model of female dominance relations in primates. *J. Theoret. Biol.*, **82**, 47–89

Clutton-Brock, T. H. & Harvey, P. H. (1976) Evolutionary rules and primate societies. In *Growing Points in Ethology*, ed. P. P. G. Bateson & R. A. Hinde. Cambridge: Cambridge University Press

Davies, N. B. & Houston, A. I. (1981) Owners and satellites: the economics of territory defence in the pied wagtail, *Motacilla alba*. *J. Anim. Ecol.*, **50**, 157–80

Garfinkel, A. (1981) *Forms of Explanation*. Yale University Press

Hammerstein, P. (1981) The role of asymmetries in animal contests. *Anim. Behav.*, **29**, 193–205

Hammerstein, P. & Parker, G. A. (1982) The asymmetric war of attrition. *J. Theoret. Biol.*, **96**, 647–82

Hausfater, G. (1975) Predatory behavior of Yellow Baboons. *Behaviour*, **56**, 44–68

Hazlett, B. A., Rubenstein, D. I. & Rittscheff, D. (1975) Starvation, aggression and energy reserves in the crayfish *Orconectes virilis*. *Crustaceana*, **28**, 11–16

Kaplan, H. & Hill, K. (1985) Food sharing among Ache Foragers: tests of explanatory hypotheses. *Curr. Anthropol.*, **26**, 223–46

Lee, R. B. (1969) Eating Christmas in the Kalahari. *Nat. Hist.*, December, 1969, 14–22, 60–63

Lee, R. B. (1979) *The !kung San*. Cambridge & New York: Cambridge University Press

Maynard-Smith, J. & Parker, G. A. (1975) The logic of asymmetric contests. *Anim. Behav.*, **24**, 159–75

Parker, G. A. (1974) Assessment strategy and the evolution of fighting behavior. *J. Theoret. Biol.*, **47**, 223–43

Parker, G. A. & Rubenstein, D. I. (1981) Role assessment, reserve strategy, and acquisition of information in asymmetric animal conflicts. *Anim. Behav.*, **29**, 221–40

Popp, J. L. & DeVore, I. (1979) Aggressive competition and social dominance theory: synopsis. In *The Great Apes: Perspectives on Human Evolution*, vol. 5, ed. D. A. Hamburg & E. R. McCown. Menlo Park, Calif.: Benjamin/Cummins Publishing Co.

Rubenstein, D. I. (1977) Population density, resource patterning, and mechanisms of competition in the Everglades pygmy sunfish. Ph.D. dissertation, Duke University. University Microfilms, Ann Arbor, Mich.

Silverin, B. (1980) Effects of long-acting testosterone treatment on free-living Pied fly catchers, *Ficedula nypoleuca*, during the breeding period. *Anim. Behav.*, **28**, 906–12

Teleki, G. (1973) *The Predatory Behavior of Wild Chimpanzees*. Lewisburg: Buchnell University Press

Tullock, G. (1983) Territorial boundaries: an economic view. *Am. Naturalist*, **121**, 440–2

Vehrencamp, S. L. (1983) A model for the evolution of despotic versus egalitarian societies. *Anim. Behav.*, **31**, 667–82

INDEX

Acacia trees 228
adolescent roles 281
age-related growth 143
age-related interactions 277
aggression,
　cross-species 307
　female 304, 338, 344
　intensity 308
　male 334, 374
agonistic buffering 205
alliances,
　male 381
　rank acquisition 219, 291
　(*see also* interventions)
allocation of care 198
Alouatta palliata 105
alpha-male role 294, 338
alternative tactics 187
anthropomorphism 61, 69
appeasement 311
assembly rules 187
asymmetry,
　in aggression 310
　in contests 216, 393
　models 394
attack coalitions 343, 345
awareness, concepts of 32

behaviour,
　and infant development 150, 250, 262
　of juveniles 283
behavioural actions 62
behavioural repertoire 58, 283
behavioural traits 4
birth,
　rank 221, 292
　seasons 232
　weights 142
body measurements 142

brain size,
　development 131
　relation to feeding 93
breast feeding 319

Callicebus moloch 79, 255
Callithrix jacchus 122
care-elicitation behaviour 193
categorical learning 36
Cercopithecus aethiops 124, 193, 228
chacma baboon 124
children 121, 147, 155
chimpanzee 32, 39, 75, 120, 124
coalitions 344, 384
cognition,
　definition 31
　development 147, 155
　feeding techniques 94
　in nature 73
　social 53
cognitive abilities 32, 157
cognitive concepts 75
cognitive maps 75, 87
communal nesting 356
communication, 161
　and consciousness 45, 47
　meshing 201
　rules 169
comparative development,
　brain 147
　motor skills 155
competition,
　and reproduction 343
　over resources 399
　(*see also* dominance)
component form gradient 165
confidence intervals 243
conflict interference 291
consciousness 44, 49
consorts 376, 392

Index

contests,
  models 394
  outcomes 396
coordination, hand 94

dayranges 88
decision making 73
development,
  cognitive 75
  constraints on 235
  human 147, 155
  rules for 187
  scaffolding 185
  social behaviour 121
demography and development 233
diet 80, 105, 228
directionality of interactions 303
dominance,
  correlations with 271, 304
  development of 219, 291
  female 291, 301
  male 207, 334, 383
  maternal 219, 291
  over resources 399
  reproductive behaviour 334, 343
  (see also competition)
dyadic interactions 214, 221, 308

ecology,
  adaptations 235, 281
  constraints on learning 114
  effects on behaviour 227
emigration, male 371, 381
energy exchange networks 286
epistemology 25
estrus (see oestrus)
evolution,
  of behavioural traits 10
  of consciousness 45
  of food-sharing 397
  of reproductive tactics 315
expectancies, in incentives 34
exploitation of infants 212
extraction of food 95, 100

familial rank 219, 293
family organisation 286
feeding selectivity 105
females,
  aggression 304, 309, 338, 344
  cooperation 355
  dominance 291, 301
  interactions 337, 345, 360
  reproduction 335, 352, 356
fertility 331
field sites,
  finances 20
  planning 13
fitness 393

foetal brain growth 133
food,
  availability 105, 228, 398
  extraction 95, 100
  sharing 397
  storage 394
foraging efficiency 83
fruit eating 111
function of behaviour 179

games theory,
  models 196, 393
  predictions 199, 396
gestation and growth 147
goals, ranging 88
Gorilla gorilla 153
grading of vocalisations 161
greetings 175, 373
grooming 59, 210
growth,
  body 142
  brain 136, 148
  foetal 133, 149

hamadryas baboon 87
harem accretion 363, 367
hierarchical distance 304
histology, brain 132
hormones,
  and behaviour 258, 329
  in humans 321, 324, 326
household analysis 281
howling monkey 105
human thinking 25
hunting 403

immatures,
  disputes 221, 291
  interactions 230, 281
immigration,
  female 364
  male 212, 364, 371
inbreeding avoidance 371
infant carrying 212, 216
infant retrieval 261
infant survivorship 210, 343, 357
infanticide 213, 347, 366
infants,
  feeding sequence 106
  interactions 248
  learning 301
  relationships with males 205, 258, 264
  relationships with mothers 152, 198, 232, 264
  social behaviour 267
infertility 335
information exchange 194
inhibition of aggression 208
innovation 113, 123